FUNDAMENTAL PHYSICS *of* ULTRASOUND

V.A. Shutilov

Translated by
Michael E. Alferieff

Gordon and Breach Science Publishers

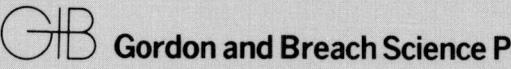

FUNDAMENTAL PHYSICS OF ULTRASOUND

FUNDAMENTAL PHYSICS OF ULTRASOUND

By

V.A. Shutilov

Translated from the Russian by
Michael E. Alferieff

GORDON AND BREACH SCIENCE PUBLISHERS
NEW YORK LONDON PARIS MONTREUX TOKYO

Gordon and Breach Science Publishers

Post Office Box 786
Cooper Station
New York, New York 10276
United States of America

Post Office Box 197
London WC2E 9PX
England

58, rue Lhomond
75005 Paris
France

Post Office Box 161
1820 Montreux 2
Switzerland

3–14–9, Okubo
Shinjuku-ku, Tokyo
Japan

Private Bag 8
Camberwell, Victoria 3124
Australia

Original version published in Russian in 1980 as *Osnovy fiziki ul'trazvuka* by Leningrad University Press

Library of Congress Cataloging in Publication Data
Main entry under title:

Shutilov, Vladimir Aleksandrovich.
Fundamental physics of ultrasound.

Translation of: Osnovy fiziki ul'trazvuka.
Bibliography: p.
Includes index.
1. Ultrasonics. I. Title.
QC244.S4713 1988 534.5′5 87–21081
ISBN 2–88124–684–2

Printed in Great Britain by Bell and Bain Ltd., Glasgow.

Contents

Foreword to the English Edition ix
Preface xi
List of Symbols xvi

I. BASIC EQUATIONS OF THE THEORY OF ELASTICITY

1. Description of the equilibrium and deformed states of a body 1
2. The stress tensor 9
3. Equation of motion 13
4. Relation between stress and strain. Generalized Hooke's Law 16
5. Energy of elastic deformation 20
6. The simplest deformations and the relation between the different elastic moduli 23

II. PROPAGATION OF ULTRASONIC WAVES IN LIQUIDS AND GASES

1. Acoustic properties of ideal liquids 28
2. The equations of hydrodynamics 30
3. Equation of state for liquids and gases 33
4. The wave equation 38
5. Plane waves 39
6. The velocity of sound 41

III. SINUSOIDAL PLANE WAVES WITH INFINITESIMAL AMPLITUDE

1. Equation of a monochromatic plane wave 47
2. Basic linear relations between the physical quantities varying in an ultrasonic wave. Characteristic impedance and acoustic impedance 49
3. Energy characteristics of the ultrasonic field. Ultrasonic intensity 54

NUMERICAL EXAMPLES. THE LOGARITHMIC SCALE OF INTENSITIES AND AMPLITUDES 58

4. Absorption of monochromatic ultrasonic waves 60
5. Shear waves in liquids. Viscous losses at the boundaries of ultrasonic beams 71

IV. FINITE-AMPLITUDE PLANE WAVES

1. Estimation of the nonlinear terms in the equations of hydrodynamics 77
2. The exact solution of the system of nonlinear equations of hydrodynamics for a nondissipative medium 80
3. The velocity of propagation of a finite-amplitude wave. Nonlinear properties of the medium 81
4. Relationship between the acoustic parameters in the second approximation 87
5. Distortion of the form of a finite-amplitude wave during propagation 88
6. Spectral analysis of a finite-amplitude wave 95
7. Intensity of distorted finite-amplitude ultrasonic waves 101
8. Absorption of finite-amplitude plane waves 103

V. STEADY FORCES ARISING IN AN ULTRASONIC FIELD

1. Radiation pressure 125
2. Radiation pressure forces on an obstacle 131
3. Steady forces acting on suspended particles in an ultrasonic field 137
4. Streaming 142

VI. ULTRASONIC CAVITATION

1. Rupture strength of liquids 149
2. Cavitation strength of a liquid 151
3. Collapse of a cavitation cavity 157
4. Dynamics of a cavitation cavity in an ultrasonic wave 163
5. Acoustic properties of a cavitating liquid 168

VII. REFLECTION, REFRACTION AND SCATTERING OF ULTRASONIC WAVES

1. Transmission and reflection of plane waves at normal incidence on the boundary between two media 172
2. Standing plane waves 180
3. Interference of oppositely traveling waves with normal reflection in an absorbing medium 185
4. Reflection and refraction of a plane wave at oblique incidence on a plane boundary between two media 188
5. Interference of plane waves at oblique incidence. Quasistanding waves 195
6. Scattering of ultrasonic waves in an inhomogeneous medium 198

VIII. TRANSMISSION OF PLANE WAVES THROUGH LAYERS. ELECTROACOUSTICAL ANALOGIES. RADIATION OF PLANE WAVES

1. Transmission of ultrasonic plane waves through a plane-parallel layer 211

2. "Antireflection" (impedance-matched) layers 217
3. Characteristic acoustic oscillations of plates 223
4. Method of electroacoustical analogies 226
5. Oscillatory systems without damping 228
6. Characteristic oscillations of electric, mechanical and acoustic oscillatory systems with damping 231
7. Forced oscillations. Resonance 237
8. Radiation of plane waves. The field of a real plane ultrasonic radiator 244

IX. SPHERICAL WAVES

1. Wave equation for spherical waves 250
2. Monochromatic spherical waves 251
3. The intensity of a spherical wave 253
4. Radiation of spherical waves from a pulsating sphere 255

X. PROPAGATION OF ULTRASOUND IN AN ISOTROPIC SOLID

1. Wave equation for an infinite solid 260
2. Reflection, refraction and transformation of ultrasonic waves at the boundaries of solids 266
3. Reflection coefficient at the boundary of a solid at oblique incidence 271
4. Rayleigh waves 285
5. Love waves 288
6. Geometric dispersion of sound in rods 291
7. Nonlinear elasticity and the origin of the nonlinear acoustics of solids 295

XI. PROPAGATION OF ULTRASOUND IN CRYSTALS

1. General acoustic equations for crystals 300
2. Relationship between the elastic moduli and the velocities of propagation of ultrasound in crystals 304
3. Cubic crystals 306
4. Crystals with lower symmetry 318
5. Influence of the piezoelectric effect on the elastic properties of crystals 335

PROBLEMS AND THEIR SOLUTIONS

Problems 341
Solutions 347
REFERENCES 363
BIBLIOGRAPHY 371
SUBJECT INDEX 373

Foreword to the English edition

At the initiative of the publishers, the English translation of "Fundamental Physics of Ultrasound" has been enhanced by the addition of problems and their solutions for each chapter, inserted at the end of the book. These additions should contribute toward a better understanding of the material, and help the reader to go beyond the limits of the book. A bibliography of monographs and manuals on ultrasonic physics has also been added.

Because of the untimely death of Professor V A Shutilov I have provided these supplements.

DOCTOR E. V. TCHARNAYA.

Preface

The term *ultrasound* in general encompasses different elastic waves with frequencies above the range of human hearing, i.e., above 15–16 kHz. Modern ultrasonic technology makes it possible to generate and detect ultrasonic oscillations with frequencies up to 10^{10}–10^{11} Hz and higher, i.e., up to frequencies approaching those of infrared radiation. At such high frequencies the wavelengths of ultrasonic waves (called hypersonic waves above $\simeq 10^9$ Hz) are comparable to intermolecular distances; however, the propagation of ultrasonic waves in different media is already sensitive to the structural characteristics of the material at the molecular, atomic, electronic and even nuclear levels. In this connection, ultrasonic methods have turned out to be very informative tools for studying the structure of materials and the different physical processes occurring in them.

On the other hand, the characteristics of ultrasonic radiation have led to extensive applications of ultrasound in the most diverse areas of the economy: sonar detection, flaw detection in various materials and structures, medical diagnostics and therapeutic action on body organs, acceleration or stimulation of different technological processes, in electronic and optical systems and in many other areas. All these applications are based on studies of physical processes occurring in ultrasonic fields in particular media. Such studies, pertaining to both purely scientific and applied problems, form an extensive body of knowledge, generally referred to as *physical acoustics* or *physical ultrasonics*. The subject matter here is so broad that within physical ultrasonics itself large independent branches have evolved: molecular acoustics, quantum acoustics, acoustoelectronics, acousto-optics, nonlinear acoustics, etc. Many books or review papers have been written on the specialized problems of physical acoustics. As a rule, they begin with an exposition of separate problems, pertaining to the foundations of the physics of ultrasound and based on the general laws of acoustics of continuous media. There are also many good books on general acoustics, such as the classic *Theory of Sound* by Lord Rayleigh,[1] *Vibrations and Sound* by P. Morse,[2] and books by Soviet authors *Lectures on the Theory of Sound* by S. N. Rzhevkin,[3] *General Acoustics* by M. A. Isakovich,[4] and the two-volume set by E. Skuchik entitled *Foundations of Acoustics*.[5] General acoustics, however, includes a wide range of problems, pertaining to primarily audible, i.e., low-frequency, sound and a number of more or less specialized subjects with their own peculiarities. These include, for example, branches of acoustics such as musical and architectural acoustics, bioacoustics, acoustics of noise and vibrations, geoacoustics, and others. Thus many problems covered in general acoustics are irrelevant to the physics of ultrasound, while others that are relevant to the physics of ultrasound are, on the other hand, inadequately covered or are omitted altogether.

The present book, which is based on lectures by the author over a period of many years to students specializing in the physics of ultrasound in the Physics Department of the Leningrad State University, can be viewed as a textbook on general ultrasonics, intended to be used prior to the study of specialized problems in this field of physics. An attempt is made in this book to single out and give a systematic exposition of the widest possible range of subject matter directly pertaining to the propagation of ultrasonic waves in media with different elastic properties and under conditions close to those encountered in scientific and practical applications of ultrasound.

The occasionally used term *ultrasonics* is not a very fortunate term, because the prefix *ultra* (as also the prefix *hyper*) refers, generally speaking, to frequencies and not to the process of propagation of elastic waves in itself. Nevertheless, the terms *ultrasonics* and *hypersonics* are now established in the scientific-technical lexicon, so that this book could equally well have been entitled *The Foundations of Ultrasonics*. In one way or another, this book is concerned with the propagation of ultrasound in different media that are regarded as being continuous. On the other hand, the propagation of ultrasound in continuous media, as already noted, conforms to the general laws of classical acoustics. As always, however, quantity (in this case – frequency) transfers to quality: high frequency, the special methods required to produce directed beams and high radiation intensities, and other characteristics of ultrasound impart certain special features to the problems of its propagation.

These special features are manifested primarily in the real and extensively used possibility of generating plane or quasi-plane waves, the special significance of the pulsed radiation regime, the action of intense ultrasound on a medium and the response of the medium to this action, the strong absorption of ultrasonic waves in gases, the possibility of propagation of shear waves in liquids, the steady forces existing in an ultrasonic field, etc. Correspondingly, the most important problems in ultrasonics are: the propagation of plane waves and their absorption, reflection, refraction, transmission through layers, focusing, and scattering; the analysis of nonlinear effects; ponderomotive forces in a plane-wave field; diffraction and interference effects in the field of real radiators of ultrasonic beams together with the analysis of deviations of the characteristics of the ultrasonic field in finite beams from the field formed by ideal plane waves; and, the propagation of different types of ultrasonic waves in unbounded and bounded solids, including crystals, etc. In this book, an attempt is made to discuss all these problems as fully as possible together with other aspects of the propagation of ultrasonic waves. Experimental data on the velocity and absorption of ultrasound in liquids and gases as well as on the velocity of sound in isotropic solids and crystals are likewise presented here. In addition to classical material, data from original sources, to which appropriate reference is made, are used.

The book is intended, aside from students, for a wide range of readers who are familiar with the foundations of higher mathematics and general physics at the technical college level. The author hopes that it will also be useful to graduate students and scientists specializing in ultrasonics or desiring to study this subject.

The author is deeply grateful to L. K. Zarembo and I. N. Kanevskii for valuable remarks about the manuscript, I. G. Mikhailov for general assistance in writing the book, N. N. Khromova for selecting the experimental data on solids, and L. D. Shutilova and B. F. Borisov for their assistance in laying out the manuscript.

V.A. Shutilov

List of Symbols

A	- work of amplitude ocillations	l	- ratio of acoustic impedances
a	- acceleration	Ma	- Mach's number
B	- nonlinear modulus of elasticity	m	- mass
		N	- quantity
C	- capacitance of capacitor	$\mathbf{n}$	- unit normal vector
c	- speed of sound	n	- polytropic index refractive index
$c_{iklj} = c_{nm}$	- moduli of elasticity		
c_P	- specific heat capacity at constant pressure	n_O	- concentration (volume)
		P	- pressure (static total)
c_V	- specific heat capacity at constant volume	P_m	- Legendre polynomial
		p	- acoustic pressure (variable)
D	- intensity		
$\mathbf{D}$	- electrical induction	Q	- Q-factor
$\bar{D}$	- cavitation index	q	- charge: gas-content factor
d	- transmission coefficient thickness	R	- radius
		R_O	- universal gas constant
E	- electromotive force Young's modulus	R_e	- ohmic resistance
		Re	- Reynolds number
$\mathbf{E}$	- intensity of electric field	$\mathbf{r}$	- radius vector
E	- effective modulus of elasticity	r	- coefficient of friction; polar coordinate
F	- force	S	- area
f	- focal length	s	- condensation
f_{ikl}	- piezoelectric constants	T	- temperature
G	- shear modulus	t	- time
I	- intensity of ultrasound current strength	U	- internal energy voltage
		$\mathbf{u}\,(\xi,\eta,\zeta)$	- displacement vector
i	- $\sqrt{(-1)}$	V	- volume
J	- Bessel function invariant	z	- specific acoustic impedance
K	- linear bulk modulus of elasticity; spring constant		
		$\tilde{z}$	- specific impedance
$\mathbf{k}$	- wave vector	α	- absorption coefficient
k	- wave number compliance factor	α_T	- coefficient of thermal expansion
L	- distance coefficient of self-induction	β	- phase: phase difference
		γ	- ratio of heat capacities

Δ – Laplacian operator
δ_0 – damping time constant
ϵ_0 – nonlinear coefficient
ϵ_{ik} – deformation dielectric constant
ζ, η – components of displacement along the z, y axes
η – coefficient of viscosity
Θ – volume expansion
θ – angle spherical coordinate
υ – damping decrement
ϱ – density reflection coefficient
σ – surface tension
σ_{ik} – mechanical stress
σ_{eff} – effective scattering cross section
v – velocity

W – energy
w – energy density
x,y,z – Cartesian coordinates
Z – total acoustic impedance
$\tilde{Z}$ – total impedance
æ – compressibility
Λ – wavelength of acoustic wave
λ_0 – thermal conductivity
λ,μ – Lamé constants
ν – frequency
ν_0 – Poisson's ratio
ξ – displacement along the x axis
τ_0 – damping time constant
φ – potential angle of rotation
Ψ – azimuthal angle potential
ω – angular frequency

I. Basic Equations of the Theory of Elasticity

§1. Description of the equilibrium and deformed states of a body

The propagation of ultrasonic waves in different media, which we shall regard as continuous, is accompanied by a periodic displacement of particles of the medium from their equilibrium positions under the action of elastic forces. Here, "particle" refers to an infinitesimal volume element which itself contains a very large number of molecules so that the medium within it can be considered to be continuous. In the normal, unperturbed state all particles in the medium are in certain equilibrium positions, determined by the balance of intermolecular forces. We shall denote the equilibrium position of a particle by a radius vector $\mathbf{r}$ (*position vector*), drawn from the origin of some system of coordinates (*laboratory system*) that is stationary relative to the given medium. For such a system we shall most often choose a Cartesian rectangular coordinate system x, y, z. In many cases, it will be more convenient to use a spherical coordinate system r, θ, ψ, which is related to the rectangular system by the relations $x = r \sin \theta \cos \psi$, $y = r \sin \theta \sin \psi$, $z = r \cos \theta$, or a cylindrical system r, θ, z in which $x = r \cos \theta$, $y = r \sin \theta$, $z = z$. We shall describe the displacement of a particle from its position of equilibrium with the help of the *displacement vector* $\mathbf{u}$. Thus the new position of the particle after displacement will be determined by the vector $\mathbf{r} + \mathbf{u}$. We shall denote the components of the displacement vector $\mathbf{u}$ along the coordinate axes by the symbols ξ, η, and ζ, respectively. The magnitude of the displacement depends on the position of the particle and, in the general dynamic case, the displacement can vary with time. Thus the components of the displacement ξ, η, and ζ are, in general, functions of the coordinates and time: $\xi = \xi(x, y, z, t)$, $\eta = \eta(x, y, z, t)$ and $\zeta = \zeta(x, y, z, t)$.

The displacement of particles from their equilibrium positions corresponds to a deformation of the medium. To describe completely the deformed state of the body as a function of time, the displacement vector **u** must obviously be represented as a function of the coordinates x, y, and z. This problem can be approached by examining successively one-, two-, and three-dimensional deformations. In so doing, keeping in mind the smallness of the deformation in an acoustic wave, we shall at first restrict our attention to infinitesimal deformations.

One-dimensional deformations. Let x denote a segment of an undeformed body lying along the x-axis between the points M and N (Fig. 1) and consider the change in this segment accompanying a

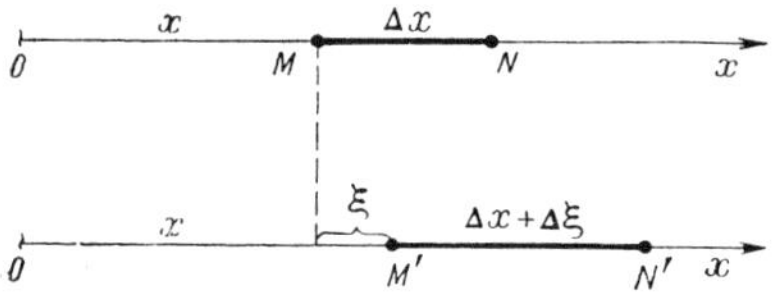

Fig. 1.

deformation of the body. The point M with coordinate x will be displaced after deformation by an amount ξ and will move into the position M' with coordinate $x + \xi$. The length of the segment MN will increase by $\Delta\xi$. The deformation of the segment MN is understood to mean the ratio formed by the increment to the length and the initial length, i.e., the quantity $\Delta\xi/\Delta x$. The deformation at the point M is defined by the expression

$$\epsilon = \lim_{\Delta x \to 0} \Delta\xi/\Delta x = d\xi/dx,$$

i.e., the deformation of an infinitesimal segment equals the derivative of the displacement with respect to the coordinate and is a dimensionless quantity. If ξ is a linear function of x, i.e., $\epsilon = const$, then such a deformation is called a *homogeneous* deformation. In this case, $d\xi/dx = \Delta\xi/\Delta x$ (homogeneous stretching of a rod). In general, $\epsilon \neq const$, i.e., the deformation is a function of the coordinate. In the dynamic case, $\epsilon = \epsilon(x, t)$.

Two-dimensional deformations. We shall now examine two-dimensional deformations. For this, we single out a segment of length Δr in the xy plane (Fig. 2) and follow its variation with the deformation of the body. Let the point M, whose radius vector before deformation is $\mathbf{r}$ with projections x and y along the coordinate axes, be displaced after deformation into the position M' with position vector $\mathbf{r} + \mathbf{u}$. The components of the displacement vector $\mathbf{u}$ are therefore ξ and η. The point N will move after deformation into the position N', and the segment singled out, which before the deformation was described by the vector $\Delta\mathbf{r}$ with components Δx and Δy, will be described after deformation by the vector $\Delta\mathbf{r} + \Delta\mathbf{u}$ with components $\Delta x + \Delta\xi$ and $\Delta y + \Delta\eta$ along the coordinate axes. The ratios $\Delta\xi/\Delta x$ and $\Delta\eta/\Delta y$ define the stretching of the projections of the segment singled out along the coordinate axes. These ratios do not, however, completely describe the deformed state, because, as is evident from Fig. 2, in addition to being stretched, the vector $\Delta\mathbf{r}$ is also rotated in the xy plane.

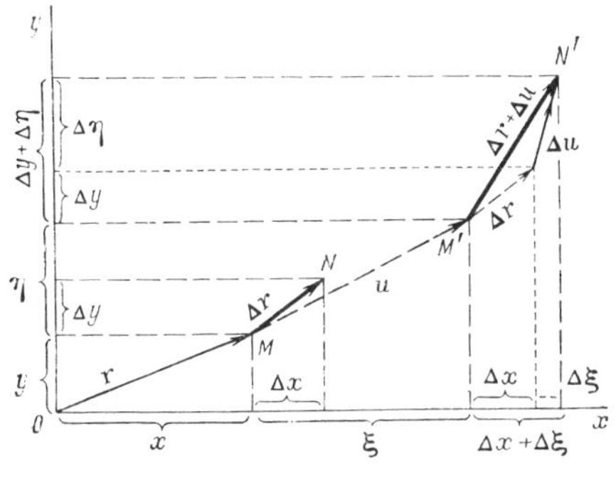

Fig. 2.

To describe this rotation, we shall examine the distortion of a rectangle constructed from the projections of the undeformed segment MN with lengths $MQ_1 = \Delta x$ and $MQ_2 = \Delta y$ (Fig. 3). After deformation, since the nonvanishing components $\Delta\xi$ and $\Delta\eta$ now appear, these projections will be stretched and displaced. As is evident from Fig. 3, the tangent of the angle of rotation of the segment $M'Q_1'$ is determined by the ratio $\tan\varphi = \Delta\eta/(\Delta x + \Delta\xi)$, while the tangent of the angle of rotation of the segment $M'Q_2'$ is determined by the ratio $\tan\varphi = \Delta\xi/(\Delta y + \Delta\eta)$. Because we are considering small deformations only, $\Delta\xi$ and $\Delta\eta$ are small compared with Δx and Δy. Letting Δx and Δy approach zero, as a measure of the displacement of the segments MQ_1 and MQ_2 in the xy plane we have

$$\varphi_{yx} = \frac{\partial\eta}{\partial x} = \epsilon'_{yx} \quad \text{and} \quad \varphi_{xy} = \frac{\partial\xi}{\partial y} = \epsilon'_{xy},$$

while stretching of the segments MQ_1 and MQ_2 is characterized by the derivatives $\partial\xi/\partial x = \epsilon_{xx}$ and $\partial\eta/\partial y = \epsilon_{yy}$.

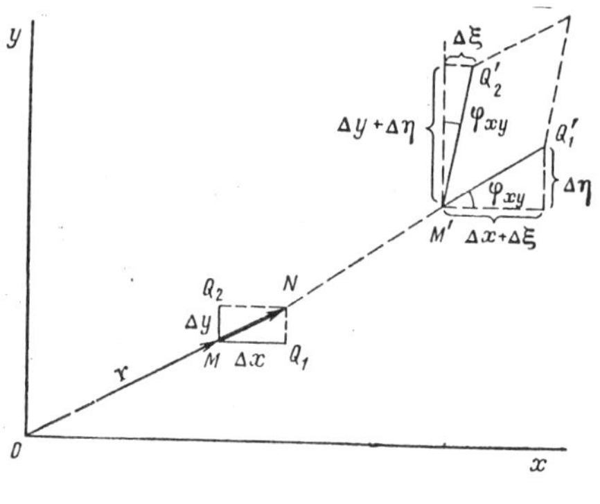

Fig. 3.

On the other hand, since the components of the displacement are functions of the coordinates, we can write

$$\Delta\xi = \frac{\partial\xi}{\partial x}\Delta x + \frac{\partial\xi}{\partial y}\Delta y = \epsilon_{xx}\Delta x + \epsilon'_{xy}\Delta y,$$

$$\Delta\eta = \frac{\partial\eta}{\partial x}\Delta x + \frac{\partial\eta}{\partial y}\Delta y = \epsilon'_{yx}\Delta x + \epsilon_{yy}\Delta y.$$

Thus the quantities ϵ_{ik} relate the components of the vector $\Delta\mathbf{u}$ to the components of the vector $\Delta\mathbf{r}$, i.e., they form a tensor of rank two, which, replacing the coordinates x, y and z by the indices 1, 2, 3, can be represented in the form

$$\epsilon'_{ik} = \begin{bmatrix} \epsilon_{11} & \epsilon'_{12} \\ \epsilon'_{21} & \epsilon_{22} \end{bmatrix}$$

It is not difficult to see that the nondiagonal components of this tensor $\epsilon'_{12} = \varphi_{xy}$ and $\epsilon'_{21} = \varphi_{xy}$, aside from shear deformations, also describe the rotation of the rectangle MQ_1NQ_2 as a whole. This is illustrated in Fig. 4, which shows the change in the posi-

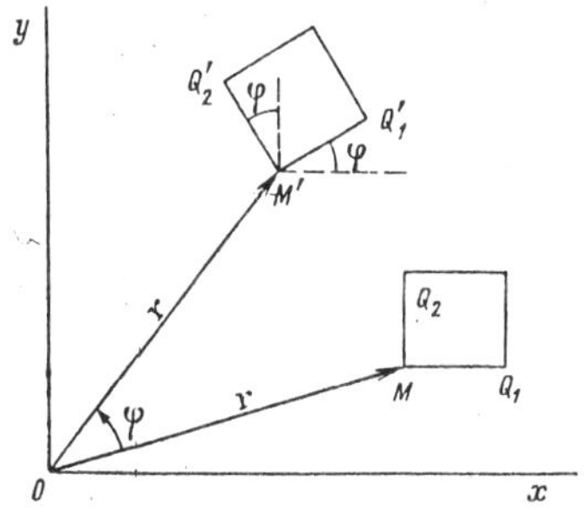

Fig. 4.

tion of this rectangle as a result of the rotation of the body by an angle φ relative to the origin of coordinates. In this case, both segments MQ_1 and MQ_2 rotate counterclockwise by an angle φ and, in accordance with the geometric meaning of ϵ'_{ik} established above, for this case we can write

$$\epsilon'_{ik} = \begin{bmatrix} 0 & -\varphi \\ -\varphi & 0 \end{bmatrix} \qquad (I.1)$$

The shape of a rectangle is not distorted in this case, but the tensor ϵ'_{ik} does not vanish. Therefore, in order to find the part of the tensor ϵ'_{ik} which describes a pure deformation, we must subtract from ϵ'_{ik} the part corresponding to the rotation of the body as a whole.

Any tensor of rank two can be represented as a sum of symmetric and antisymmetric tensors, i.e., ϵ'_{ik} can be written in the form $\epsilon'_{ik} = \epsilon_{ik} + \epsilon''_{ik}$, where $\epsilon_{ik} = (\epsilon'_{ik} + \epsilon'_{ki})/2$ and $\epsilon''_{ik} = (\epsilon'_{ik} - \epsilon'_{ki})/2$. Since $\epsilon_{ik} = (\epsilon'_{ik} + \epsilon'_{ki})/2 = (\epsilon'_{ki} + \epsilon_{ik})/2 = \epsilon_{ki}$, it is easy to see that the tensor ϵ_{ik} defined in this manner is symmetric. On the other hand, since $\epsilon''_{ik} = (\epsilon'_{ik} - \epsilon'_{ki})/2 = -(\epsilon'_{ki} - \epsilon'_{ik})/2 = -\epsilon''_{ki}$, the tensor ϵ''_{ik} is antisymmetric. According to expression (I.1), the rotation of a body is described by an antisymmetric tensor. A pure displacement is therefore described by the symmetric tensor ϵ_{ik} obtained by subtracting the antisymmetric part from ϵ'_{ik}, i.e., $\epsilon_{ik} = \epsilon'_{ik} - \epsilon''_{ik}$. This gives a symmetric tensor of rank two, called the *strain tensor*:

$$\epsilon'_{ik} = \begin{bmatrix} \epsilon_{12} & \frac{1}{2}(\epsilon'_{12} + \epsilon'_{21}) \\ \frac{1}{2}(\epsilon'_{21} + \epsilon'_{12}) & \epsilon_{22} \end{bmatrix}$$

whose diagonal components describe a stretching deformation along the coordinate axes and whose nondiagonal components equal one-half the displacement angle φ_{12} in the xy plane:

$$\epsilon_{12} = \epsilon_{21} = \frac{1}{2}(\epsilon'_{12} + \epsilon'_{21}) = \frac{1}{2}\left(\frac{\partial \xi}{\partial y} + \frac{\partial \eta}{\partial x}\right) = \frac{1}{2}(\varphi_{xy} + \varphi_{yx}) = \frac{1}{2}\varphi_{12}.$$

Three-dimensional deformations. In the three-dimensional case, by examining the deformation of a volume element in the form of a rectangular parallelepiped constructed from the projections of the segment singled out Δx, Δy, and Δz, we obtain in an analogous manner the component of the deformation $\partial \zeta/\partial z = \epsilon_{33}$, characterizing the stretching along the z axis, and the shear components, which describe the displacement in the xz and zy planes. The strain tensor in this case will have the form

$$\epsilon_{ik} = \begin{bmatrix} \epsilon_{11} & \epsilon_{12} & \epsilon_{13} \\ \epsilon_{21} & \epsilon_{22} & \epsilon_{23} \\ \epsilon_{31} & \epsilon_{32} & \epsilon_{33} \end{bmatrix}$$

where $\epsilon_{11} = \partial\xi/\partial x$, $\epsilon_{22} = \partial\eta/\partial y$, and $\epsilon_{33} = \partial\zeta/\partial z$ are the extensions along the x, y, and z axes; and,

$$\epsilon_{12} = \epsilon_{21} = \frac{1}{2}\left(\frac{\partial\xi}{\partial y} + \frac{\partial\eta}{\partial x}\right) = \frac{1}{2}\varphi_{12},$$

$$\epsilon_{23} = \epsilon_{32} = \frac{1}{2}\left(\frac{\partial\eta}{\partial z} + \frac{\partial\zeta}{\partial y}\right) = \frac{1}{2}\varphi_{23},$$

$$\epsilon_{13} = \epsilon_{31} = \frac{1}{2}\left(\frac{\partial\xi}{\partial z} + \frac{\partial\zeta}{\partial x}\right) = \frac{1}{2}\varphi_{13}$$

are equal to one-half the total displacement angles in the xy, yz, and xz planes, respectively.

Thus, for small deformations, the deformed state of the body in the vicinity of the point M with coordinates x, y, and z is completely described by six independent components of the strain tensor ϵ_{ik}, which can be represented in the general form:

$$\epsilon_{ik} = \frac{1}{2}\left(\frac{\partial u_i}{\partial x_k} + \frac{\partial u_k}{\partial x_i}\right), \tag{I.2}$$

where u_i, u_k are the components of the displacement vector and $i, k = 1, 2, 3$.

The symmetry of the tensor ϵ_{ik} (as well as of other tensors characterizing physical properties) makes it possible to write it in a simpler "matrix" form with a single index: $\epsilon_{ik} \rightarrow \epsilon_n$, where $n = 1, 2, 3, 4, 5$, and 6 and the components are read as shown in the following diagram:

$\epsilon_{ik} =$ 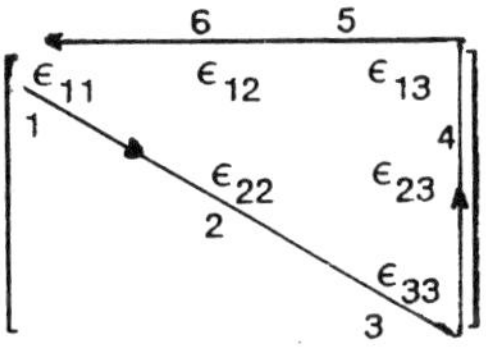
6
5
ϵ_{11}
ϵ_{12}
ϵ_{13}
1
4
ϵ_{22}
ϵ_{23}
2
ϵ_{33}
3

$$\rightarrow \epsilon_n = \begin{pmatrix} \epsilon_1 & \epsilon_6 & \epsilon_5 \\ & \epsilon_2 & \epsilon_4 \\ & & \epsilon_3 \end{pmatrix}, \tag{I.3}$$

We shall use this form often in what follows.

Due to its symmetry, the strain tensor can be referred to its principal axes. The shear components vanish in this case, and we have

$$\epsilon_{ii} = \begin{bmatrix} \epsilon_{11} & 0 & 0 \\ & \epsilon_{22} & 0 \\ & & \epsilon_{33} \end{bmatrix} = \epsilon_{ik}\delta_{ik},$$

where δ_{ik} is the *unit tensor* (the *Kronecker delta*): $\delta_{ik} = 1$ if $i = k$ and δ_{ik} = if $i \neq k$. The characteristic property of the principal axes is that they define three mutually orthogonal directions which remain mutually orthogonal under deformations of the body (they cannot rotate with the body). When a unit cube with edges parallel to the principal axes is deformed, the edges of the cube remain orthogonal to one another and their lengths become equal to $1 + \epsilon_{11}$, $1 + \epsilon_{22}$, and $1 + \epsilon_{33}$. In view of the smallness of the deformations, the change in the volume of this unit cube due to its deformation equals

$$\Theta = (1+\epsilon_{11})(1+\epsilon_{22})(1+\epsilon_{33}) - 1 \simeq \epsilon_{11} + \epsilon_{22} + \epsilon_{33}$$

Thus the *invariant* of the tensor of small strains — the trace — represents a volume expansion:

$$\Theta = \frac{\partial \xi}{\partial x} + \frac{\partial \eta}{\partial y} + \frac{\partial \zeta}{\partial z} = \operatorname{div} \mathbf{u} \qquad \text{(I.4a)}$$

Relation (I.4a) is valid when there is no discontinuity in the medium and can therefore be interpreted as the mathematical expression of continuity, i.e., it is the linearized *equation of continuity*.

In the general case of an inhomogeneous variable deformation, the volume expansion Θ is a function of the coordinates and time: $\Theta = (x, y, z, t)$. The equation of continuity with variable deformation can also be written in the form

$$d\Theta/dt = \operatorname{div} \mathbf{v}, \qquad \text{(I.4b)}$$

where $\mathbf{v} = d\mathbf{u}/dt$ is the rate-of-displacement vector.

Exact expressions for finite deformations. We shall obtain the final expressions for the components of the strain tensor immediately for the three-dimensional case by calculating the change in the distance dL between two nearby points in the body as a result of its deformation without making any approximations. The square of the distance between the two points equals $(dL)^2 = (dx_i)^2$ before the deformation and $(dL')^2 = (dx_i + du_i)^2$ after deformation. Since $du_i = (\partial u_i / \partial x_k)\, dx_k$, this expression can be written in the form

$$(dL')^2 = (dx_i)^2 + 2\frac{\partial u_i}{\partial x_k} dx_k dx_i + \frac{\partial u_i}{\partial x_k}\frac{\partial u_i}{\partial x_l} dx_k dx_l .$$

From here we obtain the following expression for the increment to the square of the distance between two nearby points: $(dL')^2 - (dL)^2 = 2\epsilon_{ik} dx_i dx_k$, where

$$\epsilon_{ik} = \frac{1}{2}\left(\frac{\partial u_i}{\partial x_k} + \frac{\partial u_k}{\partial x_i} + \frac{\partial u_l}{\partial x_i}\frac{\partial u_l}{\partial x_k}\right), \quad i, k, l = 1, 2, 3. \qquad (I.5)$$

Equation (I.5) represents the exact expression for the components of the strain tensor; for sufficiently small deformations, when the last term can be neglected as a second-order infinitesimal, it transforms into the linearized expression (1.2).

§2. The stress tensor

In a body that is not deformed, all parts of the body are in mechanical equilibrium with one another. This means that the resultant of all forces acting on a volume element singled out in the body vanishes. A deformation, however, takes the body out of the equilibrium state, as a result of which elastic forces due to intermolecular interactions appear in the body. The range of molecular forces is of the order of the distance between molecules, so that in the theory of elasticity of a continuous medium this range is assumed to equal zero. Thus, when a body is deformed, the internal forces exerted directly on an individual volume element of the body act only from the surrounding parts of the body through the surface of the element, i.e., they are surface forces, which we shall examine in what follows neglecting body forces such as gravity. Surface forces are proportional to the surface area on which they act. A force referred to unit surface area is called a *mechanical stress*.

Let dS denote an element of the surface of an arbitrary volume ΔV of a deformed body (Fig. 5) which is small enough so that the mechanical stress acting through it* may be assumed to be uniform. Let us draw the

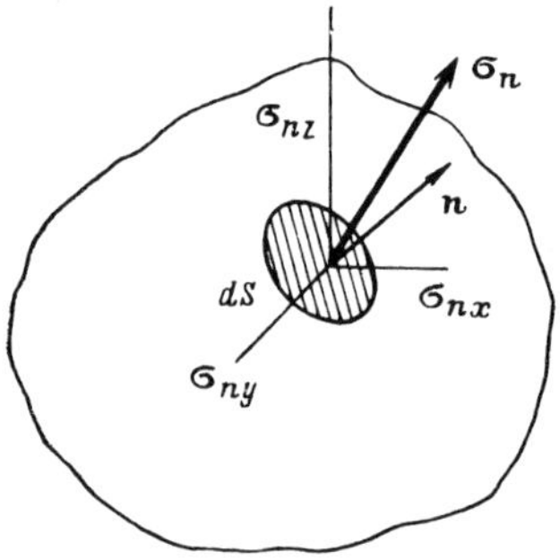

Fig. 5.

outer normal $\mathbf{n}$ to this surface. The stress acting on the surface dS is a vector whose orientation may not, in general, coincide with the normal to the surface. The sign of the stress is chosen arbitrarily. It is customary to consider a stress making an acute angle with the normal $\mathbf{n}$, i.e., a tensile stress, to be positive. The stress depends on the position and orientation of the surface element dS, so that the stress vector corresponding to a given surface element with the outer normal $\mathbf{n}$ is denoted by an index referring to this area $\sigma_\mathbf{n}$. The vector $\sigma_\mathbf{n}$ can be decomposed into the components σ_{nx}, σ_{ny}, σ_{nz} along the coordinate axes. In general, the stress $\sigma_\mathbf{n}$ and its components are functions of the coordinates and time.

---

* Below, we shall use "stress" instead of the term "mechanical stress."

To describe completely the state of stress of a body in the vicinity of some point 0, we shall construct around this point a right-angled parallelepiped with edges of length dx, dy, dz parallel to the coordinate axes which is small enough so that the stresses acting on the faces are uniform. The volume element constructed in this manner is bounded by faces with only three orientations: the outer normals lie along the coordinate axes x, y, z. We shall denote the stresses acting on these faces by $\boldsymbol{\sigma}_x$, $\boldsymbol{\sigma}_y$, and $\boldsymbol{\sigma}_z$, respectively (Fig. 6). Each of these stresses has three components along the coordinate axes: $\boldsymbol{\sigma}_x$: σ_{xx}, σ_{xy}, σ_{xz}; $\boldsymbol{\sigma}_y$: σ_{yx}, σ_{yy}, σ_{yz}; $\boldsymbol{\sigma}_z$: σ_{zx}, σ_{zy}, σ_{zz}. Here, the first index (row) indicates the face and the second (column) indicates the orientation of the projection. The nine scalar quantities σ_{ik} obtained in this manner completely describe the stress state of the body in the vicinity of a given point and form a tensor of rank two, called the *stress tensor*. This tensor is also symmetric, i.e., $\sigma_{ik} = \sigma_{ki}$, so that it contains only six independent components, and the sequence of indices is not significant. Replacing the indices x, y, z by 1, 2, 3, the stress tensor can be written in the form

$$\sigma_{ik} = \begin{bmatrix} \sigma_{11} & \sigma_{12} & \sigma_{13} \\ \sigma_{21} & \sigma_{22} & \sigma_{23} \\ \sigma_{31} & \sigma_{32} & \sigma_{33} \end{bmatrix} \qquad \text{(I.6a)}$$

or, in matrix notation,

$$\sigma_n = \begin{pmatrix} \sigma_1 & \sigma_6 & \sigma_5 \\ & \sigma_2 & \sigma_4 \\ & & \sigma_3 \end{pmatrix}, \qquad \text{(I.6b)}$$

where $n = 1, 2, 3, 4, 5$, and 6, and the matrix is read as in (I.3).

The symmetry of the stress tensor enables it to be reduced to the principal axes, in which the shear stresses vanish and only the diagonal components remain:

$$\sigma_{ii} = \begin{bmatrix} \sigma_{11} & 0 & 0 \\ 0 & \sigma_{22} & 0 \\ 0 & 0 & \sigma_{33} \end{bmatrix} = \sigma_{ik}\,\delta_{ik} \,. \tag{I.7}$$

Only tensile (compressive) stresses act on a volume element in the form of a right-angled parallelepiped with edges parallel to the principal axes.

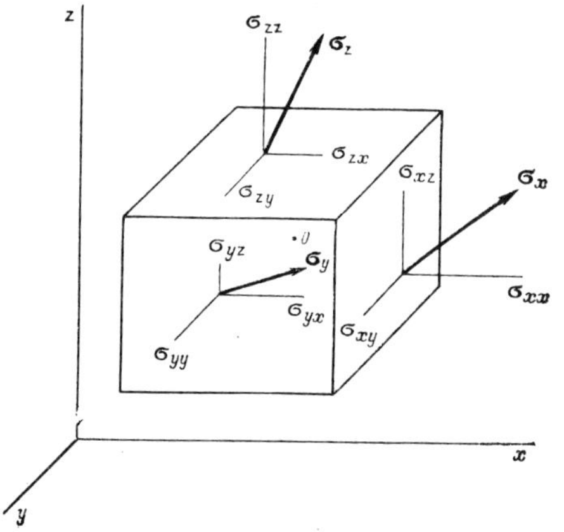

Fig. 6.

The stress tensor (I.6) describes the stressed state in the vicinity of a given point in the body. If it does not vary from point to point and does not depend on time, then it describes a homogeneous, constant (static) stress. In the general case of an inhomogeneous dynamic stress, the components of the tensor σ_{ik} are functions of the coordinates and time: $\sigma_{ik} = \sigma_{ik}(x, y, z, t)$.

§ 3. Equation of motion

In the case of inhomogeneous stress, uncompensated surface forces will act on a particle in the medium, imparting to each particle an acceleration inversely proportional to its mass. To express the resulting forces in terms of the components of the stress tensor σ_{ik}, let us examine the motion of a volume element in the form of a right-angled parallelepiped with edges dx, dy, dz, parallel to the coordinate axes (Fig. 7). Its volume is $dV = dx\,dy\,dz$, its mass is m, and its density is $\rho = m/(dV)$. Let the coordinates of the vertex M be x, y, and z. We

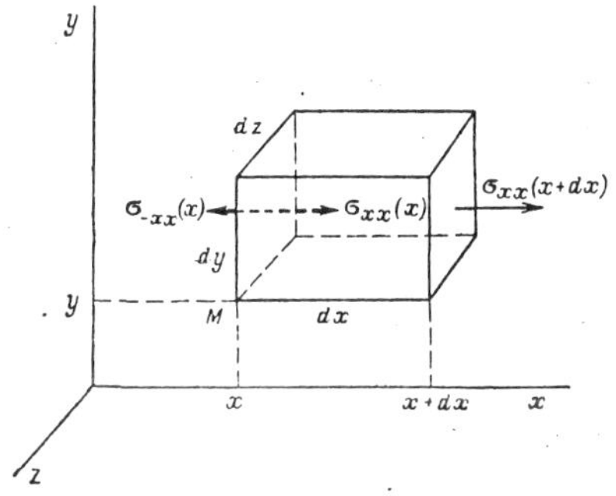

Fig. 7.

shall calculate the x-component of the resulting force acting on this volume element as a result of the difference in the stresses acting on its faces. For this, we shall first separate out the x-component of the stresses acting on the faces perpendicular to the x-axis. A stress $\sigma_{-x}(x)$ (the index of the stress by convention denotes the normal; the positive normal to the face with coordinate x is oriented in the $-x$ direction) acts on the face with coordinate x. Its x component $\sigma_{-xx}(x)$ is a scalar, denoted in Fig. 7 by an arrow in order to indicate the sign of the stress. In view of the equality of action and reaction, $|\sigma_{xx}(x)| = |\sigma_{-xx}(x)|$. The normal stress on the face with coordinate $x + dx$ is $\sigma_{xx}(x + dx)$. The resulting force acting on the faces perpendicular to the x-axis is

$$F_{xx} = [\sigma_{xx}(x + dx) - \sigma_{xx}(x)]\,dy\,dz.$$

For a sufficiently small parallelepiped, the change in stress along its edge may be assumed to be linear. Then

$$\sigma_{xx}(x + dx) = \sigma_{xx}(x) + \frac{\partial \sigma_{xx}}{\partial x} dx$$

and

$$F_{xx} = \frac{\partial \sigma_{xx}}{\partial x} dx\, dy\, dz = \frac{\partial \sigma_{xx}}{\partial x} dV.$$

Analogously, for the x-component of the forces acting on the faces perpendicular to the y and x axes we obtain:

$$F_{yx} = \frac{\partial \sigma_{yx}}{\partial y} dV; \; F_{zx} = \frac{\partial \sigma_{zx}}{\partial z} dV.$$

The total x-component of the force acting on the entire volume element is

$$F_x = \left(\frac{\partial \sigma_{xx}}{\partial x} + \frac{\partial \sigma_{xy}}{\partial y} + \frac{\partial \sigma_{xz}}{\partial z}\right) dV.$$

It imparts to the volume element an acceleration along the x-axis: $md^2\xi/dt^2$, where ξ is the displacement of the particle under examination along the x-axis. Thus the equation of motion of the particle (Newton's second law) along the x-axis is

$$\frac{\partial \sigma_{xx}}{\partial x} + \frac{\partial \sigma_{xy}}{\partial y} + \frac{\partial \sigma_{xz}}{\partial z} = \rho \frac{d^2 \xi}{dt^2}. \qquad \text{(I.8a)}$$

Analogously, for the two other axes we have:

$$\frac{\partial \sigma_{yx}}{\partial x} + \frac{\partial \sigma_{yy}}{\partial y} + \frac{\partial \sigma_{yz}}{\partial z} = \rho \frac{d^2 \xi}{dt^2}; \qquad \text{(I.8b)}$$

$$\frac{\partial \sigma_{zx}}{\partial x} + \frac{\partial \sigma_{zy}}{\partial y} + \frac{\partial \sigma_{yz}}{\partial z} = \rho \frac{d^2 \xi}{dt^2} \qquad \text{(I.8c)}$$

Replacing the indices x, y, z by 1, 2, 3 and the coordinates x, y, z by x_1, x_2, x_3, Eq. (I.8) can be combined into a single expression:

$$\partial\sigma_{ik}/\partial x_k = \rho\, dv_i/dt\ , \quad i,\ k = 1,\ 2,\ 3 \qquad \text{(I.9)}$$

where summation over repeated indices is implied. This expression represents the complete *equation of motion*, which is one of the basic equations of the dynamics of continuous media. The displacement **u** and the rate of displacement **v** are functions of the coordinates and time. For this reason, the total derivative with respect to time in Eq. (I.9) can be represented in the form

$$\frac{dv_i}{dt} = \frac{\partial v_i}{\partial t} + \frac{\partial v_i}{\partial x_k} v_k\ ,$$

where the first (local) derivative describes the change in the velocity of the particle with time at a given point in space due to the action of forces and the second term (sum of convective derivatives) represents the change in the velocity due to the displacement of the particle into neighboring points of the medium with a different velocity. For small displacements and low velocities, the convective derivatives, which are second-order infinitesimals compared with the local derivative, can be neglected, setting $dv_i/dt = \partial v_i/\partial t$.

Analogously, the instantaneous density ρ of the perturbed medium can be represented as a sum: $\rho = \rho_0 + \Delta\rho$ where ρ_0 is the equilibrium density of the unperturbed medium and $\Delta\rho$ is the change in density due to the deformation. For small deformations $\Delta\rho \ll \rho_0$, and the instantaneous density ρ can be set equal to ρ_0. Expression (I.9) then assumes the simpler form:

$$\partial\sigma_{ik}/\partial x_k = \rho_0 \partial v_i/\partial t \qquad \text{(I.10)}$$

or

$$\partial\sigma_{ik}/\partial x_k = \rho_0 \partial^2 u_i/\partial t^2\ . \qquad \text{(I.11)}$$

In this linearized form, the equation of motion is exact only for infinitesimal displacements. This is the form used in the acoustics of infinitesimal amplitudes. We shall examine the consequences of including nonlinear terms below for the example of the propagation of finite-amplitude ultrasonic waves in liquids.

§ 4. Relation between stress and strain. Generalized Hooke's Law

So far we have studied stress and strain independently of one another. Actually, however, the deformation of an elastic body involves the appearance of internal stresses in the body which strive to eliminate this deformation, i.e., to restore the equilibrium state. Thus there is a definite dependence between stress and strain, i.e.,

$$\sigma_{ik} = \sigma_{ik}(\epsilon_{jl}). \tag{1.12}$$

Experiment shows that for small strains the stress is proportional to the strain. This fact, established by Hooke for the simplest deformations, is a statement of the well-known *Hooke's law*, which is valid only for quite small strains and stresses. In application to infinitesimal-amplitude acoustics we can restrict our attention to ideal elastic media, for which the relation between stress and strain is linear. Since the stress and strain are in general defined by tensors of rank two with six independent components each, the natural generalization of Hooke's law is a linear dependence between them. The *generalized Hooke's law* can then be formulated as follows: **the components of the stress at a given point of the body are linear and homogeneous functions of all components of the strain**, i.e.,

$$\begin{aligned}
\sigma_1 &= c_{11}\epsilon_1 + c_{12}\epsilon_2 + c_{13}\epsilon_3 + c_{14}\epsilon_4 + c_{15}\epsilon_5 + c_{16}\epsilon_6;\\
\sigma_2 &= c_{21}\epsilon_1 + c_{22}\epsilon_2 + c_{23}\epsilon_3 + c_{24}\epsilon_4 + c_{25}\epsilon_5 + c_{26}\epsilon_6;\\
\sigma_3 &= c_{31}\epsilon_1 + c_{32}\epsilon_2 + c_{33}\epsilon_3 + c_{34}\epsilon_4 + c_{35}\epsilon_5 + c_{36}\epsilon_6; \qquad \text{(I.13a)}\\
\sigma_4 &= c_{41}\epsilon_1 + c_{42}\epsilon_2 + c_{43}\epsilon_3 + c_{44}\epsilon_4 + c_{45}\epsilon_5 + c_{46}\epsilon_6;\\
\sigma_5 &= c_{51}\epsilon_1 + c_{52}\epsilon_2 + c_{53}\epsilon_3 + c_{54}\epsilon_4 + c_{55}\epsilon_5 + c_{56}\epsilon_6;\\
\sigma_6 &= c_{61}\epsilon_1 + c_{62}\epsilon_2 + c_{63}\epsilon_3 + c_{64}\epsilon_4 + c_{65}\epsilon_5 + c_{66}\epsilon_6;
\end{aligned}$$

or, in general (matrix) form,

$$\sigma_n = c_{nm}\epsilon_m, \quad n, m = 1, 2, 3, 4, 5, 6 \qquad \text{(I.13b)}$$

where summation over the repeated (dummy) index (the row index) is implied. In tensor form, when two indices must be retained for the components of the stresses and strains (as, for example, in the equation of motion (I.11)), the generalized Hooke's law will have the form:

$$\sigma_{ik} = c_{iklj}\epsilon_{ij}, \qquad \text{(I.13c)}$$

The coefficients of proportionality c_{nm} are called the *linear elastic moduli* or *stiffness constants*. Their dimensions are the same as the dimensions of stress; the 36 quantities c_{nm} form a tensor of rank 4, called the *elastic modulus tensor*. In the theory of elasticity, it is shown[6,7] that this tensor is symmetric, i.e., $c_{nm} = c_{nm}$ $(c_{ikjl} = c_{jlki})$, so that it contains 21 independent constants and has the form

$$c_{nm} = \begin{bmatrix} c_{11} & c_{12} & c_{13} & c_{14} & c_{15} & c_{16} \\ c_{12} & c_{22} & c_{23} & c_{24} & c_{25} & c_{26} \\ c_{13} & c_{32} & c_{33} & c_{34} & c_{35} & c_{36} \\ c_{14} & c_{42} & c_{43} & c_{44} & c_{45} & c_{46} \\ c_{15} & c_{52} & c_{53} & c_{54} & c_{55} & c_{56} \\ c_{16} & c_{62} & c_{63} & c_{64} & c_{65} & c_{66} \end{bmatrix}.$$

In this form, the tensor c_{nm} describes the elasticity of a medium without symmetry. The existence of symmetry reduces the total number of nonzero elastic moduli and the number of independent moduli. Table 1 shows the matrices of the elastic moduli for different crystallographic systems. As is evident from this table, the elastic properties of crystals, for example, crystals of the hexagonal system, are described by only five independent elastic moduli, and for crystals

Table 1

Elastic Moduli Tensors for Different Groups of Crystals

Group	System	Class	No. in-depend. moduli	Matrix	Example
1	2	3	4	5	6
I	Triclinic	C_1, S_2	21	$\begin{matrix} c_{11} & c_{12} & c_{13} & c_{14} & c_{15} & c_{16} \\ & c_{22} & c_{23} & c_{24} & c_{25} & c_{26} \\ & & c_{33} & c_{34} & c_{35} & c_{36} \\ & & & c_{44} & c_{45} & c_{46} \\ & & & & c_{55} & c_{56} \\ & & & & & c_{66} \end{matrix}$	Copper sulfate
II	Mono-clinic	C_2, C_{2h}, C_6	13	$\begin{matrix} c_{11} & c_{12} & c_{13} & 0 & 0 & c_{16} \\ & c_{22} & c_{23} & 0 & 0 & c_{26} \\ & & c_{33} & 0 & 0 & c_{36} \\ & & & c_{44} & c_{45} & 0 \\ & & & & c_{55} & 0 \\ & & & & & c_{66} \end{matrix}$	Gypsum
III	Rhombic	$D_2 = V, C_{2v}, D_{2h}$	9	$\begin{matrix} c_{11} & c_{12} & c_{13} & 0 & 0 & 0 \\ & c_{22} & c_{23} & 0 & 0 & 0 \\ & & c_{33} & 0 & 0 & 0 \\ & & & c_{44} & 0 & 0 \\ & & & & c_{55} & 0 \\ & & & & & c_{66} \end{matrix}$	Rochelle salt
IV	Tetra-gonal	C_4, C_{4h}, C_{4v}	7	$\begin{matrix} c_{11} & c_{12} & c_{13} & 0 & 0 & c_{16} \\ & c_{11} & c_{13} & 0 & 0 & -c_{16} \\ & & c_{33} & 0 & 0 & 0 \\ & & & c_{44} & 0 & 0 \\ & & & & c_{44} & 0 \\ & & & & & c_{66} \end{matrix}$	Scheelite
V	"	S_4, D_{2d}, D_4, D_{4h}	6	$\begin{matrix} c_{11} & c_{12} & c_{13} & 0 & 0 & 0 \\ & c_{11} & c_{13} & 0 & 0 & 0 \\ & & c_{33} & 0 & 0 & 0 \\ & & & c_{44} & 0 & 0 \\ & & & & c_{44} & 0 \\ & & & & & c_{66} \end{matrix}$	Ammonium dihydrogen phosphate

Continuation of Table 1

1	2	3	4	5	6
VI	Trigonal	C_3, C_{3i}	7	$\begin{matrix} c_{11} & c_{12} & c_{13} & c_{14} & -c_{25} & 0 \\ & c_{11} & c_{13} & -c_{14} & c_{25} & 0 \\ & & c_{33} & 0 & 0 & 0 \\ & & & c_{44} & 0 & -c_{25} \\ & & & & c_{44} & c_{14} \\ & & & & & \frac{1}{2}(c_{11}-c_{12}) \end{matrix}$	Dolomite
VII	"	D_3, D_{3v}, D_{3d}	6	$\begin{matrix} c_{11} & c_{12} & c_{13} & c_{14} & 0 & 0 \\ & c_{11} & c_{13} & -c_{14} & 0 & 0 \\ & & c_{33} & 0 & 0 & 0 \\ & & & c_{44} & 0 & 0 \\ & & & & c_{44} & c_{14} \\ & & & & & \frac{1}{2}(c_{11}-c_{12}) \end{matrix}$	α-Quartz tourmaline
VIII	Hexa-gonal	C_{3h}, D_{3h}, C_6, D_6, C_{6h}, C_{6v}, D_{6h}	5	$\begin{matrix} c_{11} & c_{12} & c_{13} & 0 & 0 & 0 \\ & c_{11} & c_{13} & 0 & 0 & 0 \\ & & c_{33} & 0 & 0 & 0 \\ & & & c_{44} & 0 & 0 \\ & & & & c_{44} & 0 \\ & & & & & \frac{1}{2}(c_{11}-c_{12}) \end{matrix}$	β-Quartz cadmium sulfide
IX	Cubic	T, O, T_h, T_d, O_h	3	$\begin{matrix} c_{11} & c_{12} & c_{12} & 0 & 0 & 0 \\ & c_{11} & c_{12} & 0 & 0 & 0 \\ & & c_{11} & 0 & 0 & 0 \\ & & & c_{44} & 0 & 0 \\ & & & & c_{44} & 0 \\ & & & & & c_{44} \end{matrix}$	Alkali-halide crystals

with cubic symmetry there are only three independent moduli. It should be kept in mind here that the tables of elastic moduli presented refer to a completely determined position of the coordinate axes relative to the crystallographic axes. In an isotropic body, the elastic moduli cannot, of course, depend on the orientation of the coordinate axes, which leads to the conditions[8]

$$c_{12}=c_{13}=c_{23},\quad c_{44}=c_{55}=c_{66}=(c_{11}-c_{12})/2,\quad c_{11}=c_{22}=c_{33}, \qquad (I.14)$$

and the remaining moduli equal zero. This means that the elasticity of isotropic solids is determined by two independent moduli, for which it is customary to use the *Lamé constants* λ and μ, defined as follows: $\lambda = c_{12} = c_{13} = c_{23}$; $\mu = c_{44} = c_{55} = c_{66}$. Then, according to (I.14), $c_{11} = c_{22} = c_{33} = \lambda + 2\mu$. If the components of stresses and strains are represented by **two indices**, then Hooke's law (I.13c) for an isotropic solid can be written in the form

$$\sigma_{ik} = \lambda\Theta\delta_{ik} + 2\mu\epsilon_{ik}, \quad i, k = 1, 2, 3, \qquad \text{(I.15)}$$

where $\Theta = \epsilon_{11} + \epsilon_{22} + \epsilon_{33}$ is the volume expansion.

It should also be noted that the magnitudes of the elastic moduli c_{nm} depend on whether they are defined for adiabatic or isothermal deformation processes. In this connection, *adiabatic* and *isothermal* values of the elastic moduli are distinguished. Since the propagation of ultrasound is a nearly adiabatic process, in what follows we shall have in mind the adiabatic values of the moduli.

Since Eqs. (I.13) are linear homogeneous equations, they can be solved for the components of the strain ϵ_m. This gives a system of equations $\epsilon_m = k_{mn}\sigma_n$, relating the strains to the stresses. The coefficients of proportionality k_{mn} can be called the *elastic susceptibilities* or *elastic compliances*. They likewise form a tensor of rank four, for which the same remarks hold true as for the elastic modulus tensor. The dimensions of the compliance coefficients are the inverse of the dimensions of mechanical stress.

§ 5. Energy of elastic deformation

Let us calculate the energy of an elastically deformed body. Let the displacement vector **u** due to the deformation of the body vary by a small amount du_i. The elementary work performed in this case by the internal stresses is the product of the force $F_i = \partial\sigma_{ik}/\partial x_k$ and the displacement du_i, integrated over the entire volume of the body V: $dA = \int_V (\partial\sigma_{ik}/\partial x_k)(du_i)dV$. Integrating by parts, we obtain

$$dA = \oint \sigma_{ik}(du_i)dS - \int_V \sigma_{ik}\frac{\partial}{\partial x_k}(du_i)dV.$$

The first (surface) integral vanishes for an unbounded medium which is not deformed at infinity, since $\sigma_{ik} = 0$. The second integral, by virtue of the fact that $(\partial/\partial x_k)(du_i) = d(\partial u_i/\partial x_k)$, can be written in the form $\int \sigma_{ik} d(\partial u_i/\partial x_k)dV$. The integrand here represents the work performed by internal stresses per unit volume of the body:

$$dA' = -\sigma_{ik} d\left(\frac{\partial u_i}{\partial x_k}\right), \tag{I.16}$$

In the case of a linear-elastic deformation, taking into account the symmetry of the stress tensor σ_{ik}, we have

$$\sigma_{ik}\, d\left(\frac{\partial u_i}{\partial x_k}\right) = \sigma_{ik} d\left[\frac{1}{2}\left(\frac{\partial u_i}{\partial x_k} + \frac{\partial u_k}{\partial x_i}\right)\right] = \sigma_{ik}\, d\epsilon_{ik},$$

where ϵ_{ik} is the strain tensor. We thus obtain the following expression for the work performed by the internal stresses:

$$dA' = -\sigma_{ik} d\epsilon_{ik}. \tag{1.17}$$

For a reversible adiabatic process, this work equals minus the change in the internal energy of the body (per unit volume), i.e.,

$$dU = -dA' = \sigma_{ik} d\epsilon_{ik}. \tag{I.18}$$

From here there follows, in particular, the definition of the stress tensor in terms of the internal energy:

$$\sigma_{ik} = (\partial U/\partial\epsilon_{ik})_{\text{ad}} \tag{I.19}$$

or, in more general form, from (I.16),

$$\sigma_{ik} = \left[\frac{\partial U}{\partial(\partial u_i/\partial x_k)}\right]_{\text{ad}}. \tag{I.20}$$

Substituting into Eq. (I.18) the stresses σ_{ik} from Hooke's law (I.13c), we have $dU = c_{ikjl}\epsilon_{jl}d\epsilon_{ik}$, which after integration gives $U = c_{ikjl}\epsilon_{ik}\epsilon_{jl}/2$. This equation expresses the potential energy of an elastically deformed body in the linear approximation. The strains appear in it quadratically; for this reason, the linear elastic moduli c_{ikjl}(or c_{nm} in the two-index notation) appearing in it are called *second-order elastic moduli*. For an isotropic solid body described by by two linear elastic moduli, an expression for the internal energy can be obtained by expanding U in powers of small strains ϵ_{ik}. In so doing, it is necessary to include the fact that in the undeformed state, i.e., for $\epsilon_{ik} = 0$, the stresses must vanish, i.e., $\sigma_{ik} = 0$. Since, however, $\sigma_{ik} = \partial U/\partial\epsilon_{ik}$, it follows from this that the linear terms must drop out of the expansion of U in powers of ϵ_{ik}. In what follows, we shall be interested only in the excess energy, so that the constant term in the expansion can also be set equal to zero. As far as the quadratic (and higher-order) terms are concerned, expressions can be obtained for them based on the fact that since the internal energy is a scalar, each term in the expansion of U must also be a scalar. Two independent quadratic scalars can be formed from the components of the symmetric linearized tensor ϵ_{ik}: the square of the trace $(\epsilon_{ii})^2 = \Theta^2$ and the sum of the squares of all components ϵ_{ik}^2.[6] Thus, having expanded the internal energy in powers of ϵ_{ik}, we obtain up to quadratic terms

$$U = \lambda\Theta^2/2 + \mu\epsilon_{ik}^2, \tag{I.21}$$

where λ and μ are the Lamé constants introduced above. Differentiating this expression with respect to ϵ_{ik}, according to definition (I.19), we can find a relation between the stresses and strains which leads to Hooke's law for an isotropic solid in the form (I.15).

In general, the relation between the stresses and strains is not linear. To take this nonlinearity into account, the exact expression for the strain tensor (I.5) must be used and terms with higher powers of the strain must be retained in relations of the type (I.13). We shall examine the consequences of including the nonlinearity of elasticity in the theory of propagation of ultrasonic waves in detail below (Chaps. IV–V) for longitudinal waves in a medium described by a single elastic modulus, and we shall then briefly consider the nonlinearity of solids in Chap. X.

§6. The simplest deformations and the relation between the different elastic moduli

In accordance with Eq. (I.15), the equations of elasticity for an isotropic medium can be written in the form:

$$\sigma_{11} = (\lambda + 2\mu)\epsilon_{11} + \lambda\epsilon_{22} + \lambda\epsilon_{33} = \lambda\Theta + 2\mu\epsilon_{11};$$

$$\sigma_{22} = \lambda\Theta + 2\mu\epsilon_{22}; \quad \sigma_{33} = \lambda\Theta + 2\mu\epsilon_{33}; \tag{I.22}$$

$$\sigma_{32} = \sigma_{23} = 2\mu\epsilon_{32}; \quad \sigma_{13} = \sigma_{31} = 2\mu\epsilon_{13}; \quad \sigma_{12} = \sigma_{21} = 2\mu\epsilon_{12}.$$

These equations can be solved for the components of the strain, which gives

$$\epsilon_{11} = \frac{2(\lambda + \mu)\sigma_{11} - \lambda\sigma_{22} - \lambda\sigma_{33}}{2\mu(3\lambda + 2\mu)};$$

$$\epsilon_{22} = \frac{-\lambda\sigma_{11} + 2(\lambda + \mu)\sigma_{22} - \lambda\sigma_{33}}{2\mu(3\lambda + 2\mu)}; \tag{I.23}$$

$$\epsilon_{33} = \frac{-\lambda\sigma_{11} - \lambda\sigma_{22} + 2(\lambda + \mu)\sigma_{33}}{2\mu(3\lambda + 2\mu)}.$$

Analyzing Eqs. (I.22) and (I.23) we can single out several of the simplest cases of deformations which, in the dynamic regime, can propagate in an isotropic body in the form of the corresponding elastic waves.

One-dimensional stress (extension of a rod). Let $\sigma_{11} = \sigma_{xx} = \sigma$ be the only nonvanishing components of the stress tensor. From Eq. (I.23), for this case, we have:

$$\epsilon_{11} = \frac{(\lambda + \mu)\sigma}{\mu(3\lambda + 2\mu)}; \quad \epsilon_{22} = \epsilon_{33} = -\frac{\lambda\sigma}{2\mu(3\lambda + 2\mu)}. \tag{I.24}$$

Thus a positive normal stress acting along the x-axis gives rise to extension in this direction and isotropic compression in the transverse directions (all elastic moduli, including the Lamé constants, are positive). Since strains along the y and z axes in a continuous medium must be accompanied by corresponding stresses, the initial condition of one-dimensional stress can be satisfied only in the presence of free lateral surfaces. Therefore, the case examined is realized for extension of a rod oriented along the x-axis.

The coefficient in front of the stress in the first of equations (I.24), by definition, represents the coefficient of extension of the rod being stretched and the inverse quantity is the effective modulus of elasticity, which in this case is called *Young's modulus*:

$$E = \frac{(3\lambda + 2\mu)\mu}{\lambda + \mu}. \tag{I.25}$$

Substituting (I.25), the first equation of (I.24) assumes the form $\epsilon_{11} = \sigma/E$. Thus Young's modulus characterizes the rigidity of the rod relative to longitudinal extension (compression) and determines the mechanical stress with which the magnitude of the strain must equal 1, i.e., the length of the rod changes by a factor of two (of course, with Hooke's law remaining valid). The values of Young's modulus E for some isotropic bodies are presented in Table 2.

Table 2

Values of Young's Modulus, Poisson's Ratio, and the Shear Modulus for Some Isotropic Solids

Material	$E \cdot 10^{-10}$, N/m^2	ν_0	$G \cdot 10^{-10}$, N/m^2
Tungsten	36.0	0.27	13.3
Steel 3	22–24	0.30	8.5–8.8
Iron	21	0.28	8.2
Copper	12.0	0.35	4.6
Brass	9–10	0.35	3.0–3.7
Gold	8.0	0.41	2.9
Aluminum	7.0	0.34	2.6
Tin	5.4	0.33	2.0
Lead	1.6	0.44	0.6
Fused quartz	7.4	0.18	3.2
Glass crown	7.2	0.25	2.9
flint	5.5	0.23	2.4
Porcelain	6.0	0.23	2.4
Ice	1.0	0.33	0.4
Plexiglass	0.5	0.35	0.15

The absolute ratio of the transverse deformation of the rod to the longitudinal deformation, i.e., the relative elongation due to the longitudinal stress, is called *Poisson's ratio* (ν_0):

$$\nu_0 = \left|\frac{\epsilon_{22}}{\epsilon_{11}}\right| = \left|\frac{\epsilon_{33}}{\epsilon_{11}}\right| = \epsilon_{22}\frac{E}{\sigma} = \frac{\lambda}{2(\lambda + \mu)}. \quad \text{(I.26)}$$

Thus $\epsilon_{22} = \epsilon_{33} = -\nu_0\sigma/E$, i.e., the transverse contraction of the rod accompanying longitudinal stretching is characterized by the stiffness E/ν_0. The values of ν_0 for different media fall in the range 0.2 – 0.5. Solving Eqs. (I.25) and (I.26) for the Lamé constants λ and μ, we find an expression for these coefficients in terms of E and ν_0:

$$\lambda = \nu_0 E[(1 + \nu_0)(1 - 2\nu_0)]^{-1}, \quad \text{(I.27)}$$

$$\mu = E[2(1 + \nu_0)]^{-1}. \quad \text{(I.28)}$$

One-dimensional stretching. Assume, on the contrary, that only a longitudinal stretching deformation, for example along the x-axis, is possible, i.e., $\epsilon_{11} = \epsilon_{xx} \neq 0$, and the remaining components of the strain tensor ϵ_{ik} equal zero. This situation is realized, in particular, in a longitudinal ultrasonic plane wave propagating in the bulk of an isotropic solid, which is unbounded along the y and z axes. In this case, in accordance with Hooke's law (I.22),

$$\sigma_{11} = (\lambda + 2\mu)\epsilon_{11}, \quad \sigma_{22} = \sigma_{33} = \lambda\epsilon_{11}. \quad \text{(I.29)}$$

Thus, in this case, transverse stresses arise on the boundary of the deformed part of the medium; the stiffness of the medium, on the other hand, with respect to its longitudinal stretching is characterized by the modulus

$$c_{11} = \lambda + 2\mu. \quad \text{(I.30)}$$

Expressing the Lamé constants in terms of E and ν_0, with the help of Eqs. (I.27) and (I.28) we obtain $c_{11} = E(1 - \nu_0)[(1 + \nu_0)(1 - \nu_0)]^{-1}$. It follows from here that for any real value of ν_0, the modulus $E < c_{11}$.

Physically, this means that the absence of transverse contraction inhibits the stretching of the medium, which corresponds to a high effective stiffness of the medium with one-dimensional stretching.

Pure shear. Assume, for example, that a shear (tangential) force $\sigma_{12} = \sigma_T$ acts in the xy plane; the remaining components of the stress tensor equal zero. In this case, from Eq. (I.22) we have: $\epsilon_{12} = \epsilon_{21} = \sigma_T/(2\mu)$. According to definition (I.2), the component of the strain tensor ϵ_{12} equals one-half the displacement angle in the xy plane: $\epsilon_{12} = \varphi_{12}/2$. Therefore, the total displacement angle in this plane $\varphi = \sigma_T/\mu = \sigma_T/G$.

Thus the Lame constant μ is a *shear modulus* G, which determines the magnitude of the displacement angle φ for a fixed tangential stress σ_T. The relation of this modulus to Young's modulus E and Poisson's ratio ν_0 is given by relation (I.28), from which it follows that the shear modulus is a factor of 2.5 – 3 smaller than Young's modulus. The numerical values of the shear modulus for different isotropic solids are also presented in Table 2.

Hydrostatic compression. Assume that identical compressive stresses $-\sigma_{11} = -\sigma_{22} = -\sigma_{33} = P$, act on a volume element in the form of a cube, whose edges are oriented parallel to the coordinate axes x, y, z, and that there are no tangential stresses. In this case, Eqs. (I.22) assume the following form:

$$-P = \lambda\Theta + 2\mu\epsilon_{11}, \quad -P = \lambda\Theta + 2\mu\epsilon_{22},$$

$$-P = \lambda\Theta + 2\mu\epsilon_{33}, \quad \epsilon_{12} = \epsilon_{23} = \epsilon_{13} = 0.$$

Adding these equations, we obtain

$$P = -(\lambda + \frac{2}{3}\mu)\Theta. \qquad \text{(I.31)}$$

Expression (I.31) represents Hooke's law for hydrostatic compression. The quantity

$$K = \lambda + (2/3)\mu \qquad \text{(I.32)}$$

is called the *modulus of hydrostatic compression* or the *bulk modulus of elasticity*. Substituting into (I.32) the expressions for λ and μ (I.27) and (I.28), we obtain a relation between the bulk modulus of elasticity

K and Young's modulus E and the Poisson ratio ν_0: $K = E[3(1 - 2\nu_0]^{-1}$. It follows from here, in particular, that the limiting value of ν_0 for an incompressible medium ($K = \infty$) is 0.5. Comparing expressions (I.32) and (I.30), we find another relation between the moduli c_{11} and K:

$$c_{11} = K + (4/3)\mu. \tag{I.33}$$

A compressive (negative) stress P is called a *positive pressure*. Thus the sign of the pressure is defined in an inverse manner: the pressure is positive when it is oriented into the volume under examination. A positive pressure, according to (I.31), corresponds to negative volume expansion Θ $(K > 0)$. If the density of the medium within the element under examination with volume V_0 is $\rho_0 = m/V_0$ before deformation and after deformation it increases to the value $\rho = \rho_0 + \Delta\rho$, then the relative change in density $\Delta\rho/\rho_0 = -\Delta V/V_0$, where $\Delta V = V - V_0$ and V is the volume of the deformed element. The quantity $s = \Delta\rho/\rho$ is called the *condensation*. Hooke's law for the condensation can be represented in the form

$$s = P/K. \tag{I.34}$$

A medium without shear elasticity. Ideal fluidity of a medium (ideal liquid or gas) corresponds to the absence of shear elasticity ("elasticity of shape"), i.e., the shear modulus $G = 0$ for such a medium. Therefore, the elasticity of an ideally flowing isotropic medium is characterized by only one elastic constant λ, which in this case, according to expression (I.32), equals the modulus of hydrostatic compression K. Due to the absence of shear stresses, only a normal stress (pressure), which is a scalar, acts on any surface element singled out within the flowing medium. Accordingly, any volume element in such a medium is subjected only to hydrostatic compression. In the following chapters we shall examine the propagation of ultrasonic waves precisely in such media, proceeding from there to media characterized by a larger number of linear elastic moduli.

II. Propagation of Ultrasonic Waves in Liquids and Gases

§ 1. Acoustic properties of ideal liquids

In subsequent chapters we shall examine the propagation of ultrasonic waves in an unbounded medium that exhibits only volume elasticity and not elasticity of shape and viscosity, i.e., it is an ideal liquid. According to §I.6, in such a medium, to which we ascribe the properties of an ideal compressible liquid, only elastic hydrostatic compression deformations are possible and, therefore, only one type of elastic wave -- a compression (rarefaction) wave -- can propagate in it. This considerably simplifies the analysis of perturbations and at the same time enables obtaining the basic acoustic relations for the most common types of waves which can exist in both liquids (and gases) and solids. In solids, as we have seen, other elastic deformations, corresponding to different types of waves which will be examined below, can also exist. However, the relations that we shall obtain for compression waves in an ideal liquid will also be valid for other waves, so that their basic features are common to different types of waves in different media. Real liquids have some elasticity of shape. Such elasticity is significant only at very high deformation rates, greatly exceeding the velocity of ultrasonic waves at the highest frequencies with which they can propagate in the liquid without significant attenuation. This provides justification for considering the deformation rate in an ultrasonic wave to be low enough that the shear elasticity of real liquids can be completely neglected.

In a viscous liquid, the formation of shear waves is also possible. As will be shown below, however, such waves are attenuated over negligibly small distances from the source, so that they can be neglected. The presence of viscosity in a real liquid, as well as other mechanisms responsible for the loss of energy of elastic oscillations, also lead to the attenuation of ultrasonic compression waves as they propagate in such a dissipative medium. This attenuation is not, however, as high as that of viscous waves, and at the first stage it can also be neglected and included later in the results obtained for an ideal liquid.

Due to the absence of shear stresses in an ideal liquid, the stresses (pressure) existing in it always act perpendicular to any surface area singled out in the liquid, and the force of the pressure applied to the volume element passes through its center of mass, producing only translational motion of the particles. Thus the motion of particles in an ideal liquid must be irrotational, which is expressed mathematically by the condition

$$\text{curl } \mathbf{v} = 0, \tag{II.1}$$

where $\mathbf{v}$ is the displacement velocity of the particles.

We shall take the density ρ, the pressure P, and the displacement of particles from their equilibrium position $\mathbf{u}$ or their displacement rate $\mathbf{v} = d\mathbf{u}/dt$ as the basic acoustic parameters of the liquid. We shall also assume that each quantity named above consists of a constant component and a finite increment, which varies in an acoustic wave, i.e., it depends on the coordinates and time thus:

$$\begin{aligned} P &= P_0 + p(x, y, z, t), \\ \rho &= \rho_0 + \Delta\rho(x, y, z, t), \\ \mathbf{V} &= \mathbf{v}_0 + \mathbf{v}(x, y, z, t), \end{aligned} \tag{II.2}$$

where P_0 is the static pressure (for example, atmospheric pressure in a gas or internal pressure in a liquid); ρ_0 is the density of the unperturbed medium, corresponding to the pressure P_0; $\mathbf{v}_0$ is the velocity of steady flow, which, below, we shall assume equals zero, setting $\mathrm{V} = \mathbf{v}$. Since in what follows $P_0 = const$ and $\rho_0 = const$, we have $dP = dp$ and $d\rho = d(\Delta\rho)$, so that in differentiating the quantities P, ρ, and $\mathbf{v}$ their total and variable values need not be distinguished. When compression waves propagate in the medium, the temperature also oscillates, suggesting that the temperature of the medium T should be introduced as a fourth variable acoustic parameter. However, assuming that the propagation of ultrasonic waves is an adiabatic process, using the corresponding adiabatic values of the elastic moduli, and neglecting losses due to the finite thermal conductivity of the medium, the temperature need not be introduced as an independent parameter. The increment to the temperature in an acoustic wave ΔT can be found with the help of the well-known thermodynamic relation for an adiabatic compression process:

$$\frac{T_0 + \Delta T}{T_0} = \left(\frac{p}{p_0}\right)^{\frac{\gamma - 1}{\gamma}}, \qquad \text{(II.3)}$$

where T_0 is the equilibrium temperature and $\gamma = c_p/c_V$ is the ratio of the specific heat capacities.

Thus any problem in the acoustics of an ideal liquid reduces to finding the parameters p, ρ, and $\mathbf{v}$ as functions of the coordinates and time. These parameters are related by the equations of motion, continuity, and elasticity, presented in Chap. I for the general case of anisotropic media exhibiting elasticity of shape. In the particular form applicable to flowing media, these equations form the system of equations of hydrodynamics (written in Euler's notation) which comprise the basic system of acoustic equations for liquids and gases.

§2. The equations of hydrodynamics

Equation of motion. Substituting the stress tensor $\sigma_{ik} = -p\delta_{ik}$ into Eq. (I.9), we obtain the equation of motion of an ideal liquid (gas) in the form

$$-\frac{\partial p}{\partial x_k}\delta_{ik} = \rho\frac{dv_i}{dt} = \rho\left(\frac{\partial v_i}{\partial t} + \frac{\partial v_i}{\partial x_k}v_k\right) \qquad \text{(II.4)}$$

or in vector form

$$-\nabla p = \rho\frac{d\mathbf{v}}{dt} = \rho\frac{\partial \mathbf{v}}{\partial t} + (\mathbf{v}\cdot\nabla)\mathbf{v},$$

where ρ is the total (instantaneous) density: $\rho = \rho_0 + \Delta\rho$; the minus sign corresponds to the definition of positive pressure.

These equations can be linearized by neglecting the convective derivatives and setting $\rho \simeq \rho_0$ ($\Delta\rho/\rho_0 \ll 1$); we then have, by analogy with (I.10),

$$-\nabla p = \rho_0 \frac{\partial \mathbf{v}}{\partial t}. \qquad \text{(II.5)}$$

In this linearized form the equation of motion can be considered to be exact only for hypothetical infinitesimal perturbations.

Velocity potential. The irrotational nature of the motion of an ideal liquid enables us to introduce a more convenient scalar parameter instead of the velocity vector $\mathbf{v}$. The condition for there to be no particle rotation (II.1) in projections along the coordinate axes has the form

$$\frac{\partial v_x}{\partial y} - \frac{\partial v_y}{\partial x} = 0, \quad \frac{\partial v_y}{\partial z} - \frac{\partial v_z}{\partial y} = 0, \quad \frac{\partial v_z}{\partial x} - \frac{\partial v_x}{\partial z} = 0. \qquad \text{(II.6)}$$

This condition enables us to introduce a scalar function $\varphi(x, y, z, t)$ defined as

$$\mathbf{v} = -\nabla\varphi, \qquad \text{(II.7)}$$

for which the left sides of Eq. (II.6) indeed vanish. By analogy with the potential energy, whose derivatives with respect to the coordinates determine the magnitude of the acting force, the function $\varphi(x,y,z,t)$ is called the *velocity potential*. According to definition (II.7), the components of the velocity vector $\mathbf{v}$ are related to this function by the relations:

$$v_x = -\partial\varphi/\partial x, \quad v_y = -\partial\varphi/\partial y, \quad v_z = -\partial\varphi/\partial z. \qquad \text{(II.8)}$$

Replacing the velocity $\mathbf{v}$ in the equation of motion (II.5) by its expression in terms of the velocity potential (II.7), we obtain

$$\nabla p = \rho_0 \frac{\partial}{\partial t}(\nabla\varphi) = \nabla\left(\rho_0 \frac{\partial\varphi}{\partial t}\right),$$

whence follows another form of the equation of motion in terms of the velocity potential:

$$p = \rho_0 \partial\varphi/\partial t. \qquad \text{(II.9)}$$

Equation (II.9) also determines the variable pressure in terms of the velocity potential φ. The exact equation of motion with the velocity potential has the form: $p = \rho d\varphi/dt$.

The **equation of continuity** in the form (I.4a) or (I.4b), which is the mathematical expression of the continuity of the medium, has the same form for all media. Replacing in (I.4b) the volume expansion by the condensation $s = -\Theta$, we obtain the equation of continuity in the form

$$-\frac{1}{\rho}\frac{d\rho}{dt} = \text{div } \mathbf{v}. \qquad \text{(II.10)}$$

Since $\rho = \rho(x_i, t)$, the total derivative $d\rho/dt = \partial\rho/\partial t + v_i\partial\rho/\partial x_i$. The exact equation of continuity can therefore also be written in the form:

$$-\frac{\partial \rho}{\partial t} = \frac{\partial \rho}{\partial x_i}v_i + \rho\frac{\partial v_i}{\partial x_i} = \frac{\partial}{\partial x_i}(\rho v_i) \qquad \text{(II.11)}$$

or

$$-\partial\rho/\partial t = \nabla(\rho\mathbf{v}).$$

If $\Delta\rho \ll \rho_0$, then to a first approximation we can set $\rho \simeq \rho_0$ and $d\rho/dt \simeq \partial\rho/\partial t$. Then we obtain the linearized equation of continuity $-\partial\rho/\partial t = \rho_0 \text{ div } \mathbf{v}$, which is valid for infinitesimal perturbations.

Replacing in the last expression the velocity vector $\mathbf{v}$ by the velocity potential φ, according to its definition (II.7), we obtain another form of the linearized equation of continuity:

$$\frac{\partial \rho}{\partial t} = \rho_0 \text{div}(\nabla\varphi) = \rho_0\Delta\varphi, \qquad \text{(II.12)}$$

where Δ is the Laplacian operator, i.e., the sum of second derivatives with respect to the coordinates.

Equation of conservation of momentum. The equation of motion and the equation of continuity can be combined into a single equation, which is often used in hydrodynamic and acoustic calculations. Multiplying (II.11) by v_i and adding the i-th component of Eq. (II.4), we obtain

$$\frac{\partial}{\partial t}(\rho v_i) = -\frac{\partial}{\partial x_k}(\rho v_i v_k + p\delta_{ik}), \qquad \text{(II.13)}$$

where δ_{ik} is the unit tensor. Equation (II.13) expresses the law of conservation of momentum of a unit volume of an ideal medium in differential form. The significance of this equation becomes clear when it is integrated over some fixed region with volume V and surface area S. The integration gives

$$\frac{\partial}{\partial t}\int_V \rho v_i dV = -\oint_S (\rho v_i v_k + p\delta_{ik}) n_k dS, \qquad \text{(II.14)}$$

where **n** is the unit outer normal vector of the surface S. The left side of this equation expresses the change in the momentum of a fixed volume of the stationary space, and the right side gives the momentum flux through the surface enclosing this volume. Since the change in momentum determines the force acting on the surface of the volume element singled out, the surface integral on the right side of Eq. (II.14) determines the components of this force: $F_i = -\oint_S \Pi_{ik} n_k dS$, where the tensor

$$\Pi_{ik} = \rho v_i v_k + p\delta_{ik} \qquad \text{(II.15)}$$

can be called the *tensile stress tensor*. We shall need this particular form of Eqs. (II.13)–(II.15) below for calculating the pressure of ultrasonic radiation.

In linearized form the equation of conservation of momentum (II.13) assumes the form: $\partial(\rho_0 v_i)/\partial t = -\partial p/\partial x_i$.

§ 3. Equation of state for liquids and gases

We obtained the equation of elasticity for an ideal compressible liquid (gas) in §1.6 in the form of Hooke's law for hydrostatic compression (I.34):

$$p = Ks, \qquad \text{(II.16)}$$

where K is the linear bulk modulus of elasticity, $s = \Delta\rho/\rho_0$, and $p = P - P_0$ is the excess pressure. The linearized equation (II.16) is valid only for deformations which are so small that the relation between the pressure and density of the liquid (gas) can be assumed to be linear. In general, however, the dependence between the pressure and density is nonlinear. The exact equation of elasticity can then be written in terms of the function

$$P = P(\rho), \tag{II.17}$$

where $\rho = \rho_0 + \Delta\rho$. Since the inequality $\Delta\rho \ll \rho_0$ is satisfied in all practical cases encountered in acoustics, the function (II.17) can be expanded in a Taylor series in powers of the increment to the density $\Delta\rho$ about the value $\rho = \rho_0$ $(\Delta\rho = 0)$:

$$P = P_0 + \left(\frac{dP}{d\rho}\right)_{\rho=\rho_0} \Delta\rho + \frac{1}{2!}\left(\frac{d^2P}{d\rho^2}\right)_{\rho=\rho_0} \Delta\rho^2 + \ldots \tag{II.18}$$

or

$$P - P_0 = \rho_0\left(\frac{dP}{d\rho}\right)_{\rho=\rho_0} s + \frac{\rho_0^2}{2}\left(\frac{d^2P}{d\rho^2}\right)_{\rho=\rho_0} s^2 + \ldots, \tag{II.19}$$

where P_0 is the constant pressure corresponding to the density ρ_0. The coefficient in front of the first power of the condensation represents the linear bulk modulus K, introduced above, whose exact value is defined as the derivative:

$$K = \rho_0\left(\frac{dP}{d\rho}\right)_{\rho=\rho_0}. \tag{II.20}$$

Correspondingly, the coefficient in front of s^2

$$\rho_0^2\left(\frac{d^2P}{d\rho^2}\right)_{\rho=\rho_0} = B, \tag{II.21}$$

which also has the dimensions of pressure, can be called the *nonlinear bulk modulus of elasticity* ("third" order modulus).

For small condensations the quadratic terms in Eq. (II.18) and (II.19) can be neglected and we can set

$$P \simeq P_0 + \left(\frac{dP}{d\rho}\right)_{\rho=\rho_0} \Delta\rho, \text{ i.e. } P-P_0 = p \simeq \rho_0\left(\frac{dP}{d\rho}\right)_{\rho=\rho_0} s \simeq Ks. \tag{II.22}$$

This linearization of Eq. (II.17) is equivalent to replacing the derivative $dP/d\rho$ by the ratio of the finite pressure and density increments:

$$\frac{P-P_0}{\rho-\rho_0} \simeq \left(\frac{dP}{d\rho}\right)_{\rho=\rho_0} = \tan\alpha,$$

where α is the slope angle of the tangent to the curve $P(\rho)$ at the point $\rho = \rho_0$. (Fig. 8), and can be used for small deviations of the density $\Delta\rho$. Thus the linear equation (II.16) is formally exact only for infinitesimal perturbations.

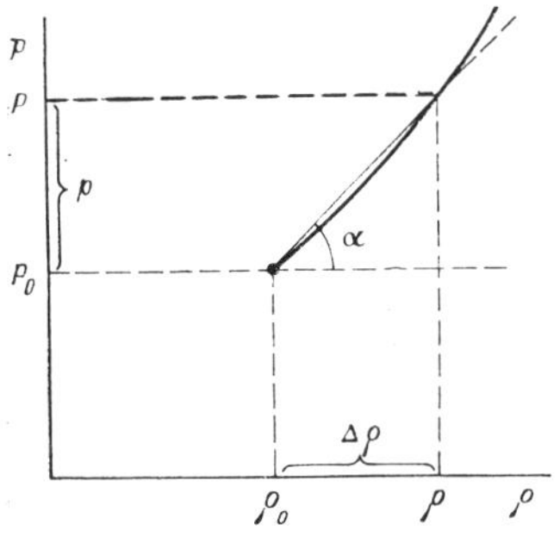

Fig. 8.

Equation (II.17) is the rheological *equation of state*. Its explicit form depends on the nature of the compression process and the properties of the medium itself. For **isothermal** compression of an ideal gas the relation between pressure and density is given by the Boyle–Mariotte law:

$$P/P_0 = \rho/\rho_0, \qquad \text{(II.23)}$$

i.e., it is linear. The bulk compression modulus (isothermal) in this case equals $K_{is} = \rho_0 (dP/d\rho)_{T\ =\ const} = P_0$, where P_0 is the static pressure.

As already noted, the propagation of sound (ultrasound) is a nearly **adiabatic** process. For this reason, Poisson's equation must be used as the rheological equation of state of the gas:

$$P/P_0 = (\rho/\rho_0)^{\nu}, \qquad \text{(II.24)}$$

where $\gamma = c_p/c_v$ is the ratio of the specific heat capacities at constant pressure and constant volume. This equation is nonlinear. We find the magnitude of the linear bulk modulus (the adiabatic modulus in this case) from relation (II.20):

$$K_{ad} = \rho_0 \left(\frac{dP}{d\rho}\right)_{\rho=\rho_0} = \nu P_0. \tag{II.25}$$

Therefore, the stiffness of the gas under adiabatic compression is a factor of γ larger than the stiffness under isothermal compression. Thus the linearized adiabatic equation of state of the gas assumes the following form:

$$P = \nu P_0 s. \tag{II.26}$$

For liquids the equation of state cannot be presented in an explicit form. By analogy with gases it can be written in the form

$$P/P_0 = (\rho/\rho_0)^n, \tag{II.27}$$

called *Tait's equation*, where n is an empirical parameter, characterizing the nonlinearity of the elasticity of this liquid.

The adiabatic equation of state of a liquid can be represented in the following linearized form:

$$P = K_{ad} s, \tag{II.28}$$

where $K_{ad} = \rho_0 (dP/d\rho)^{ad}_{\rho=\rho_0}$ is the adiabatic linear bulk modulus, which is related to the isothermal modulus K_{is} by the well-known thermodynamic relation:

$$K_{ad} = K_{is}\left(1 + \frac{\alpha_T^2 T K_{is}}{\rho_0 c_V}\right), \tag{II.29}$$

where T is the absolute temperature, and α_T is the coefficient of thermal expansion.

We shall use below the adiabatic equation of state, using the adiabatic value of the bulk modulus of elasticity and its inverse

$$\chi_{ad} = K_{ad}^{-1} = \rho_0^{-1} (d\rho/dP)^{ad}_{\rho\to 0},$$

called the compression coefficient or the *adiabatic compressibility* of the medium (dropping the index "ad"). The adiabatic modulus (or compressibility) characterizes the elasticity of the medium under dynamic compression, whereas the isothermal modulus characterizes the elasticity of the medium under static (infinitely slow) compression. For this reason, the adiabatic modulus (compressibility) is sometimes called the *dynamic* modulus (compressibility), whereas the isothermal modulus is called the *static* modulus. The adiabatic value of the linear bulk modulus of elasticity or compressibility of the medium is determined directly from measurements of the velocity of propagation of ultrasonic waves in the medium. Values of the adiabatic modulus K_{ad}, adiabatic compressibility κ_{ad}, internal pressure P_0 and the parameter n for several pure liquids are presented in Table 3. The nonlinearity of elasticity is manifested much more strongly in liquids (due to their close packing) than in gases, for which the value of the parameter $n = \gamma = c_p/c_V$ is approximately 1.3–1.5. Nevertheless, for the time being we shall neglect this nonlinearity, restricting our attention as before to small (in the limit, infinitesimal) deformations, for which the linear equation (II.28) is valid.

Table 3

Elastic Properties of Liquids at 20°C

Liquid	$P_0 \cdot 10^{-8}$, N/m²	$\kappa_{ad} \cdot 10^{10}$, m²/N	$K_{ad} \cdot 10^{-10}$, N/m²	n
Distilled water	3.2	4.5	0.22	7.6
Ethyl alcohol	1.0	9.1	0.11	10.6
Carbon tetrachloride	1.3	7.1	1.4	11.8
Benzene	1.9	6.3	0.16	9.4
Dichloroethane	2.3	7.5	0.13	8.7
Glycerin	4.8	2.0	0.5	10.4
Mercury	24.5	0.33	3.0	12.0

§ 4. The wave equation

The equation of motion $p = \rho_0 \partial\varphi/\partial\tau$, the equation of continuity $\Delta\varphi = (\partial\rho/\partial t)/\rho_0$, and the equation of state $p = Ks$ obtained above form a closed system of linear equations for the three acoustic variables p, ρ, and φ. Since the density ρ is a function of the pressure p, the derivative $\partial\rho/\partial t$ in the equation of continuity can be represented in the form

$$\partial\rho/\partial t = (\partial\rho/\partial p)_0 \partial p/\partial t. \qquad \text{(II.30)}$$

We shall find the derivative $\partial p/\partial t$ by differentiating the equation of motion:

$$\partial p/\partial t = \rho_0 \partial^2\varphi/\partial t^2. \qquad \text{(II.31)}$$

Substituting expression (II.30) together with (II.31) into the equation of continuity (II.12), we obtain

$$\Delta\varphi = (1/c_0^2)(\partial^2\varphi/\partial t^2). \qquad \text{(II.32)}$$

where the quantity

$$c_0^2 = (dp/d\rho)_{\rho=\rho_0} \qquad \text{(II.33)}$$

has the dimensions of velocity squared, so that c_0 is the velocity of propagation of an infinitesimal deformation ($\Delta\rho \to 0$) or the velocity of sound in the zeroth-order approximation. We find the value of c_0 by differentiating the equation of state (II.22):

$$c_0 = \sqrt{K/\rho_0} = \sqrt{1/\kappa\rho_0} \qquad \text{(II.34)}$$

where K and κ are the bulk modulus and the compressibility for an adiabatic process.

Equation (II.32) is called the *wave equation*. It has the same form for different types of perturbations and is the fundamental equation of acoustics. The physical interpretation of the wave equation follows from its derivation: $\Delta\varphi$ is the rate of volume compression of the medium; the density changes with time, according to (II.30), as a result of the change in pressure, which satisfies the equation of motion, i.e., Newton's second law (II.9); finally, the relation between the pressure and density is given by Hooke's law, i.e., the equation of state (II.22), which determines the velocity of propagation of condensation in an elastic medium.

The Laplacian of the velocity potential has the following forms in the general three-dimensional case. In rectangular (Cartesian) coordinates

$$\Delta\varphi = \frac{\partial^2\varphi}{\partial x^2} + \frac{\partial^2\varphi}{\partial y^2} + \frac{\partial^2\varphi}{\partial z^2}, \qquad \text{(II.35)}$$

in spherical coordinates (r, θ, ψ)

$$\Delta\varphi = \frac{1}{r^2}\frac{\partial}{\partial r}\left(r^2\frac{\partial\varphi}{\partial r}\right) + \frac{1}{r^2\sin\theta}\frac{\partial}{\partial\theta}\left(\sin\theta\frac{\partial\varphi}{\partial\theta}\right) + \frac{1}{r^2\sin^2\theta}\frac{\partial^2\varphi}{\partial\varphi^2} \qquad \text{(II.36)}$$

and, in cylindrical coordinates (r, θ, z)

$$\Delta\varphi = \frac{1}{r}\frac{\partial}{\partial r}\left(r\frac{\partial\varphi}{\partial r}\right) + \frac{1}{r^2}\frac{\partial^2\varphi}{\partial\theta^2} + \frac{\partial^2\varphi}{\partial z^2}.$$

If the velocity potential (and, together with it, the remaining acoustic parameters) depends on only one coordinate, then this corresponds to the one-dimensional case; if this coordinate is one of the Cartesian coordinates, then we have one-dimensional plane waves. Plane acoustic waves are realized in practice only at ultrasonic frequencies. In this respect, they comprise the well-known basic elements of ultrasonics, so that below we shall be mainly concerned with the problems of the propagation of ideal plane waves, taking into account later the limits of applicability of the results obtained in the field of a real planar radiator of ultrasound.

§ 5. **Plane waves**

In the one-dimensional planar case, when the potential depends on only one Cartesian coordinate and time $\varphi = \varphi(x, t)$, the wave equation (II.32) assumes the form

$$\frac{\partial^2\varphi}{\partial x^2} = \frac{1}{c_0^2}\frac{\partial^2\varphi}{\partial t^2}. \qquad \text{(II.37)}$$

Equation (II.37) is a second-order linear differential equation. Its solution can be obtained by substituting the variables ξ and η for the variables x and t (d'Alambert's method):

$$\xi = x - c_0 t, \quad \eta = x + c_0 t, \tag{II.38}$$

which are related to the previous variables x and t by the relations $x = (\xi + \eta)/2$ and $t = (\eta - \xi)/2$.

Assuming that the potential φ depends on x and t through the new variables ξ and η, i.e., $\varphi = \varphi(\xi, \eta)$, where $\xi = \xi(x,t)$ and $\eta = \eta(x,t)$ are given in the form (II.38), we find the derivatives entering into Eq. (II.37) from the rule for differentiating composite functions:

$$\frac{\partial^2 \gamma}{\partial x^2} = \frac{\partial}{\partial x}\left(\frac{\partial \varphi}{\partial x}\right) = \frac{\partial}{\partial x}\left(\frac{\partial \varphi}{\partial \xi}\frac{\partial \xi}{\partial x} + \frac{\partial \varphi}{\partial \eta}\frac{\partial \eta}{\partial x}\right) = \frac{\partial}{\partial x}\left(\frac{\partial \varphi}{\partial \xi} + \frac{\partial \varphi}{\partial \eta}\right) =$$

$$= \frac{\partial^2 \varphi}{\partial \xi^2} + 2\frac{\partial^2 \varphi}{\partial \xi \partial \eta} + \frac{\partial^2 \varphi}{\partial \eta^2};$$

and analogously

$$\frac{\partial^2 \varphi}{\partial t^2} = \frac{\partial}{\partial t}\left(\frac{\partial \varphi}{\partial t}\right) = c_0^2\left(\frac{\partial^2 \varphi}{\partial \xi^2} - 2\frac{\partial^2 \varphi}{\partial \xi \partial \eta} + \frac{\partial^2 \varphi}{\partial \eta^2}\right).$$

Substituting these results into (II.37), we obtain $4\partial^2\varphi/\partial\xi\partial\eta = 0$, i.e.,

$$\frac{\partial^2 \varphi}{\partial \xi \partial \eta} = \frac{\partial}{\partial \eta}\left(\frac{\partial \varphi}{\partial \xi}\right) = 0. \tag{II.39}$$

It follows from here that the derivative $\partial\varphi/\partial\xi$ is independent of η and is a function of the variable ξ only, i.e., $\partial\varphi/\partial\xi = f(\xi)$ Integrating this expression with respect to ξ, we obtain

$$\varphi = \int f(\xi)d\xi + f'(\eta). \tag{II.40}$$

The first term in the solution (II.40) is a function of ξ only, and the second term is a function of η only. Denoting these terms by $f_1(\xi)$ and $f_2(\eta)$, we obtain $\varphi = f_1(\xi) + f_2(\eta)$ or, in terms of the previous variables x and t:

$$\varphi = f_1(x - c_0 t) + f_2(x + c_0 t). \tag{II.41}$$

This solution describes two plane waves: a forward wave, i.e., a wave propagating along the positive x-axis with velocity c_0, and a backward wave, propagating in the opposite direction with the same velocity. Indeed, if the function f_1 initially (at $t = 0$) has the value $f_1(x_0)$ at the point $x = x_0$, then at time t the disturbance described by this function reaches the coordinate $x = x_0 + c_0t$. Since $x - c_0t = x_0$, $f_1(x - c_0t) = f_1(x_0)$, i.e., the initial value of the disturbance $f_1(x_0)$ propagates in the direction of the positive x-axis with velocity c_0. The same can be said for the backward wave also, which, for the time being, we shall neglect below, since it differs from the forward wave only because of the arbitrary choice of the positive orientation of the x-axis.

Thus the solution of the wave equation in the form (II.41) or in the form of a single function describing the forward wave

$$\varphi = f(x - c_0t), \tag{II.42}$$

is distinctive not because of the form of the function, but because of the form of the argument, which corresponds to a one-dimensional acoustic plane wave. The form of the wave, however, can be arbitrary in accordance with the arbitrary form of the function f, which, however, must satisfy the Dirchlet condition. In this case, the complex function $f(x - c_0t)$ can be expanded in a Fourier series, i.e., it can be represented as a sum of harmonic components. By virtue of the principle of superposition, however, which is valid for linear differential equations, which the wave equation (II.32) or (II.37) is, each of these components will be a particular solution of this equation, in exactly the same way as any sum of particular solutions, including also the sum of the forward and backward waves. Thus any complex perturbation can be represented as a superposition of harmonic oscillations, and the analysis of a complicated disturbance can be reduced to the analysis of the propagation of sinusoidal (monochromatic) waves, with which we shall be concerned in the following chapters.

§ 6. The velocity of sound

The quantity c_0, entering into the wave equation (II.37) and its solutions (II.41) or (II.42), represents the velocity with which elastic deformation waves, in this case compression (rarefaction) waves, propagate. The process of propagation of such waves is what constitutes the concept of sound (or ultrasound), so that c_0 is the *velocity of sound* (*ultrasound*). Its magnitude is determined by Eq. (II.34): $c_0 = \sqrt{(K/\rho_0)}$, which is exact only for infinitesimal perturbations (sound waves with infinitesimal amplitude). The inclusion of the nonlinearity of elasticity for real finite-amplitude waves leads to a correction to the magnitude of the velocity. As we shall see below, however, this correction is small, so that the velocity of sound is practically constant over a very wide range of amplitudes, as confirmed by direct experiments.[9,10]

Furthermore, since the velocity of sound c_0 is determined by the adiabatic modulus K, in an ideal medium it does not depend on frequency, i.e., there is no dispersion of sound. Dispersion does exist in a real medium because different relaxation processes are always present. However, it occurs in a comparatively narrow range of frequencies, forming a dispersion step, not exceeding several percent, on the curve of the velocity versus the frequency. For this reason, even in real media the quantity c_0 can be considered to be practically independent of frequency (at least, over a wide frequency range) and there is no need to distinguish between the velocity of sound and the velocity of ultrasound.

The velocity of sound in gases can be calculated using the expression for the adiabatic bulk modulus of the gas (II.25), i.e., from the equation

$$c_0 = \sqrt{\nu P_0/\rho_0}. \tag{II.43}$$

For atmospheric air at 0 °C ($P_0 = 1$ atm $= 1.016 \cdot 10^6$ dynes/cm^2, $\rho_0 = 1.29 \cdot 10^{-3}$ g/cm^3, $\gamma = 1.41$) a calculation gives $c_0 = 3.33 \cdot 10^4$ cm/s. The corresponding value for an isothermal process ($\gamma = 1$) under the same conditions is $c_0 = 1.8 \cdot 10^4$ cm/s. Experiment, on the other hand, gives the value $c_0 = 3.32 \cdot 10^4$ cm/s, which indicates the adiabatic nature of the propagation of sound. For other gases, the measured values of the velocity of sound likewise agree very well with the value computed from the molecular-kinetic theory. Using Clapeyron's equation $P_0/\rho_0 = R_0T$, where R_0 is the universal gas constant per gram of gas and T is the absolute temperature, from (I.43) we obtain: $c_0 = \sqrt{(\gamma R_0 T)}$, i.e., the velocity of sound in an ideal gas increases with temperature as $T^{1/2}$ (approximately by 60 cm/(s·deg) near room temperature) and is independent of the static pressure. The latter behavior is explained by the fact that the velocity of sound is determined by the ratio of the static pressure to the density, which under static compression, by virtue of the Boyle--Mariotte law, remains constant. The increase in c_0 with temperature, on the other hand, is related to the fact that because of momentum transport the elasticity of the gas increases with increasing temperature.

The velocity of sound in a liquid cannot be calculated with the same accuracy because a satisfactory model, which would enable the calculation of the magnitude of the bulk modulus theoretically, does not exist for liquids. For this reason, c_0 for liquids can be calculated from the experimental data or the isothermal modulus K_{is} (measured by static methods), which is related to the adiabatic modulus K by the relation (II.29), or directly from the adiabatic modulus, which, in its turn, is determined from acoustic measurements and the equation $K = \rho_0 c_0^2$. The value of c_0 for distilled water at 20 °C is $1.49 \cdot 10^3$ m/s. In other liquids at the same temperature the velocity of sound varies from $\simeq 0.9 \cdot 10^3$ m/s to $\simeq 2.0 \cdot 10^3$ m/s. In some liquid metals it attains $3 \cdot 10^3$ m/s. Values of the velocity of sound for a number of liquids and gases are presented in Table 4, which also shows their densities ρ_0 and the products of the density and the velocity of sound $\rho_0 c_0$, called the *specific characteristic impedance* (see below).

In contrast to gases, the velocity of sound in almost all liquids decreases monotonically and quite considerably (by 2 – 6 m/(s·deg)) with temperature. Water and some liquid metals (antimony, tellurium) are the only exceptions. The velocity of sound in water at low temperatures increases with a temperature coefficient $dc_0/dT \simeq 2.5$ m/(s·deg), attaining a maximum value of 1550 m/s at 67 °C, and then decreases as in normal liquids (Fig. 9). The well-known anomalous properties of water, which are related to its structural peculiarities, which in their turn cause the packing density of water molecules to increase with temperature, are manifested in this behavior of the velocity of sound velocity. Liquid antimony and tellurium, in which the velocity of sound exhibits an analogous temperature dependence,

Table 4

Acoustic Properties of Some Liquids and Gases at Normal Pressure

Substance	Chemical formula	T, °C	$\rho_0 \cdot 10^{-3}$ kg/m^2	c_0, m/s	$\rho_0 c_0 \cdot 10^{-4}$, kg/(m$^3 \cdot$ s)
1	2	3	4	5	6
Nitrogen	N_2	—197	0.815	869	71
		20	1.17	351	0.04
Aniline	C_3H_4O	20	1.022	1656	170
Argon	Ar	—189	1.424	863	123
Acetone	CH_3CHCH_3	20	0.792	1192	94
Benzene	C_6H_6	20	0.878	1326	116
Bromoform	$CHBr_3$	20	2.890	928	268
Bromobenzene	C_6H_5Br	50	1.454	1074	156
Water	H_2O	20	0.998	1490	150
Hydrogen	H_2	—252.7	0.355	1127	40
		20	0.10	1284	0.013
Air	...	20	1.29	343	0.045
Helium	He	—269.1	0.125	180	2.3
		0	0.18	965	0.017
Hexane	C_6H_{14}	20	0.654	1083	71
Glycerin	$C_3H_8O_3$	20	1.260	1923	242
Diacetyl	$C_4H_6O_2$	25	0.990	1236	122
Dioxane	$C_4H_8O_2$	20	1.033	1389	143
Dichloroethane	$C_2H_4Cl_2$	20	1.250	1240	156
Diethyl phthalate	$C_6H_4(C_3O_2H_5)_2$	25	1.121	1470	165
Isopentane	C_5H_{12}	0	0.641	950	61
Indium	In	156	7.033	2215	1558
Potassium	K	75	0.824	1882	155
Kerosene	...	34	0.825	1295	107
Xylene	C_8H_{10}	20	0.860	1330	114
Oxygen	O_2	—183.6	1.143	911	104
		20	1.33	328	0.044
Acid sulfuric	H_2SO_4	15	1.84	1440	257
formic	HCOOH	20	1.216	1287	156
acetic	CH_3COOH	20	1.050	1150	121
Oil spindle	...	25	0.866	1431	124
linoleic	...	31	0.922	1772	163
olive	...	32	0.904	1381	125
trans-former	...	25	0.865	1415	122
Nitrobenzene	$C_6H_5NO_2$	20	1.207	1473	178
Octane	C_8H_{18}	20	0.703	1197	84
Tin	Sn	230	6.96	2462	1720

Continuation of Table 4

1	2	3	4	5	6
par(acet)-aldehyde	$C_6H_{12}O_3$	20	0.994	1204	120
Pentane	C_5H_{12}	20	1.263	1158	146
Pyridine	$C_6H_{15}N$	20	0.982	1445	142
Mercury	Hg	20	13.59	1451	1972
Carbon disulfide	CS_2	20	1.263	1158	146
Alcohol-amyl	$C_5H_{11}OH$	20	0.816	1294	106
benzyl	C_7H_7OH	20	1.045	1540	161
butyl	C_4H_9OH	20	0.810	1268	103
methyl	CH_3OH	20	0.792	1123	89
propyl	C_3H_7OH	20	0.804	1223	98
ethyl	C_2H_5OH	20	0.789	1180	93
Toluene	C_7H_8	20	0.866	1328	115
Carbon dioxide	CO_2	20	1.85	268	0.052
Acetic anhydride	$(CH_3CO)_2O$	24	1.075	1384	149
Formamide	$HCONH_2$	20	1.139	1550	177
Chlorobenzene	C_6H_5Cl	20	1.107	1291	143

apparently also have similar structural anomalies. In some liquid metals structural rearrangement leads to a quite complicated temperature dependence, which is illustrated in Fig. 10 for liquid bismuth.[11]

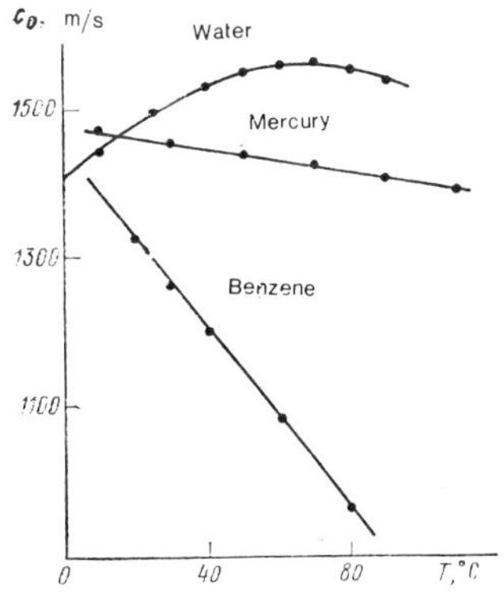

Fig. 9.

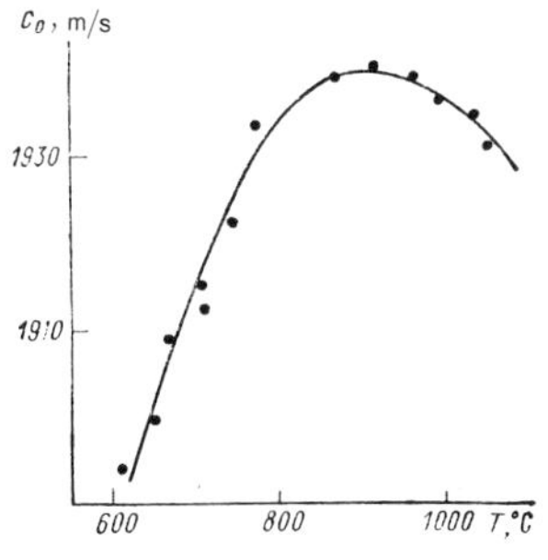

Fig. 10

The velocity of sound in all liquids increases nearly linearly with the static pressure up to pressures of several thousands of atmospheres. As an example, Fig. 11 shows the pressure dependence of the velocity

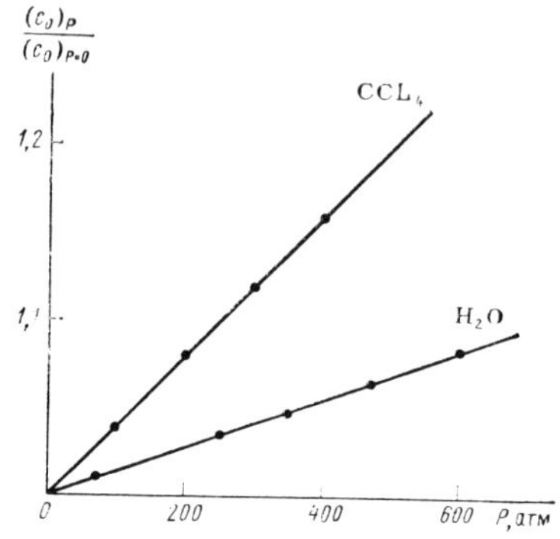

Fig. 11.

of sound for water and carbon tetrachloride at 20 °C. It follows from these data that the velocity of sound in water increases approximately by 0.1 m/(s · atm); for carbon tetrachloride and other organic liquids, the coefficient dc_0/dp is of the order of 0.3 – 0.4 m/(s · atm).[12]

III. Sinusoidal Plane Waves with Infinitesimal Amplitude

§ 1. Equation of a monochromatic plane wave

Following § II.5, we shall now study the most interesting case, when the source of ultrasonic waves oscillates harmonically with frequency ω. The velocity potential $\varphi(x,t)$ can then be represented in the form

$$\varphi(x, t) = \Psi(x) \sin(\omega t + \beta), \tag{III.1}$$

where β is an arbitrary initial phase of the oscillations. The equivalent complex form of Eq. (III.1) is often used:

$$\varphi(x, t) = \mathrm{Re}[\Psi(x) e^{i(\omega t+\beta)}] = \mathrm{Re}[\tilde{\Psi}(x) e^{i\omega t}] \tag{III.2}$$

or

$$\varphi(x, t) = \mathrm{Im}[\tilde{\Psi}(x) e^{i\omega t}], \tag{III.3}$$

where, in order to take into account the initial phase β, the modulus $\tilde{\Psi}(x)$ is in general complex: $\tilde{\Psi}(x) = \Psi(x) e^{i\beta}$. Equations (III.1)–(III.3) are entirely equivalent, since the real and imaginary parts of the function $\exp(ia)$ ($\sin a$ and $\cos a$) differ only by a constant initial phase, which is included in the modulus $\tilde{\Psi}(x)$ and is not important when examining one oscillation only. The complex notation, on the other hand, greatly simplifies the calculations, so that we shall employ it often, omitting, as is customary, the symbol Re (or Im).

Substituting (III.2) or (III.1) into the wave equation (II.37), we obtain

$$\frac{\partial^2 \tilde{\Psi}(x)}{\partial x^2} + k^2 \tilde{\Psi}(x) = 0, \tag{III.4}$$

where $k = \omega/c_0$ is the *wave number*. In the general case of an arbitrarily directed wave the wave number is multiplied by a unit vector **n** normal to the wave front. Then, the *wave vector* $\mathbf{k} = k\mathbf{n}$ also determines the direction of propagation of the wave. In the present case, $\mathbf{k} = k_x = k$. The homogeneous differential equation (III.4) obtained above is called *Helmholtz's equation*.* The general solution of this equation has the form $\tilde{\Psi}(x) = \tilde{A}e^{-ikx} + \tilde{B}e^{+ikx}$ where the arbitrary constants (complex amplitudes) A and B must be found from the boundary conditions.

Thus we finally obtain:

$$\varphi(x,\ t) = \tilde{A}e^{i(\omega t - kx)} + \tilde{B}e^{i(\omega t + kx)}. \qquad \text{(III.5)}$$

This solution describes two monochromatic waves: a forward and backward wave. Examining only the forward wave and setting the initial phase equal to zero, we obtain

$$\varphi(x,\ t) = \varphi_{max}e^{i(\omega t - kx)} \qquad \text{(III.6)}$$

or

$$\varphi(x,\ t) = \varphi_{max}\sin(\omega t - kx), \qquad \text{(III.7)}$$

where φ_{max} is the amplitude of the wave (in this case, the velocity potential) and x is the distance traversed by the wave from the origin of coordinates. Using the fact that the angular frequency $\omega = 2\pi\nu$, where ν is the directly measured frequency of the oscillations, we shall write Eq. (III.7) in the form

$$\varphi(x,\ t) = \varphi_{max}\sin 2\pi\left(\nu t - \frac{\nu x}{c_0}\right).$$

It is evident from this notation that at $x = const$, the phase of the oscillations changes by 2π over a time $T = \nu^{-1}$ called the *period of a complete oscillation*, and the phase changes by 2π in space ($t = const$) over a distance $x = c_0/\nu = \Lambda$ called the *wavelength*.

---

*Helmholtz's equation has the following form in the general three-dimensional case: $\Delta\tilde{\Psi} + k^2\tilde{\Psi} = 0$.

§ 2. Basic linear relations between the physical quantities varying in an ultrasonic wave. Characteristic impedance and acoustic impedance

We shall now find the relations between the quantities varying in the field of ultrasonic waves with infinitesimal amplitude, i.e., in the linear approximation.

Let the velocity potential be given for the forward wave in the form (III.7). We shall find the variable (sound) pressure in the wave from (II.9) by differentiating expression (III.7) with respect to time and multiplying by ρ_0:

$$p = \rho_0 \frac{\partial \varphi}{\partial t} = \rho_0 \omega \varphi_{max} \cos(\omega t - kx) = p_{max} \cos(\omega t - kx), \quad \text{(III.8)}$$

where $p_{max} = \rho_0 \omega \varphi_{max}$ is the amplitude of the pressure. The displacement velocity of the particles ("particle velocity") is determined from (II.8) by differentiating φ with respect to the coordinate:

$$v_x = v = -\frac{\partial \varphi}{\partial x} = k\varphi_{max} \cos(\omega t - kx) = v_{max} \cos(\omega t - kx), \quad \text{(III.9)}$$

where $v_{max} = k\varphi_{max}$ is the amplitude of the velocity. Comparing Eqs. (III.8) and (III.9), we see that the pressure and particle velocity in the forward wave have the same phase and are related to one another by the relation

$$p = \rho_0 c_0 v = v_{max} \rho_0 c_0 \cos(\omega t - kx). \quad \text{(III.10)}$$

For the backward wave, $\varphi = \varphi_{max} \sin(\omega t + kx)$, we obtain analogously

$$\begin{aligned} p &= -\rho_0 c_0 v = -\rho_0 c_0 v_{max} \cos(\omega t - kx) = \\ &= \rho_0 c_0 v_{max} \cos(\omega t - kx + \pi), \end{aligned} \quad \text{(III.11)}$$

i.e., the pressure and velocity in the backward wave oscillate with opposite phase. Thus the compression phase ($p > 0$) in the forward wave corresponds to a positive value of the particle velocity, whose sign, of course, is determined by the corresponding choice of the positive orientation of the coordinate axis; in the rarefaction phase ($p < 0$) the velocity is negative. In the backward wave the opposite relation holds between the sign of the pressure and the sign of the particle velocity.

It is not difficult to show that the relation (III.10) is satisfied for any form of the wave profile (infinitesimal amplitude). Indeed, let the forward wave be given in the general form (II.42). Introducing the argument $x - c_0 t \equiv \xi$, from the rule for differentiating composite functions we obtain:

$$p = \rho_0 \frac{\partial \varphi(\xi)}{\partial t} = \rho_0 \frac{\partial \varphi(\xi)}{\partial \xi}\frac{\partial \xi}{\partial t} = -\rho_0 c_0 \frac{\partial \varphi(\xi)}{\partial \xi},$$

$$v = -\frac{\partial \varphi(\xi)}{\partial x} = -\frac{\partial \varphi(\xi)}{\partial \xi}\frac{\partial \xi}{\partial x} = -\frac{\partial \varphi(\xi)}{\partial \xi},$$

i.e., $p = \rho_0 c_0 v$, which agrees with the results (III.10) obtained for the forward sinusoidal wave. In relations (III.10) and (III.11), p and v are any **local** values of the acoustic pressure and particle velocity. For the **amplitude** values in the forward wave we have, correspondingly,

$$p_{max} = \rho_0 c_0 v_{max}. \qquad \text{(III.12)}$$

The quantity $\rho_0 c_0 \equiv z_0$ is called the *specific characteristic (acoustic) impedance* of the medium. This terminology comes from the fact that the coefficient $\rho_0 c_0$ in Eq. (III.10) and (III.11) determines the magnitude of the particle velocity with fixed acoustic pressure. The pressure force acting on an area S equals $F_p = \rho_0 c_0 S v$. Correspondingly, the quantity $\rho_0 c_0 S$ can be called the *total acoustic impedance* of the medium on the surface S.

The values of the specific acoustic impedances for different liquids and gases are presented in the last column of Table 4. It is evident from this table that the acoustic impedance of liquids is three to four orders of magnitude higher than the acoustic impedance of gases. This means that for the same amplitude of the pressure the amplitude of the acoustic velocity of fluid particles in a liquid will be 10^3–10^4 times smaller than in a gas. On the other hand, for a fixed amplitude of the particle velocity, the pressure in a liquid exceeds the pressure in the gas by three to four orders of magnitude. Since, however, the particle velocity in the medium is determined by the oscillations of the surface of the source (under the condition of continuity), this means that the same source emitting ultrasound into a liquid and a gas while oscillating with the same velocity amplitude creates a variable pressure that is

10^3–10^4 times higher in the liquid than in the gas. In solids ($\rho_0 \simeq 10^4$ kg/m^3), for compression waves ($c_0 \simeq 5 \cdot 10^3$ m/s), $z_0 \simeq \rho_0 c_0 \simeq 5 \cdot 10^7$ $kg/(m \cdot s)$, i.e., approximately ten times higher than for liquids. Therefore, the pressure (normal stress) created by the same source emitting ultrasound into a solid is greater by the same factor than the pressure created in a liquid. Of course, in this case, media with different acoustic impedances will create different to the source, damping its oscillations. The analysis of this problem is not, however, the subject of the present chapter, which is concerned only with the propagation of ultrasound.

Thus the pressure and particle velocity in the forward plane wave have the same phase, and their ratio is characterized by a real quantity: the specific acoustic resistance $\rho_0 c_0$. In general, the pressure and velocity can differ in phase, as occurs, for example, in the backward plane wave. For this reason, in general, the ratio of the pressure to the particle velocity is characterized by a complex number, called the *specific acoustic impedance*: $p/v = \tilde{z} = \tilde{z}_0 + iy$. The imaginary part of this number determines the magnitude of the phase shift between p and v. Multiplication of the specific impedance by the area S on which the pressure p acts gives, correspondingly, the *total impedance*: $\tilde{Z} = \tilde{z}S$.

We shall encounter below the problem of determining and calculating impedances many times. Here, as an example, we shall also find the ratio of the pressure to the particle velocity in a field created by superposed forward and backward plane waves with equal amplitude. According to (III.5) the total pressure in this field is

$$p(x, t) = p_{max}[e^{i(\omega t - kx)} + e^{i(\omega t + kx)}];$$

for the total particle velocity, based on (III.5) together with expressions (III.10) and (III.11), we obtain

$$v(x, t) = [p_{max}/(\rho_0 c_0)]\,[e^{i(\omega t - kx)} - e^{i(\omega t + kx)}].$$

The specific acoustic impedance, in this case, equals

$$z = \frac{p(x, t)}{v(x, t)} = \rho_0 c_0 \frac{e^{i(\omega t - kx)} + e^{i(\omega t + kx)}}{e^{i(\omega t - kx)} - e^{i(\omega t + kx)}}.$$

which

$$s = \Delta\rho/\rho_0 = p/K = (p_{max}/K)\sin(\omega t - kx),$$

where K is the bulk modulus of elasticity. Therefore,

$$\Delta\rho = (\rho_0 p_{max}/K)\sin(\omega t - kx), \qquad \text{(III.15)}$$

i.e., the density oscillates with an amplitude

$$\rho_{max} = \rho_0 p_{max}/K = p_{max}/c_0^2. \qquad \text{(III.16)}$$

in phase with the pressure.

The optical refractive index n of the medium varies with the density of the medium. A relation can be found between the density and the refractive index from the well-known Lorenz–Lorentz relation, according to which

$$\frac{n^2-1}{n^2+2}\,\frac{1}{\rho} = \frac{n_0^2-1}{n_0^2+2}\,\frac{1}{\rho_0}, \qquad \text{(III.17)}$$

where n_0 is the refractive index of the unperturbed medium. Setting in (III.17) $n = n_0 + \Delta n$ and $\rho = \rho_0 + \Delta\rho$ and neglecting terms which are quadratic in Δn, we obtain after some simple calculations

$$\Delta n \simeq \frac{(n_0^2-1)(n_0^2+2)}{6n_0\rho_0}\Delta\rho = \frac{N_0}{\rho_0}\rho_{max}\sin(\omega t - kx),$$

i.e., the refractive index oscillates in phase with the density (and pressure) with amplitude $n_{max} = (\rho_{max}/\rho_0)N_0$, where $N_0 \equiv (n_0^2-1)(n_0^2+2)/(6n_0)$.

Table 5 summarizes the equations relating all of the enumerated characteristics of ultrasonic waves. These equations enable us to calculate the amplitude value of any parameter of the ultrasonic field, if the quantities ω, ρ_0, and c_0, which are easily measured, are known together with one of the following: A, v_{max}, a_{max}, p_{max}, ρ_{max}, s_{max}, or n_{max}.

Returning to the forward monochromatic plane wave and using the relationship obtained for it between the pressure and the velocity in the form (III.10) or (III.11), we shall find a relation between these parameters and the other acoustic variables.

The particles of the medium, oscillating in the ultrasonic wave with velocity*

$$v = v_{max} \sin(\omega t - kx), \qquad \text{(III.13)}$$

will undergo an acceleration

$$a = \partial v/\partial t = \omega v_{max} \cos(\omega t - kx) = \omega v_{max} \sin(\omega t - kx + \pi/2),$$

which leads the velocity in phase by $\pi/2$, with amplitude

$$a_{max} = \omega v_{max}. \qquad \text{(III.14)}$$

We shall find the displacement of particles from their equilibrium position ξ by integrating expression (III.13), which gives

$$\xi = \int v\, dt = \frac{v_{max}}{\omega} \cos(\omega t - kx) = \frac{v_{max}}{\omega} \sin(\omega t - kx - \pi/2).$$

The displacement therefore lags the velocity in phase by 90° and the acceleration by 180°. The amplitude of the displacement $\xi_{max} \equiv A$, which we shall refer to below as the *amplitude of oscillations* in an ultrasonic wave, is related to the *amplitude of the particle velocity* by the relation $\xi_{max} = A = v_{max}/\omega$. Substituting $v_{max} = \omega A$ into expression (III.14), we obtain $a_{max} = \omega^2 A$. Thus the acceleration of the oscillating particles with $A = const$ is proportional to the square of the frequency of the oscillations. At ultrasonic frequencies it can be several hundreds of thousands of times greater than the acceleration of gravity.

We shall also present expressions for the relative deformation (compression) and the variable density in the ultrasonic wave. In a wave with infinitesimal amplitude the condensation s is related to the acoustic pressure p by the linear equation of state (II.22), according to

*For a single wave sin and cos are equivalent.

Table 5

Linear Relations Between the Amplitude Values of the

	A	v_{max}	a_{max}
A	A	$\frac{v_{max}}{\omega}$	$\frac{d_{max}}{\omega^2}$
v_{max}	ωA	v_{max}	$\frac{a_{max}}{\omega}$
a_{max}	$\omega^2 A$	ωv_{max}	a_{max}
p_{max}	$\rho_0 c_0 \omega A$	$\rho_0 c_0 v_{max}$	$\frac{\rho_0 c_0 a_{max}}{\omega}$
ρ_{max}	$\frac{\omega \rho_0 A}{c_0}$	$\frac{\rho_0 v_{max}}{c_0}$	$\frac{\rho_0 a_{max}}{\omega c_0}$
s_{max}	$\frac{\omega A}{c_0}$	$\frac{v_{max}}{c_0}$	$\frac{a_{max}}{\omega c_0}$
n_{max}	$\frac{\omega A N_0}{c_0}$	$\frac{N_0}{c_0} v_{max}$	$\frac{N_0 a_{max}}{\omega c_0}$

§ 3. Energy characteristics of the ultrasonic field. Ultrasonic intensity

During the propagation of an ultrasonic wave each particle of the medium undergoes an oscillatory motion about its position of equilibrium with a velocity v, which is accompanied by a periodic change in density and pressure in the vicinity of the particle. In addition, as we have seen, in a plane wave the pressure and velocity have the same phase. This means that the pressure forces do positive work. In the absence of

Continuation of Table 5

Parameters of an Ultrasonic Field of Plane Waves

p_{max}	ρ_{max}	s_{max}	n_{max}
$\frac{p_{max}}{\omega\rho_0 c_0}$	$\frac{\rho_{max} c_0}{\omega\rho_0}$	$\frac{c_0}{\omega} s_{max}$	$\frac{c_0}{\omega N_0} n_{max}$
$\frac{p_{max}}{\rho_0 c_0}$	$\frac{\rho_{max} c_0}{\rho_0}$	$c_0 s_{max}$	$\frac{c_0}{N_0} n_{max}$
$\frac{\omega p_{max}}{\rho_0 c_0}$	$\frac{\omega c_0 \rho_{max}}{\rho_0}$	$\omega c_0 s_{max}$	$\frac{\omega^2 c_0}{N_0} n_{max}$
p_{max}	$c_0^2 \rho_{max}$	$\rho_0 c_0^2 s_{max}$	$\frac{\rho_0 c_0^2}{N_0} n_{max}$
$\frac{p_{max}}{c_0^2}$	ρ_{max}	$\rho_0 s_{max}$	$\frac{\rho_0}{N_0} n_{max}$
$\frac{p_{max}}{K}$	$\frac{\rho_{max}}{\rho_0}$	s_{max}	$\frac{n_{max}}{N_0}$
$\frac{N_0 p_{max}}{K}$	$\frac{N_0}{\rho_0} \rho_{max}$	$N_0 s_{max}$	n_{max}

absorption this work cannot be transformed into heat, but must remain in the form of the energy of the oscillatory motion of the particles of the elastic medium, i.e., in the form of "acoustic" energy. Thus, when an oscillating source emits ultrasound, its energy is transmitted to the adjacent medium in the form of acoustic energy, which propagates in the medium with the velocity of sound, filling an increasingly larger space called the *ultrasonic field*. The energy of each volume element in this field represents the sum of the kinetic energy of the oscillating particles and the potential energy of elastic deformation. The kinetic energy of a particle with volume V_0 and density ρ_0 equals

$$W_{kin} = \frac{1}{2}\rho_0 V_0 v^2 = \frac{1}{2}\rho_0 V_0 v_{max}^2 \sin^2(\omega t - kx)\,. \qquad \text{(III.18)}$$

The potential energy of this particle W_{pot} equals the work that must be performed in order to change the volume from V_0 to V. The relative change in volume due to an infinitesimal change in condensation from s to $s + ds$, equals $d\Theta = -ds$. The absolute change in volume corresponding to this change in the condensation is $dV = -V_0 ds$. The work performed in so doing $-pdV = pV_0 ds$, according to (II.28), equals $dA = pV_0 ds = KsV_0 ds$. We shall find the total work performed in changing the volume from V_0 to V by integrating dA over the range of condensation from 0 to s:

$$A = W_{pot} = \int_0^s V_0 Ks\, ds = V_0 Ks^2/2.$$

Making the substitutions $K = \rho_0 c_0^2$ and $s = v/c_0$, we obtain

$$W_{pot} = (\rho_0 V_0/2)\, v_{max}^2 \sin^2(\omega t - kx)\,. \qquad \text{(III.19)}$$

We note that the potential and kinetic energies are equal and vary in phase, i.e., potential energy is not transferred into kinetic energy. Adding expressions (III.18) and (III.19), we obtain the total energy in the volume V_0:

$$W = W_{kin} + W_{pot} = \rho_0 V_0 v_{max}^2 \sin^2(\omega t - kx)\,. \qquad \text{(III.20)}$$

Dividing this expression by V_0 we obtain the *instantaneous energy density*, i.e., the energy per unit volume of the medium $w = W/V_0 = \rho_0 v_{max}^2 \sin^2(\omega t - kx)$. The *average energy density* $\overline{w} = \rho_0 v_{max}^2/2 = p_{max}^2/(2\rho_0 c_0^2) = p_{max}^2/(2K)$ is one of the basic energy characteristics of the ultrasonic field.

It should be noted that the result obtained is valid for the case of an **unbounded** ultrasonic field examined here, as well as for other particular cases in which the amount of matter in the ultrasonic field remains constant, i.e., the average density of a volume element of the medium remains constant. If this condition is not satisfied, then, as will be shown in Chap. IV, the average kinetic energy density does not equal the average potential energy density.

Instead of the amplitude values of the pressure and velocity we can introduce their effective values $p_{eff} = p_{max}/\sqrt{2}$, $v_{eff} = v_{max}/\sqrt{2}$. Then the expressions for the average energy density will assume the form $\overline{w} = \rho_0 v_{eff}^2 = p_{eff}^2/(\rho_0 c_0^2)$.

As already noted, in an ultrasonic wave of the type (III.7) energy is transported from the source in the direction of propagation of the wave. For the energy characteristic of the radiation, we introduce the concept of the *energy flux density* or the *ultrasonic intensity*. **The ultrasonic intensity is equal to the energy crossing a unit surface area perpendicular to the direction of propagation of the ultrasonic wave per unit time.** Since the acoustic energy propagates with the velocity of sound c_0, the intensity is determined by multiplying the energy density $\overline{w}$ by c_0, which gives

$$I = wc_0 = \frac{v_{max}^2}{2}\rho_0 c_0 = \frac{p_{max}^2}{2}\,\frac{1}{\rho_0 c_0} = \frac{v_{max} p_{max}}{2}, \quad \text{(III.21)}$$

or in terms of effective values: $I = v_{eff}^2 \rho_0 c_0 = p_{eff}^2/(\rho_0 c_0) = v_{eff} p_{eff}$. The intensity, in contrast to the energy density, is a vector quantity characterizing the directed flow of energy. Thus, if the energy densities of the forward and backward waves add on superposition, then the intensities are subtracted, so that, for example, the total intensity in the field of two oppositely traveling waves with identical amplitude equals zero. In addition to the intensity, we can introduce the concept of the total flux of acoustic energy or the *acoustic radiation power* through the surface S defined as

$$D = IS = \frac{v_{max}^2}{2}\rho_0 c_0 S = \frac{p_{max}^2}{2}\,\frac{S}{\rho_0 c_0} = \frac{(p_{max} S)\, v_{max}}{2}. \quad \text{(III.22a)}$$

This equation assumes that the intensity is constant over the surface S. The acoustic power can in general be determined by integrating over the area:

$$D = \int_S [\mathbf{I}(S) \cdot \mathbf{n}]\, dS. \quad \text{(III.22b)}$$

Thus the radiation intensity represents the specific power, i.e., the power per unit surface area. If the power is measured in watts and the unit surface area is 1 cm^2, then the unit of measurement of intensity is 1 W/cm^2, which is the most commonly employed unit.

We note that the equations for the ultrasonic intensity (III.21) or the acoustic power (III.22a) are analogous to the equations for the ac power dissipated as Joule heat in an ohmic resistance R_e:

$$D = \frac{I_{max}^2}{2} R_e = \frac{U_{max}^2}{2R_e} = \frac{I_{max} U_{max}}{2} = I_{eff} U_{eff}.$$

The analog of the current is the particle velocity v, the analog of the electrical voltage U is the acoustic pressure force $F = pS$, and the analog of the ohmic resistance R_e is the acoustic resistance $\rho_0 c_0 S$. In addition, just as the quantity R_e determines the irreversible losses to Joule heat liberated in the resistive element, the acoustic resistance characterizes the irreversible "losses" of the acoustic power in the form of radiation into the adjacent medium. For this reason, the acoustic resistance is also called the *radiation resistance*.

NUMERICAL EXAMPLES. THE LOGARITHMIC SCALE OF INTENSITIES AND AMPLITUDES

We shall present numerical estimates for the characteristic sonic and ultrasonic intensities.

The **sensitivity** of the human ear at a frequency $\nu = 1000$ Hz (the region of maximum sensitivity) corresponds to an acoustic pressure amplitude of $p_{max} \simeq 10^{-4}$ $N/m^2 \simeq 10^{-9}$ atm. When such sound propagates in air, the amplitudes of the particle velocity, the condensation, and the oscillations are $v_{max} = p_{max}/(\rho_0 c_0) \simeq 3 \cdot 10^{-2}$ cm/s, $s_{max} = v_{max}/c_0 \simeq 10^{-9}$, and $A = v_{max}/\omega \simeq 5 \cdot 10^{-9}$ $cm = 0.5$ Å, respectively, and the intensity is $I = 0.5\, \rho_0 c_0 v_{max}^2 \simeq 20 \cdot 10^{-16}$ W/cm^2.

The **pain threshold** of the human ear at the same frequency (the acoustic oscillations are now perceived as a painful sensation) corresponds to a pressure amplitude $p_{max} \simeq 10^2$ $N/m^2 = 10^{-3}$ atm. In this case, the amplitudes of the particle velocity, the condensation, and the displacement and the intensity in air are, respectively, $v_{max} \simeq 30$ cm/s, $s_{max} = 10^{-3}$, $A \simeq 5 \cdot 10^{-3}$ $cm = 50$ μm, and $I \simeq 2 \cdot 10^{-3}$ W/cm^2.

Ultrasonic oscillations in water with a frequency $\nu = 1$ $MHz = 10^6$ Hz and an easily realizable intensity of 1 W/cm^2. We find the corresponding pressure amplitude from the equation $p_{max} = \sqrt{(2I\rho_0 c_0)}$.

For water ($\rho_0 c_0 = 1.5 \cdot 10^5\ g/(cm^2 \cdot s)$) this gives $p_{max} \approx \sqrt{3}\ atm \approx 1.7\ atm \approx 17 \cdot 10^4\ Pa$. In this case, the corresponding amplitudes of the particle velocity, displacement, and condensation are $10\ cm/s$, $2 \cdot 10^{-6}\ cm = 200\ \mathring{A}$, and 10^{-4}, respectively.

Modern ultrasonic technology enables realizing in liquids intensities of the order of several hundreds of watts per square centimeter in plane wave fields and up to several thousands and tens of thousands of watts per square centimeter at a frequency of several megahertz in focused fields. We shall assume that the limiting intensity of a plane radiator is $I = 1000\ W/cm^2$. In water the corresponding amplitudes of the pressure, particle velocity, condensation, and oscillations at 1 MHz are $p_{max} \approx 55\ atm$, $v_{max} \approx 4\ m/s$, $s_{max} \approx 3 \cdot 10^{-3}$, and $A \approx 1\ \mu m$, respectively. We note that even at such enormous intensities the condensation does not exceed 10^{-3} and therefore the condition, adopted below, that the deformations in an acoustic wave are small holds in liquids up to intensities of several tens of watts per square centimeter.

Comparison of the numerical examples presented above shows that the range of acoustic intensities is extremely wide: it encompasses about 20 orders of magnitude. For this reason, a *logarithmic scale*, in which the ratio of two intensities I_1 and I_2 is defined as the difference of the acoustic levels by the equation $\Delta\beta = 10 \log (I_1/I_2)$ and is measured in decibels, is often used in acoustics and ultrasonics. Thus a ten-fold difference between intensities I_1 and I_2 corresponds to a 10 dB difference of the levels. If $I_1/I_2 = 100$, then $\Delta\beta = 20\ dB$; if $I_1/I_2 = 1000$, then $\Delta\beta = 30\ dB$; etc. The entire range of acoustic intensities from 10^{-16} to $10^4\ W/cm^2$ on a logarithmic scale fits into a level difference of ~200 dB. For a fixed difference of levels in decibels, the ratio of the intensities can be found from the equation $I_1/I_2 = 10^{\Delta\beta/10}$. Therefore, a difference of $\Delta\beta = 1\ dB$ corresponds to an intensity ratio $I_1/I_2 = 10^{0.1} \approx 1.26$, i.e., the intensities differ by approximately 25%.

Since the intensity is proportional to the square of the amplitude, we have the following expression for the ratio of the amplitudes of two acoustic waves, for example the pressure amplitudes:

$$\Delta\beta = 20 \log \frac{p_{max\ 1}}{p_{max\ 2}} \quad \text{and} \quad \frac{p_{max\ 1}}{p_{max\ 2}} = 10^{\Delta\beta/20}, \qquad \text{(III.23)}$$

so that a difference in levels equal, for example, to 40 *dB* corresponds to an amplitude ratio $p_{max\ 1}/p_{max\ 2} = 100$; 60 *dB* corresponds to 1000; 80 *dB* corresponds to 10^4; 100 *dB* corresponds to 10^5; etc.

§4. Absorption of monochromatic ultrasonic waves

So far we have been studying the propagation of ultrasonic waves in an ideal medium without energy losses. In a real medium, however, due to the presence of dissipative processes, part of the energy of the ultrasonic wave is transformed into heat. At the same time the intensity and amplitude of the ultrasonic wave continuously decrease as it propagates and the wave is attenuated. This attenuation of the wave, which involves the transformation of part of its energy into heat, is called *absorption*.* In most materials ultrasonic absorption is caused primarily by internal friction (viscosity). Acoustic energy can also be dissipated due to heat conduction in the medium as well as different molecular processes, whose analysis falls outside the scope of classical (i.e., continuous medium) acoustics and forms the foundation of so-called molecular acoustics.[13]

Ultrasonic absorption due to internal friction can be easily calculated by introducing the coefficient of viscosity of the medium η, and taking into account the fact that viscous stresses are functions of the gradient of displacement velocity of the particles of the medium. In this case, to a first approximation viscous stresses may be assumed to be proportional to the first power of the **deformation velocity** (Newton's law for internal friction forces). We shall restrict our attention, as we have done above, to plane waves propagating along the x axis. Adding to the elastic stress σ for a one-dimensional deformation $\partial\xi/\partial x$ (including shear elasticity) a viscous stress proportional to the rate of this deformation, $\eta\partial^2\xi/\partial x\partial t = \eta\partial v/\partial x$, we obtain the one-dimensional rheological equation of state in the form

$$\sigma = c_{11}\frac{\partial\xi}{\partial x} + \eta\frac{\partial v}{\partial x}, \tag{III.24}$$

*Nondissipative processes such as diffraction, scattering by inhomogeneities in the medium, etc. can also attenuate ultrasound. We shall use the term absorption to mean attenuation due to dissipative losses only.

where the effective modulus c_{11} is defined by relation (I.33): $c_{11} = K + (4/3)\,G$. By analogy with this modulus, the coefficient of viscosity η can be represented as a sum:

$$\eta = \eta_b + (4/3)\eta_s. \tag{III.25}$$

In expression (I.33), the modulus K characterizes the elasticity of the medium relative to its volume compression, while the modulus G characterizes the elasticity relative to shear. Shear elasticity in liquids and gases can be neglected compared with volume elasticity, setting $c_{11} = K$ in (III.24). Analogously, the term η_b in (III.25) characterizes the viscosity of the medium relative to the volume compression and it can be called the *bulk viscosity*, while η_s is the usual *coefficient of shear viscosity*, which characterizes viscous losses acccompanying shear deformation. In most simple liquids these losses are much higher than the losses accompanying volume deformation, so that bulk viscosity can be neglected in them* by setting

$$\eta = (4/3)\eta_s. \tag{III.26}$$

Substituting expression (III.24) into the linearized equation of motion (I.11) we obtain the following wave equation in one dimension for the displacement ξ:

$$K\frac{\partial^2\xi}{\partial x^2} + \eta\frac{\partial^2\xi}{\partial x^2\partial t} = \rho_0\frac{\partial^2\xi}{\partial t^2}$$

or for the displacement velocity $v = \partial\xi/\partial t$ along the x-axis

$$K\frac{\partial^2 v}{\partial x^2} + \eta\frac{\partial^3 v}{\partial x^2\partial t} = \rho_0\frac{\partial^2 v}{\partial t^2}. \tag{III.27}$$

For a sinusoidal disturbance $v(x,t) = \tilde{v}(x)\exp i\omega t$, we obtain

$$(K + i\omega\eta)\frac{\partial^2\tilde{v}(x)}{\partial x^2} + \rho_0\omega^2\tilde{v}(x) = 0. \tag{III.28}$$

---

*For information on the role of bulk viscosity in ultrasonic absorption the reader is referred to the references concerning molecular acoustics cited in the bibliography.

A comparison of this equation with Helmholtz's equation for the velocity potential* in a nonviscous medium (III.4) reveals the simplest, though formal, method for solving this equation. For this, we introduce the complex modulus

$$\tilde{E} = K + i\omega\eta, \qquad \text{(III.29)}$$

the complex velocity of sound $\tilde{c}_0 = \sqrt{(\tilde{E}/\rho_0)}$, and the corresponding complex wave number

$$\tilde{k} = \omega/\tilde{c}_0 = \omega/\sqrt{\tilde{E}/\rho_0}\,. \qquad \text{(III.30)}$$

Then, Eq. (III.28) assumes the following form:

$$\frac{\partial^2 \tilde{v}(x)}{\partial x^2} + \tilde{k}^2 \tilde{v}(x) = 0, \qquad \text{(III.31)}$$

i.e., it has the same form as Eq. (III.4). Therefore, the function $\tilde{v}(x) = \tilde{A}e^{-i\tilde{k}x} + Be^{i\tilde{k}x}$, which describes two plane waves whose initial phases are taken into account by the complex coefficients $\tilde{A}$ and $\tilde{B}$, is a solution of Eq. (III.31). Examining as before only the forward wave and assuming that its initial phase equals zero, i.e., $\tilde{v}(x) = v(x)$ and $\tilde{A} = v_{max\,0} = v_{max}(x = 0)$, we obtain the solution of (III.27) in the form

$$v(x,\ t) = v_{max} e^{i\omega t} e^{-i\tilde{k}x}. \qquad \text{(III.32)}$$

To clarify the physical significance of this equation, let us separate the complex wave number $\tilde{k}$ into real and imaginary parts. According to expressions (III.30) and (III.29), we have

$$\tilde{k} = \frac{\omega\sqrt{\rho_0}}{\sqrt{K + i\omega\eta}} = \frac{k}{\sqrt{1 + i\omega\eta/K}}, \qquad \text{(III.33)}$$

where $k = \omega(\rho_0/K)^{1/2} = \omega/c_0$ is the real wave number, equal to the ratio of the frequency to the velocity of sound.

As is evident from Eq. (III.28), the coefficient $\omega\eta$ characterizes the magnitude of the viscous stress at frequency ω, and the modulus K characterizes the elastic stresses. In most cases of practical interest

---

*As already noted, any acoustic variable, including the particle velocity v, can enter into the velocity potential in Eq. (III.4).

viscous forces are much smaller than elastic forces, so that we can assume $\omega\eta \ll 1$. Then according to (III.33), we have

$$\tilde{k} = k\left(1 + i\frac{\omega\eta}{K}\right)^{-1/2} \simeq k - i\frac{\omega\eta k}{2K} = k - i\frac{\omega^2\eta}{2\rho_0 c_0^3}.$$

Introducing the notation Im $(\tilde{k}) = \alpha_0$ and substituting the result obtained

$$k = k - i\alpha_0 \qquad \text{(III.34)}$$

into Eq. (III.32), we obtain finally

$$v(x, t) = v_{max\,0}e^{i(\omega t - \tilde{k}x)} = v_{max\,0}e^{-\alpha_0 x}e^{i(\omega t - kx)}. \qquad \text{(III.35)}$$

Relation (III.35) describes a forward plane wave (propagating in the $+x$ direction), whose amplitude decreases exponentially according to the law

$$v_{max} = v_{max\,0}e^{-\alpha_0 x} \qquad \text{(III.36)}$$

from $v_{max} = v_{max\,0}$ at $x = 0$ to the value v_{max} exp$(-\alpha_0)$ at a distance x from the origin. Therefore, the coefficient

$$\alpha_0 = \omega^2\eta/(2\rho_0 c_0^3) = 2\pi\nu^2\eta/(\rho_0 c_0^3) \qquad \text{(III.37)}$$

represents the ultrasonic absorption coefficient determined by the viscous losses. It indicates the distance at which the amplitude of the wave decreases by a factor of $e = 3.14$, namely, $x = \alpha_0^{-1}$, and has the dimensions of inverse length, i.e., it is measured in cm^{-1} or m^{-1}. The zero index indicates that this absorption coefficient is calculated based on the linear relations of hydrodynamics, so that the quantity α_0 obtained characterizes the absorption of **sinusoidal** waves of infinitesimal amplitude. Strictly speaking, the wave described by expression (III.35) is no longer monochromatic. In real situations, however, for α_0 not too large the deviation from monochromaticity over a distance of several wavelengths can be neglected, making the assumption that the wave (III.35) is sinusoidal and taking into account the weak exponential decrease of its amplitude according to (III.36).

As is evident from expression (III.37), in a given medium the ultrasonic absorption coefficient increases with frequency in proportion to ν^2. For this reason, it is the ratio of the absorption coefficient to the square of the ultrasonic frequency rather than the absorption coefficient α_0 that must be taken as the parameter characterizing the absorptivity of a given medium:

$$\alpha_0/\nu^2 = 2\pi^2\eta/(\rho_0 c_0^3). \qquad \text{(III.38)}$$

If, under the assumption that the attenuation of the ultrasonic wave is due only to shear viscosity, the bulk viscosity η_v in the expression for η (III.25) is neglected, then we obtain the following expression for the absorption coefficient

$$\alpha_0/\nu^2 = 8\pi^2\eta_s/(3\rho_0 c_0^3), \qquad \text{(III.39)}$$

which enables calculating the quantity α_0/ν^2 for a given medium from the known value of η_s, determined by viscosimetric methods. Equation (III.39) was first obtained by Stokes, and the values of α_0/ν^2 calculated using this equation are customarily called *Stokes* values. Experience shows that the experimental value of α_0/ν^2 is always higher than the Stokes value. This is related to the neglect of the bulk viscosity (whose contribution in some organic liquids exceeds the contribution of shear viscosity by two to three orders of magnitude) as well as other loss mechanisms, in particular losses due to heat conduction in the medium.

Due to the finite heat conduction of real media between regions of compression and rarefaction, heat transfer occurs in the ultrasonic wave, which destroys the adiabaticity of the process and leads to additional energy losses. Taking into account heat conduction, the expression for α_0/ν^2, obtained by Kirchhoff, and called the *Stokes–Kirchhoff equation*, has the form

$$\frac{\alpha_0}{\nu^2} = \frac{2\pi^2}{\rho_0 c_0^3}\left(\eta + \frac{\gamma - 1}{c_p}\lambda_0\right), \qquad \text{(III.40)}$$

where c_p is the specific heat capacity at constant pressure, $\gamma = c_p/c_V$ is the ratio of the specific heat capacities, and λ_0 is the thermal conductivity of the medium. The magnitude of the correction term in (III.40) is determined primarily by the thermal conductivity λ_0. Losses due to heat conduction are significant in gases and in liquid metals; in other liquids they do not exceed several percent of the viscous losses and can be neglected, and the difference between the Stokes and experimental values of α_0/ν^2 can be ascribed to the bulk viscosity.

As far as the frequency dependence of the ultrasonic absorption coefficient is concerned, experience shows that the ratio α_0/ν^2 is indeed constant, at least over a wide frequency range. In addition, as a result of the relaxation (delay) of different molecular processes, in a comparatively narrow range of frequencies, the characteristic of a given medium, the curve of α_0/ν^2 versus the frequency and the dispersion curves exhibit relaxation steps, after which the quantity α_0/ν^2 drops to a new constant value approximating, in the limit, the Stokes value. As an example illustrating the magnitude of ultrasonic absorption in different media, Table 6 presents the experimental values of α_0/ν^2 at frequencies in the megahertz range for several liquids and gases at normal pressure and room temperature, as well as the values of α_0/ν^2 calculated for the same media using Eq. (III.39).* As is evident from the table, the theoretical and experimental data are almost identical for monatomic liquids and gases, whereas for substances with complex molecules, α_0/ν^2 at megahertz frequencies can exceed by several orders of magnitude the values calculated using the Stokes–Kirchhoff equation, approaching them only in the gigahertz range (see the data for benzene, toluene, and acetone). For solutions of polymers, which have an enormous "macroscopic" shear viscosity, the Stokes values of α_0/ν^2 calculated taking this viscosity into account, are several orders of magnitude higher than the experimental values. This is explained by the fact that ultrasonic absorption in such solutions is determined primarily by their "microviscosity," which is close to the viscosity of the solvent, whereas the "macroviscosity" of these solutions is due to the interaction of the polymer chains, which practically do not participate in ultrasonic absorption. As the temperature increases, ultrasonic absorption in simple liquids outside the region of relaxation, as a rule, decreases due to the decrease in the shear viscosity. As the

---

*More detailed information on the velocity and absorption of ultrasound in liquids and gases can be found in the review literature on molecular acoustics (see, for example, Ref. 12).

Table 6

Absorption of ultrasonic waves in some liquids and gases

Medium	T, °C	$\eta_s \cdot 10^4$ g/(cm · s)	$\alpha_0/\nu^2 \cdot 10^{17}$, s^2/cm Eq. (III.39)	exp.	Frequency range, MHz
Water	20	100	8.5	25	7—250
Alcohol methyl	20	60	15	34	1—250
ethyl	20	120	22	54	1—220
Acetone	20	35	7	30	5—70
Toluene	20	60	7.8	80	1—70
Benzene	20	64	8.7	900	1—170
Xylene	25	70	8.4	78	1—15
Cyclohexane	21	66	10	77	15
Nitrobenzene	25	200	14	80	1—15
Carbon tetra-chloride	20	96	20	500	1—100
Carbon disulfide	20	37	5	6000	1—10
Acetic acid	18	122	17	90000	0.5
Fthyl acetate	25	45	8.3	500	1.0
Glycerin	20—25	$14 \cdot 10^4$	250	2500	0.1 —10^4
Oil olive	21—25	$0.8 \cdot 10^4$	1100	1200	1—4
castor	21	$10 \cdot 10^4$	8400	7800	1—4
Gelatin in water (1.5%)	20	10^6	$8 \cdot 10^5$	48	5
Polystyrene in benzene (4%)	20	$17 \cdot 10^3$	1400	83	5
Mercury	20	155	5.4	6	20—50
Zinc	420	130	33	3.7	20—50
Bismuth	300	170	6.2	9.3	20—50
Helium	—269		204	231	15
	+ 18	2.0	5200	$3 \cdot 10^4$	0.6
Argon	—188		10.1	10.5	44
	+ 20	2.2	$19 \cdot 10^2$	$19 \cdot 10^3$	0.4
Hydrogen	—256		5.6	5.6	44
	+ 20	0.9	1700	35800	0.6
Oxygen	—186		7.3	8.6	44
	+ 20	1.9	1800	19000	0.6
Air	20	1.8	12400	$(2—3) \cdot 10^4$	0.2—0.1
Carbon dioxide	16.6	1.4	$13 \cdot 10^3$	$3 \cdot 10^4$	0.3

pressure increases, the absorption increases. In gases the opposite relationship is mainly observed: ultrasonic absorption in gases increases with temperature and decreases with increasing pressure.

As already noted, dissipative processes are not the only factors responsible for the decrease in the amplitude of an ultrasonic wave. For this reason, the decrease in amplitude must be characterized by an "attenuation coefficient," which in general can be represented in the form of a sum $\alpha_0 = \sum \alpha_{0i}$, where α_{0i} are the attenuation coefficients which describe, for example, the attenuation due to shear or bulk viscosity, heat conduction, and other absorption mechanisms, as well as scattering by inhomogeneities in the medium, diffraction of the wave, etc. In general, if the dependence of the attenuation of the wave amplitude on distance is given $v_{max} = v_{max}(x)$ then the attenuation coefficient α_0, which by definition determines the relative decrease of the amplitude of the wave per unit distance, can be calculated from the following equation:

$$\alpha_0 = (-1/v_{max\,0})\,(dv_{max}/dx)\,. \qquad \text{(III.41)}$$

Thus the quantity α_0 characterizes the damping of the wave with distance, i.e., in space, and for this reason can be called the *spatial attenuation coefficient*. A wave propagating with velocity c_0 traverses a distance x over a time $t = x/c_0$. Substituting $c = x_0 t$ into Eq. (III.36), we obtain the law governing the decrease of the wave amplitude with time:

$$v_{max} = v_{max\,0}e^{-\alpha_0 t c_0} = v_{max\,0}e^{-t/\tau_0},$$

where the following notation has been introduced:

$$\tau_0 = (\alpha_0 c_0)^{-1}. \qquad \text{(III.42)}$$

The coefficient τ_0, which has the dimensions of time, is called the *damping time constant*. It characterizes the damping of the wave with time: according to (III.42), over a time $t = \tau_0$ the amplitude of the wave decreases by a factor e. The quantity inverse to τ_0, i.e.,

$$\delta_0 = \tau_0^{-1} = \alpha_0 c_0 \qquad \text{(III.43)}$$

can be called the *temporal attenuation coefficient*. If the damping is due to Stokes absorption, then according to the definition of δ_0 Stokes' equation (III.39) can be written in the form $\delta_0 = (8/3)[\pi^2\nu^2\eta_s/(\rho_0 c_0^2)]$. The *logarithmic damping decrement*, defined as the natural logarithm of the ratio of the amplitudes of two successive oscillations, i.e., two neighboring waves (Fig. 12), $\upsilon = \ln(v_{max1}/v_{max2})$, can also be introduced as a characteristic of the damping of an ultrasonic wave. Let $v_{max1} = v_{max0}\exp(-\delta_0 t)$. Then $v_{max2} = v_{max0}\exp(-\delta_0(t + T)$, where T is the period of a single oscillation. Therefore,

$$\upsilon = \ln(e^{\delta_0 T}) = \delta_0 T = \delta_0/\nu = \alpha_0 c_0/\nu = \alpha_0\Lambda, \qquad \text{(III.44)}$$

where $\Lambda = c_0/\nu$ is the wavelength of the ultrasonic wave. Thus the dimensionless parameter $\alpha_0\Lambda$ (the "decrement") characterizes the decrease in the amplitude of the ultrasonic wave over a single period of the oscillations. The deviation of the wave from monochromaticity due to damping can obviously be neglected if $\alpha_0\Lambda \ll 1$.

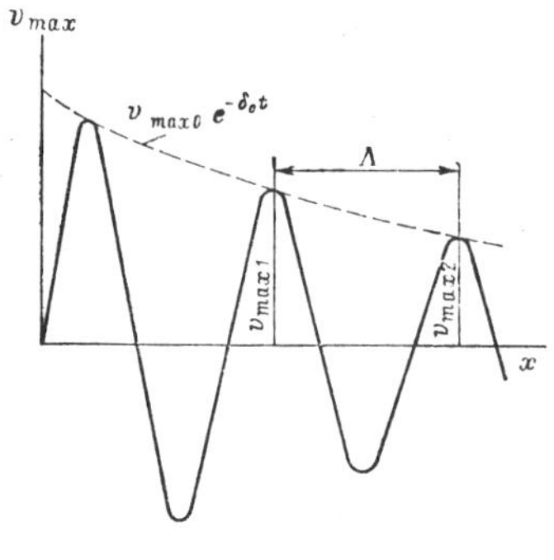

Fig. 12.

The attenuation coefficient α_0 introduced here, by virtue of its definition, following from Eq. (III.36) and (III.41), characterizes the damping of the amplitude of the ultrasonic wave and, for this reason, can be called the *amplitude attenuation coefficient*. Since the amplitude characteristics are related by linear relations (see Table 5), the exponential law of attenuation (III.36) with coefficient α_0 is valid for any acoustic parameter, for example, for the pressure amplitude

$$p_{max} = p_{max\,0}e^{-\alpha_0 x}, \qquad \text{(III.45)}$$

for the displacement amplitude $A = A_0\exp(-\alpha_0 x)$, etc.

The energy of an ultrasonic wave, however, is proportional to the square of the wave amplitude. For this reason, the law of attenuation of, for example, the ultrasonic intensity with distance from the source can be written in the form

$$I = I_0 e^{-2\alpha_0 x} = I_0 e^{-\alpha_0' x}, \qquad \text{(III.46)}$$

where

$$\alpha_0' = 2\alpha_0 \qquad \text{(III.47)}$$

is the *energy attenuation coefficient*, which gives the distance from the source $x = 1/\alpha_0'$ at which the energy of an ultrasonic wave with infinitesimal amplitude decreases by a factor of e.

The coefficient α_0' can also be defined in a different manner. By analogy with (III.46) we have for the energy density

$$\overline{w} = \overline{w}_0 e^{-\alpha_0' x}, \qquad \text{(III.48)}$$

where $\overline{w}_0$ is the energy density at $x = 0$. Making the substitution $x = c_0 t$ we obtain $\overline{w} = \overline{w}_0\exp(-\alpha_0' c_0 t)$. Over a unit time interval the energy density of the wave, i.e., the energy per unit volume, decreases to the quantity $\overline{w} = \overline{w}_0\exp(-\alpha_0' c_0)$. If the damping is small, i.e., $\alpha_0' c_0 \ll 1$, then this expression can be approximately written as a series expansion, retaining only the linear term in the expansion:

$$\overline{w} \simeq \overline{w}_0(1 - \alpha_0' c_0) = \overline{w}_0 - \overline{w}_0\alpha_0' c_0. \qquad \text{(III.49)}$$

The quantity $\overline{w}_0 - \overline{w} = \Delta\overline{w}_{abs}$ can be interpreted as the average energy absorbed per unit volume of the medium per unit time. From (III.49) we obtain $\Delta\overline{w}_{abs} \simeq \overline{w}_0 c_0 \alpha_0'$. But, according to (III.21) $\overline{w}_0 c_0$ is the ultrasonic intensity I_0 at $x = 0$ (i.e., $t = 0$). Therefore, for the coefficient α_0' we obtain

$$\alpha_0' \simeq \Delta\overline{w}_{abs}/I_0, \qquad \text{(III.50)}$$

i.e., the ultrasonic energy absorption coefficient can be defined as the ratio of the average energy absorbed per unit volume of the medium per unit time to the ultrasonic intensity, i.e., to the total energy flowing **into this volume** per unit time. For the amplitude coefficient α_0, according to (III.47), we have correspondingly: $\alpha_0 = \overline{\Delta w}_{abs}/(2\, I_0)$. This definition of α_0' and α_0 is in many cases more convenient because it enables the calculation of the ultrasonic absorption coefficient without the calculation of the complex wave number.

If the law of attenuation of the energy of the wave with distance is known, then the energy attenuation coefficient can be found from the equation

$$\alpha_0' = -\frac{1}{\overline{w}_0}\frac{d\overline{w}(x)}{dx} = -\frac{1}{I_0}\frac{dI(x)}{dx}. \tag{III.51}$$

The temporal attenuation coefficients are calculated analogously from the known time dependence of the amplitude of the wave or its energy, i.e., for example,

$$\delta_0 = -\frac{1}{v_{\max 0}}\frac{dv_{\max}(t)}{dt} = -\frac{1}{2I_0}\frac{dI(x)}{dt}, \tag{III.52}$$

where the zero index corresponds to time $t = 0$.

We note that for exponential attenuation, expressed as (III.45) or as (III.48), characteristic for the absorption of plane waves with infinitesimal amplitude, Eqs. (III.51), (III.52), and (III.41) give a constant attenuation coefficient. In general, these coefficients may depend on distance (time).

The ultrasonic attenuation is often described on a logarithmic scale (in decibels). To relate the absorption coefficient measured in this scale to the coefficient α_0 measured in inverse centimeters we shall use the relations for the pressure amplitudes (III.45) and (III.23). Setting in them $x = 1$ cm we obtain $p_{\max 0}/p_{\max} = 10^{\Delta\beta/20}$, whence $\Delta\beta = 20\alpha_0 \log e = 8.68\alpha_0 = 4.34\alpha_0'$, where α_0' is the intensity absorption coefficient measured in inverse centimeters. The change in the average ultrasonic intensity with time can be measured in decibels per second. If the temporal amplitude attenuation coefficient δ_0 or intensity attenuation coefficient δ_0' (in inverse seconds) is known, we can calculate the attenuation in decibels per second from the equation $\Delta\beta' = 8.68\delta_0 = 4.34\delta_0'$.

We shall give some numerical examples, using the experimental data presented in Table 6.

For air under normal conditions with $\nu = 1$ MHz, we have $\alpha_0 = (3 \cdot 10^{-13}) \cdot 10^{12}$ $cm^{-1} = 0.3$ $cm^{-1} = 2.6$ dB/cm. This means that the amplitude of the acoustic pressure in air at a frequency of 1 MHz decreases e-fold over a distance of $\simeq 0.3$ cm. The wavelength of the ultrasonic wave in air at this frequency is $\Lambda = c_0/\nu \simeq 0.03$ cm. In addition, the decrement $\upsilon = \alpha_0\Lambda = 0.01$. Therefore, the ratio of the amplitudes of two neighboring waves is $\exp(\alpha_0\Lambda) = 1.01$, i.e., over a distance of one wavelength the amplitude decreases by $\simeq 1\%$, so that the deviation of the wave from monochromaticity, even as a result of such large absorption, can be neglected.

At the same frequency of 10^6 Hz, the ultrasonic absorption coefficient in water is $\alpha_0 = 25 \cdot 10^{-5}$ cm^{-1} $(2.2 \cdot 10^{-3}$ $dB/cm)$, i.e., the amplitude of the ultrasonic wave decreases e-fold over a distance of approximately 40 m. The decrement in this case equals $\upsilon = \alpha_0 c_0/\nu \simeq 4 \cdot 10^{-5}$, i.e., the decrease in amplitude over the wavelength is negligibly small.

Let us now consider a strongly absorbing fluid, for example, glycerin. For glycerin, at the same frequency $(\alpha_0/\nu^2 = 25 \cdot 10^{-25}$ $cm^{-1})$ we have $\alpha_0 = 0.025$ cm^{-1} $(0.22$ $dB/cm)$ and $\alpha_0\Lambda = 5 \cdot 10^{-3}$. The damping decrement increases with frequency (in proportion to the frequency). However, from the estimates made above it follows that, at least in small volumes of a low-viscosity medium, the propagation of monochromatic ultrasonic waves can be studied neglecting absorption, taking it into account as an additional factor in cases when the effect of absorption plays a significant role in the phenomenon under study.

§ 5. Shear waves in liquids. Viscous losses at the boundaries of ultrasonic beams

Above, we characterized an ideal medium by the absence of shear stresses in it, making the assumption that it exhibits only bulk elasticity described by the modulus of hydrostatic compression K. In real liquids, however, in which shear elasticity can likewise be neglected, at least in the megahertz frequency band, shear stresses can appear as a result of the finite shear viscosity η_s ("viscous stresses"). Therefore, shear (transverse) waves excited by a tangentially oscillating plane can also propagate in a real liquid. These waves must necessarily be damped, since the absorption of the longitudinal wave examined above is due precisely to the presence of shear in it.

The equation of motion for a plane shear wave propagating along the x-axis with displacement ζ along the z-axis can be written in the form

$$\frac{\partial^2 \zeta}{\partial t^2} = \frac{G^*}{\rho_0} \frac{\partial^2 \zeta}{\partial x^2}, \quad \text{(III.53)}$$

where G^* is the effective shear modulus. According to the general definition, it is equal to the ratio of the tangential stress σ_{zx} to the corresponding deformation $\epsilon_{zx} = \partial\zeta/\partial x$:

$$G^* = \sigma_{zx}/\epsilon_{zx}. \quad \text{(III.54)}$$

The viscous stress σ_{zx} in a Newtonian fluid (for which $\eta_s = const$) is proportional to the first power of the deformation $\sigma_{zx} = \eta_s(\partial\epsilon_{zx}/\partial t) = \eta_s(\partial v_z/\partial z)$. For a sinusoidal process with frequency ω, $\partial/\partial t = i\omega$ so that $\sigma_{zx} = i\omega\eta_s\epsilon_{zx}$, where according to (III.54) $G^* = i\omega\eta_s$. Since in this case $\partial\zeta/\partial t = v = i\omega\zeta$, Eq. (III.53) assumes the form of the well-known "diffusion equation":

$$\partial v/\partial t = (\eta_s/\rho_0)(\partial^2 v/\partial x^2). \quad \text{(III.55)}$$

To find the absorption and velocity of the viscous shear wave we shall seek the solution of this equation in the form of a damped forward plane wave of the type (III.35), i.e.,

$$v = v_{max\,0} \exp i(\omega t - kx) \quad \text{(III.56)}$$

with complex wave number $\tilde{k}$ defined by relation (III.34):

$$k = k^* + i\alpha_s, \quad \text{(III.57)}$$

where $k^* = \omega/c^*$ is a real wave number which is equal to the ratio of the frequency to the velocity of propagation of the wave sought; α_s is the absorption coefficient. Substituting expression (III.56) into Eq. (III.55) we obtain $-i\omega = \eta_s \tilde{k}^2/\rho_0$, whence, since $\sqrt{(-i)} = (1 - i)\cdot\sqrt{(1/2)}$,

$$k = i\sqrt{i\omega\rho_0/\eta_s} = \sqrt{\pi\nu\rho_0}(1-i)/\eta_s, \qquad \text{(III.58)}$$

where $\nu = \omega/(2\pi)$ is the frequency. Thus the equation of the wave sought (III.56) assumes the form

$$v = v_{\max 0}\exp\, i\,[\omega t - \sqrt{\pi\nu\rho_0/\eta_s}\,(1-i)x]\,. \qquad \text{(III.59)}$$

The amplitude of this wave decays exponentially

$$v_{\max} = v_{\max 0}\exp(-\sqrt{\pi\nu\rho_0/\eta_s}\cdot x) \qquad \text{(III.60)}$$

with an attenuation coefficient

$$\alpha_s = \mathrm{Im}\,\tilde{k} = \sqrt{\pi\nu\rho_0/\eta_s}. \qquad \text{(III.61)}$$

We find the velocity of propagation of this wave c^* from the ratio $c^* = \omega/k^*$, where k^* is the real part of the complex wave number (III.57), i.e., according to (III.58) $k^* = \alpha_s = (\sqrt{\pi\nu\rho_0/\eta_s})$. Thus

$$c^* = \omega\sqrt{\eta_s}/\sqrt{\pi\nu\rho_0} = 2\sqrt{\pi\nu\eta_s/\rho_0}, \qquad \text{(III.62)}$$

i.e., the velocity of propagation of the viscous shear wave depends strongly on the ultrasonic frequency. For example, at a frequency of $\nu = 1\ MHz$ at room temperature $c^* = 400\ cm/s$ in water ($\eta_s = 0.01\ P$, $\rho_0 = 1\ g/cm^3$) and $c^* = 10^4\ cm/s$ in glycerin ($\eta_s = 14\ P$, $\rho_0 = 1.26\ g/cm^3$).

The absorption of shear waves, according to (III.61), is characterized by the quantity $\alpha_s/\sqrt{\nu} = \sqrt{(\pi\rho_0/\eta_s)}$. At room temperature it equals approximately 18 cm/s for water and approximately 0.5 cm/s for glycerin. Thus at the lowest ultrasonic frequency $\nu = 20\ kHz = 2\cdot 10^4\ s^{-1}$ the absorption coefficient of a shear wave is $\alpha_s = 2\cdot 10^3\ cm^{-1}$ in water, $\alpha_s = 50\ cm^{-1}$ in glycerin, and $\alpha_s \simeq 400\ cm^{-1}$ in air ($\rho_0 \simeq 10^{-3}\ g/cm^3$, $\eta_s = 10^{-4}\ P$), i.e., less than in water. For longitudinal waves with the same frequency the absorption coefficient is $10^{-7}\ cm^{-1}$ in water and $10^{-5}\ cm^{-1}$ in glycerin (see Table 6). At a frequency of $\nu = 2\ MHz$ the absorption coefficient of a shear wave increases by an order of magnitude; the absorption of longitudinal waves, however, increases by four orders of magnitude (since $\alpha_0 \propto \nu^2$). Due to this difference in the frequency dependences the values of α_s and α_0 may be equal at some frequency. For water, for example, this occurs at a frequency of the order of 10^{12} *Hz* and for glycerin at $\simeq 10^{10}$ *Hz*. At such high frequencies the shear **elasticity** of liquids, which increases with the rate of shear deformation, i.e., frequency, can no longer be neglected.[13, 14] In the megahertz frequency band, however, the absorption coefficient of shear waves in liquids, whose viscosity is not too high, is many orders of magnitude higher than the absorption coefficient of longitudinal waves. From the estimates presented it is evident that a shear wave decays over a very short distance from the source: its amplitude decreases *e*-fold over a distance $\Delta = = \alpha_s^{-1} = \sqrt{[\eta_s/(\pi\nu\rho_0)]}$, called the *penetration depth*. For the examples presented above this quantity is only $2 \cdot 10^{-2}$ *cm* for glycerin and $2 \cdot 10^{-3}$ *cm* for water.

We shall also find the damping decrement of a shear wave from its definition (III.44), which gives, $\upsilon = \alpha_s c^*/\nu = 2\pi$. Thus the damping decrement of a viscous shear wave (which determines the logarithm of the ratio of neighboring amplitudes) does not depend on frequency and is equal to a very large constant number, which shows that the shear wave in a liquid, for all practical purposes, decays over a distance equal to the wavelength. For this reason only viscous stresses existing near the surface of a tangentially oscillating source and propagating in a thin boundary layer of the liquid need be considered. These stresses can appear in the reaction on the source, the transmission of a shear wave by elastic bodies through a thin layer of liquid, the formation of vortex flows in the layer of liquid near a wall, additional losses on reflection of a longitudinal wave incident obliquely on a solid boundary in a viscous medium [15], and other similar effects when the appearance of viscous stresses must be taken into account in the calculation.

Viscous losses, in particular, can arise at the boundary of a real ultrasonic beam surrounded by unperturbed liquid, since under the condition of continuity at the boundary the oscillating particles in the liquid in the boundary layer of the beam will create viscous stresses in the unperturbed medium.[16] Here, part of the beam energy will be transformed into viscous waves, decaying over a distance Δ from the boundary of the beam, i.e., additional energy losses will occur in a finite beam. An approximate calculation of the corresponding loss factors in the form of the ratio of the average energy flux dissipated through the boundary of the beam to the intensity of the beam gives[16]

$$\gamma^* = c^*/(8c_0) = [1/(4c_0)]\sqrt{\eta_s \pi \nu/\rho_0}, \qquad \text{(III.63)}$$

which is determined by the velocity of the viscous wave c^* (III.62) and the velocity of sound c^* in the medium. Since the velocity c^* increases with the ultrasonic frequency and $c_0 \simeq const$, the viscous losses at the boundary of the beam also increase with the frequency. As the frequency increases, however, the absorption of volume waves also increases, which is determined by the absorption coefficient α_0 (III.38). It is therefore interesting to compare the viscous energy losses at the boundary of the beam ΔI_b with the losses in the volume of the beam ΔI_v. For a square beam with side a, using the relations (III.63) and (III.39) and neglecting the bulk viscosity ($\eta = \eta_s$), this gives:

$$\frac{\Delta I_b}{\Delta I_v} = \frac{2\gamma^*}{a\alpha_0} = \frac{c_0^2}{4\pi^2 a}\sqrt{\pi\rho_0/\eta_s\,\nu^3} = \frac{\alpha_s}{k^2 a}, \qquad \text{(III.64)}$$

where $k = 2\pi/\Lambda$ is the wave number for volume waves.

Thus the viscous losses at the boundary of the beam decrease with increasing ultrasonic frequency, with increasing transverse dimensions of the beam, as well as with increasing coefficient of viscosity. They become comparable ($\Delta I_b = \Delta I_v$) when $k^2 a = \alpha_s$. For water, for example, for $ka = 10$, according to (III.64) this equality is satisfied at a frequency of $\nu^* \simeq 10^8$ Hz. At lower frequencies the losses at the boundary even exceed the losses in the volume of the beam. The condition $ka = 10$ ($a \simeq 1.5\Lambda$) is not, however, compatible with the condition of directivity of the beam $ka >> 1$ ($a >> \Lambda$). As the parameter ka increases, the frequency ν^* decreases rapidly. Thus for $ka = 100$, for water, it already consitutes only about 1 MHz. At higher frequencies the losses at the boundary of the beam drop below the losses in the volume of beam; at the same time, at high frequencies the condition $ka >> 1$ becomes stronger. It is also necessary to take into account the fact that in a real beam the velocity gradients at the boundary of the beam are smoothed by the diffraction spreading of the boundary.

Thus at high ultrasonic frequencies the viscous losses at the boundary of a beam can be neglected, thereby preserving the results obtained for longitudinal plane waves and for real ultrasonic beams whose transverse dimensions are much larger than the wavelength in a large volume of a real liquid.

IV. Finite-Amplitude Plane Waves

§ 1. Estimation of the nonlinear terms in the equations of hydrodynamics

As we have already repeatedly noted, the linearized equations of hydrodynamics (II.5), (II.12), and (II.16) are exact only in the limit of infinitesimal perturbations and, therefore, a solution of the linear wave equation of the type

$$v = v_{max} \sin \omega (t - x/c_0) \qquad \text{(IV.1)}$$

strictly speaking describes a hypothetical wave with an infinitesimal amplitude. Any real ultrasonic wave, however, has a finite amplitude and in order to describe such a wave accurately the exact (nonlinear) equations of hydrodynamics must be used from the start. In the one-dimensional case, according to Eqs. (II.4) and (II.11), for a nondissipative medium these equations have the following form in Eulerian variables:

equation of motion:

$$\frac{\partial v}{\partial t} + v\frac{\partial v}{\partial x} = -\frac{1}{\rho}\frac{\partial p}{\partial x} \qquad \text{(IV.2)}$$

equation of continuity:

$$\frac{\partial \rho}{\partial t} + v\frac{\partial \rho}{\partial x} = -\rho\frac{\partial v}{\partial x} \qquad \text{(IV.3)}$$

where $\rho = \rho_0 + \Delta\rho_0 (x, t)$ is the total local density of the medium, which is related to the pressure p by the adiabatic equation of state

$$p = p(\rho), \qquad \text{(IV.4)}$$

which, in general, is likewise nonlinear and can be represented as an infinite Taylor series in powers of $\Delta\rho$

$$p = P - P_0 = \left(\frac{dp}{d\rho}\right)_{\rho=\rho_0} \Delta\rho + \frac{1}{2!}\left(\frac{d^2p}{d\rho^2}\right)_{\rho=\rho_0} \Delta\rho^2 + \frac{1}{3!}\left(\frac{d^3p}{d\rho^3}\right)_{\rho=\rho_0} \Delta\rho^3 + \ldots$$

or in powers of the condensation $\Delta\rho/\rho_0 = s$

$$p = \rho_0\left(\frac{dp}{d\rho}\right)_{\rho=\rho_0} s + \frac{\rho_0^2}{2!}\left(\frac{d^2p}{d\rho^2}\right)_{\rho=\rho_0} s^2 + \frac{\rho_0^3}{3!}\left(\frac{d^3p}{d\rho^3}\right)_{\rho=\rho_0} s^3 + \ldots \quad \text{(IV.5)}$$

We shall estimate the nonlinear terms in Eq. (IV.2)–(IV.4), using the relations of linear acoustics obtained in the preceding sections. Assuming a sinusoidal disturbance of the form (IV.1) we have:

$$\left|v\frac{\partial v}{\partial x}\right|_{max} \equiv b = \frac{\omega v_{max}^2}{c_0}; \quad \left|\frac{\partial v}{\partial t}\right|_{max} \equiv a = \omega v_{max}; \quad \frac{b}{a} = \frac{v_{max}}{c_0};$$

$$\left|v\frac{\partial \rho}{\partial x}\right|_{max} \equiv b' = \frac{\omega v_{max}\rho_{max}}{c_0}; \quad \left|\frac{\partial \rho}{\partial t}\right|_{max} \equiv a' = \omega\rho_{max}; \quad \frac{b'}{a'} = \frac{v_{max}}{c_0}.$$

Thus the ratio of the nonlinear to the linear terms, whose maximum value equals the ratio of the amplitude of the particle velocity v_{max} to the velocity of sound c_0, is the same for both Euler equations. Borrowing the terminology of hydrodynamics, in which the ratio of the flow velocity to the velocity of sound is called the *Mach number*, the quantity v_{max}/c_0 can be called the *acoustic Mach number*. According to the relations presented in Table 5, the acoustic Mach number

$$Ma = \frac{v_{max}}{c_0} = s_{max} = \frac{(\Delta\rho)_{max}}{\rho_0} = \frac{2\pi A}{\Lambda} = \frac{p_{max}}{K} \approx \frac{p_{max}}{P_0}. \quad \text{(IV.6)}$$

Therefore, the ratio of the quadratic term to the nonlinear term in the equation of state (IV.5) is also of the same order of magnitude.

The ultrasonic intensity used in the laboratory for physical measurements usually does not exceed 0.01– 0.1 W/cm^2. Such intensities correspond to Mach numbers of the order of 10^{-5}–10^{-6}. Thus the condition for the amplitudes of acoustical parameters to be small compared to their mean values, i.e., the condition

$$Ma = s_{max} \ll 1 \qquad \text{(IV.7)}$$

is well satisfied. This enables us to neglect the nonlinear terms in Eqs. (IV.2)– (IV.5) for describing real small-amplitude waves, which indeed conform quite accurately to the laws of linear acoustics.

According to the estimates made in § III.4, in water the comparatively high ultrasonic intensity $I = 1$ W/cm^2 ($v_{max} \simeq 0.1$ m/s, $c_0 \simeq 1500$ m/s) corresponds to $Ma = 6 \cdot 10^5$. For very high intensities ($\sim$ 100 W/cm^2) the acoustic Mach number attains a value of 10^{-3}, which may be viewed as a limit for ultrasonic plane waves in liquids and solids. In gases, due to their low density, such Mach numbers are achieved with much lower amplitudes, but the low efficiency of emission of ultrasound in gases makes it impossible to realize high Mach numbers at ultrasonic frequencies in them. Thus inequality (IV.7) holds for practically all ultrasonic waves. Nevertheless, at sufficiently high ultrasonic intensities the effects of finite amplitude ("nonlinear effects") become quite apparent, and in order to analyze the propagation of such waves the nonlinear terms, despite their smallness, must be included in Eqs. (IV.2)–(IV.5).

In this connection, it should be noted that the term "finite-amplitude wave" encountered in the literature is not completely satisfactory from the **experimental** point of view, because any **real** wave has a finite amplitude. Nonlinear effects, however, are not manifested in all real waves, but only in waves with sufficiently large amplitude; the precise value of this amplitude depends on the sensitivity of the apparatus and the method used to record the particular nonlinear effect of interest. From the theoretical point of view this term has a completely well-defined meaning: it indicates that nonlinear terms must be included in the equations of hydrodynamics and the consequences of doing so must be taken into account. It is in this sense that this term will be used in this book. We shall, however, call a real ultrasonic wave in which nonlinear effects are actually manifested simply a "large-amplitude wave," excluding in so doing strong shock waves (produced, for example, by explosions and discharges), which correspond to Mach numbers close to one and obey different laws of propagation (see, for example, Ref. 17).

We shall now study the consequences of including the nonlinear terms in the equations of hydrodynamics, neglecting for the time being dissipative processes, whose role we shall clarify later.

§ 2. The exact solution of the system of nonlinear equations of hydrodynamics for a nondissipative medium

The system of nonlinear differential equations (IV.2) and (IV.3) was solved by a number of authors using different methods in the middle of the 19th century. The most complete solution was obtained by Riemann, whose method is based on the assumption that the pressure and density are related by a general relationship of the form (IV.4). Introducing the previous notation

$$c^2 = dp/d\rho, \qquad \text{(IV.8)}$$

we shall rewrite Eqs. (IV.2) and (IV.3) in the form

$$\frac{\partial v}{\partial t} + v\frac{\partial v}{\partial x} + c^2\frac{\partial(\ln\rho)}{\partial x} = 0,$$

$$\frac{\partial(\ln\rho)}{\partial t} + v\frac{\partial(\ln\rho)}{\partial x} + \frac{\partial v}{\partial x} = 0.$$

We multiply the second of these equations by $+c$ and $-c$ and add the results to the first equation. Combining like terms we obtain:

$$\frac{\partial}{\partial t}(v + c\ln\rho) + (v + c)\frac{\partial}{\partial x}(v + c\ln\rho) = 0,$$

$$\frac{\partial}{\partial t}(v - c\ln\rho) + (v - c)\frac{\partial}{\partial x}(v - c\ln\rho) = 0. \qquad \text{(IV.9)}$$

We introduce the new variables:

$$v + c\ln\rho \equiv 2g, \quad v - c\ln\rho \equiv 2h; \qquad \text{(IV.10)}$$

$$\xi \equiv x - t(v + c), \quad \eta \equiv x + t(v - c). \qquad \text{(IV.11)}$$

Performing the differentiation operation in (IV.9), taking into account the fact that the quantities g and h depend on x and t through the variables ξ and η, we obtain the simple system of equations: $c\partial g/\partial\eta = 0$; $c\partial h/\partial\xi = 0$. Since $c \neq 0$, $\partial g/\partial\eta = \partial h/\partial\xi = 0$, from which it follows that g does not depend on η and h does not depend on ξ, i.e., $g = f_1(\xi)$ and $h = f_2(\eta)$. Since according to (IV.10) $g + h = v$, we have $v = f_1(\xi) + f_2(\eta)$. Substituting here the values of ξ and η from the identities (IV.11) we obtain the exact solution: $v(x,t) = f_1[x - t(c + v)] + f_2[x + t(c - v)]$, which represents two plane waves propagating in opposite directions along the x-axis. For a harmonically oscillating source, $v(x = 0) = v_{max} \sin \omega t$ we have the particular solution

$$v(x, t) = v_{max} \sin \omega\left(t - \frac{x}{c+v}\right), \qquad \text{(IV.12)}$$

which differs from the equation for the infinitesimal-amplitude wave (IV.1) only by the phase velocity.

§ 3. The velocity of propagation of a finite-amplitude wave. Nonlinear properties of the medium

Riemann's exact solution (IV.12) describes a plane wave propagating in a nondissipative medium along the positive x-axis with a phase velocity

$$c' = c + v(x, t), \qquad \text{(IV.13)}$$

which now depends on v, i.e., on x and t, and can therefore be called the *local velocity*. Thus the local velocity is the velocity of propagation of a given phase of the wave, characterized by a definite value of the particle velocity v. The quantity c in expression (IV.13), sometimes called the "local" velocity of sound, is defined by relation (IV.8) and can be found by differentiating the adiabatic equation of state, represented in the form of the series (IV.5), with respect to the density. Because the acoustic Mach numbers, i.e., the condensations $s = \Delta\rho/\rho_0$, in an ultrasonic field are small, this series can be terminated — with adequate accuracy — at the quadratic term, and the equation can be written in the form of the following approximately equality:

$$p \approx Ks + (B/2)s^2, \qquad \text{(IV.14)}$$

where the coefficients K (the adiabatic bulk modulus of elasticity) and B (the "nonlinear" modulus) are defined by the relations (II.20) and

(II.21).

Differentiating expression (IV.14) with respect to the density, based on Eq. (IV.8) we obtain

$$c = \left(\frac{dp}{d\rho}\right)^{1/2} = \left(\frac{K}{\rho_0} + \frac{Bs}{\rho_0}\right)^{1/2} = c_0\left(1 + \frac{B}{2K}s\right); \qquad \text{(IV.15)}$$

where $c_0 = (K/\rho_0)^{1/2}$, according to (II.34), is the velocity of propagation of an infinitesimal-amplitude wave and the factor enclosed in parentheses gives the largest correction to this velocity contributed by the quadratic term in the equation of state (IV.14). Using the linear relation between the condensation and the particle velocity and substituting it into (IV.15), we obtain

$$c = c_0 + (B/2K)v. \qquad \text{(IV.16)}$$

Thus by including the quadratic term in the equation of state, we find that the local velocity c now depends on the variable quantity v. This dependence is determined solely by the elastic nonlinearity of the medium, which, according to (IV.16), is determined by the ratio of the coefficients in front of the quadratic and linear terms of the adiabatic equation of state (IV.14). By virtue of this property the ratio B/K is customarily called the *nonlinearity parameter of the medium*.

Substituting now Eq. (IV.16) into Eq. (IV.13) we obtain an equation for the local velocity:

$$c' = c_0 + (B/2K)v + v = c_0 + \epsilon_0 v(x, t), \qquad \text{(IV.17)}$$

where $\epsilon_0 = (B/K + 2)/2$.

Thus the local velocity c' with which different phases of the finite-amplitude wave (IV.12) propagate exceeds the local velocity by an amount v. This increment is determined solely by the convective derivatives in Euler's equations, i.e., the **nonlinearity of the hydrodynamic equations** (IV.2) and (IV.3). The **elastic nonlinearity of the medium**, however, increases this correction by a factor of ϵ_0. The coefficient ϵ_0 in (IV.17) is therefore also a well-defined characteristic of the nonlinearity of the elastic properties of the medium and can thus be called the *nonlinear coefficient*.

The nonlinearity parameter B/K or the coeffcient ϵ_0 can be calculated if the equation of state of the medium is given in an explicit form.

For an isothermal process in an ideal gas the equation of state is given by the Boyle – Mariotte law:

$$P/P_0 = \rho/\rho_0.$$

In this case $c = (dP/d\rho)^{1/2} = (P_0/\rho_0)^{1/2} = c_0$, $B = \rho_0^2(d^2P/d\rho^2) = 0$, $B/K = 0$, $\epsilon_0 = 1$, $c' = c_0 + v$. Thus, due to the nonlinearity of the hydrodynamic equations (IV.2) and (IV.3), the local velocity differs from c_0 even if the equation of state is linear.

For an adiabatic process in the gas Poisson's equation (II.24) can serve as the equation of state:

$$P = P_0(\rho/\rho_0)^{\gamma}. \qquad \text{(IV.18)}$$

Differentiating this equation twice with respect to the density at the point $\rho = \rho_0$ and multiplying by ρ_0^2, according to (II.21) we obtain:

$$B = \rho_0^2(d^2P/d\rho^2)_{\rho=\rho_0} = \gamma(\gamma - 1)P_0.$$

In this case the adiabatic linear bulk modulus of elasticity $K = \rho_0(dP/d\rho)_{\rho=\rho_0} = \gamma P_0$. From here

$$B/K = \gamma - 1 \qquad \text{(IV.19)}$$

and

$$\epsilon_0 = (\gamma + 1)/2 \qquad \text{(IV.20)}$$

According to (IV.16), (IV.17), (IV.19), and (IV.20) we obtain

$$c = c_0 + \frac{\gamma - 1}{2}v; \quad c' = c_0 + \frac{\gamma + 1}{2}v. \qquad \text{(IV.21)}$$

It should be emphasized here that the expressions (IV.20) and (IV.21), which are based on approximate equalities, are actually exact equations for the case of an adiabatic process studied here, for which the relation between the pressure and density is given by a power-law function of the form (IV.18). This happened because we used the approximate relations twice: the equation of state in the form (IV.14) and the linear relation between the condensation and the particle velocity, which in the second approximation has a more complicated form (see the next section).

The propagation of ultrasound in liquids is likewise an adiabatic process, for which a theoretically well-founded equation of state does not yet exist in an explicit form. Experiments on the compressibility of simple liquids and isotropic solids show, however, that the adiabatic equation of state for these media can be approximately represented by an equation analogous to (IV.18), called the *empirical Tait equation*:

$$P/P_0 = (\rho/\rho_0)^n, \tag{IV.22}$$

in which the index of the isentrope n is equivalent to the parameter γ in Poisson's equation. This empirical parameter, according to Eqs. (IV.19) and (IV.20), is related to the nonlinearity parameter B/K and the nonlinear coefficient by the following relations: $n = (B/K) + 1 = 2\epsilon_0 - 1$.

The nonlinearity parameter B/K for liquids can be measured by different methods based on the study of the propagation of large-amplitude ultrasonic waves.[19,20] Table 7 presents the experimental values of B/K for a number of liquids, obtained in Refs. 21 and 22, as well as the magnitudes of these parameters for some gases, corresponding to their empirical specific heat capacity ratios γ. It is evident from Table 7 that the nonlinearity of liquids greatly exceeds that of gases. Nonlinear effects accompanying the propagation of large-amplitude ultrasonic waves are therefore strongest in liquids, even though the high "internal" pressure precludes large Mach numbers in them.

The nonlinearity parameter of liquids B/K or the coefficient ϵ_0 can also be approximately calculated, based on thermodynamic considerations, from the dependence of the velocity of sound on the temperature and pressure. Indeed, the fact that the local velocity is different at different points of the wave profile can be explained by the dependence of the velocity of sound on the pressure and temperature, which at these points have different values uniquely related to the particle velocity v and the thermodynamic characteristics of the given medium. Thus the increment to the local velocity $(B/2K)$ in expression (IV.16) can be represented in the form

$$\Delta c = \left(\frac{\partial c}{\partial P}\right)_T \Delta P + \left(\frac{\partial c}{\partial T}\right)_P \Delta T, \tag{IV.23}$$

Table 7

Values of the Nonlinearity Parameter B/K for Liquids and Gases

Medium	T, °C	B/K (exp.)	B/K* (theor.)
Liquid nitrogen	−195	3.1	—
Amyl acetate	20	5.1	—
Acetone	20	8.6	—
A-70 gasoline	20	10.2	—
Benzene	20	8.4	6.8
Water	20	6.6	5.2
Glycerin	20	9.4	8.8
Dichloroethane	20	7.6	—
Xylol	20	8.7	—
Transformer oil	20	6.5	—
Mercury	20	—	10.5
Terpentine	20	9.5	—
Alcohol ethyl	20	9.6	—
methyl	20	8.0	7.6
propyl	20	8.9	8.0
butyl	20	8.6	8.4
hexyl	20	9.7	—
Carbon disulfide	20	—	5.4
Toluene	20	9.4	7.9
Carbon tetrachloride	20	10.8	9.2
Chloroform	20	10.6	—
Ethyl acetate	20	5.0	—
Ethyl ether	20	3.1	—
Nitrogen gas	−195	—	0.40
Ammonia	20–40	—	0.40
Argon	0	—	0.67
Hydrogen	0–17	—	0.40
Air	0–100	—	0.40
Water vapor	100	—	0.33
Helium	18	—	0.63
Oxygen	13–200	—	0.40
Carbon dioxide	4–11	—	0.36

*The value of B/K for liquids is calculated from Eq. (IV.24) and for gases from the definition $B/K = \gamma - 1$, where γ is the empirical ratio of the specific heat capacities.

where ΔP and ΔT are the increments to the pressure and temperature at points with particle velocity v. The dependence of the temperature on pressure in an adiabatic process is given by the well-known thermodynamic relation $\Delta T = \Delta P\, T_0 \alpha_{T\,\mathrm{isob}} / (\rho_0 c_p)$, where $\alpha_{T\,\mathrm{isob}} = (1/V_0)(dV/dT)_p$ is the isobaric coefficient of thermal expansion; T_0 is the equilibrium temperature of the medium; and, c_p is the specific heat capacity at constant pressure. The pressure in the sound wave is related to the particle velocity (to a first approximation) by the relation $\Delta P \equiv p = \rho_0 c_0 v$. Substituting these values of ΔP and ΔT into the equality (IV.23) we obtain

$$\frac{B}{K} = 2\rho_0 c_0 \left(\frac{\partial c}{\partial P}\right)_T + \frac{2 c_0 T_0 \alpha_{T\,\mathrm{isob}}}{c_p} \left(\frac{\partial c}{\partial T}\right)_P . \qquad \text{(IV.24)}$$

The values of B/K calculated in this manner for a number of liquids at room temperature are also presented in Table 7. Although these values are somewhat lower than the measured values, the differences do not exceed the spread in the measurements of B/K performed by different methods.

Since all quantities entering into Eq. (IV.24) are measured with high accuracy, this equation enables following details such as the change in the nonlinearity parameter B/K with temperature, pressure, concentration of dissolved substances, etc. This, however, requires that, aside from data on acoustic measurements of the quantities c_0, $(\partial c/\partial P)_T$, and $(\partial c/\partial T)_P$, the corresponding values of ρ_0, α_T, and c_p also be known. For fluids for which these data are available, a calculation using Eq. (IV.24) enables the construction of graphs of the temperature dependence of B/K. Figure 13 illustrates, as an example, the general behavior of the nonlinearity parameter for water as a function of the temperature. Analogously, it is possible to observe the increase in the parameter B/K with hydrostatic pressure, as well as with the concentration of solutions of ionic salts in water.[23]

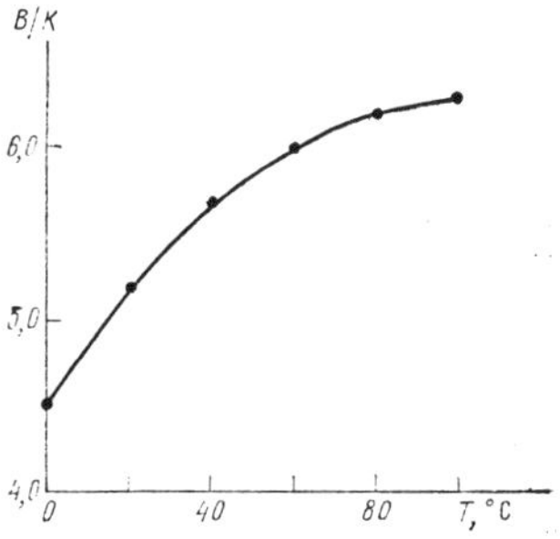

Fig. 13.

An estimate of the magnitude of the terms in Eq. (IV.24) shows that the second term is only several percent of the first term. Therefore, for known values of α_T and c_p the nonlinearity parameter B/K can be roughly calculated from measurements of the isothermal pressure dependence of the velocity of sound, making use of the approximate equation $B/K \approx 2\rho_0 c_0 (\partial c/\partial P)_T$.

§4. Relationship between the acoustic parameters in the second approximation

The addition of nonlinear terms to the equations of hydrodynamics leads to more complicated dependences between the parameters of the ultrasonic field than the simple equations presented in Table 5. Nonlinear corrections to these equations can be easily calculated from relations (IV.13) and (IV.17):

$$c' = c + v = c_0 + \epsilon_0 v, \qquad \text{(IV.25)}$$

where $c = \sqrt{(dP/d\rho)}$. Taking the adiabatic relationship between P and ρ in the form (IV.22), i.e.,

$$P/P_0 = (\rho/\rho_0)^n = (\rho/\rho_0)^{2\epsilon_0 - 1}, \qquad \text{(IV.26)}$$

where $P = P_0 + p$, $\rho = \rho_0 + \Delta\rho$, we have

$$c = c_0(\rho/\rho_0)^{\epsilon_0 - 1} = c_0(P/P_0)^{(\epsilon_0 - 1)/(2\epsilon_0 - 1)}, \qquad \text{(IV.27)}$$

while from Eq. (IV.25)

$$c = c_0 + (\epsilon_0 - 1)v, \qquad \text{(IV.28)}$$

where $c_0 = (dP/d\rho)^{1/2}_{\rho=\rho_0}$. Equating the right sides of relations (IV.27) and (IV.28), we obtain

$$v = \frac{c_0}{\epsilon_0 - 1}\left[\left(\frac{\rho}{\rho_0}\right)^{\epsilon_0 - 1} - 1\right] = \frac{c_0}{\epsilon_0 - 1}\left[\left(\frac{P}{P_0}\right)^{\frac{\epsilon_0 - 1}{2\epsilon_0 - 1}} - 1\right]. \qquad \text{(IV.29)}$$

Solving these equations exactly up to quadratic terms we find in the second-order approximation a relation between the particle velocity v in the forward wave and the variable density $\Delta\rho$ and pressure p:

$$v = \frac{c_0}{\rho_0}\Delta\rho + \frac{\epsilon_0 - 2}{2} c_0 \frac{\Delta\rho^2}{\rho_0^2}, \qquad \text{(IV.30)}$$

$$v = \frac{p}{\rho_0 c_0} - \frac{\epsilon_0}{2(2\epsilon_0 - 1)} \frac{p^2}{P_0 \rho_0 c_0}. \qquad \text{(IV.31)}$$

In the backward wave all signs are reversed. Solving these equations for p and $\Delta\rho$ we have for the forward wave in the second approximation:

$$p = \rho_0 c_0 v + (\epsilon_0/2)\rho_0 v^2; \qquad \text{(IV.32)}$$

$$\Delta\rho = \frac{\rho_0}{c_0} v - \frac{\epsilon_0 - 2}{2} \frac{\rho_0}{c_0^2} v^2. \qquad \text{(IV.33)}$$

Thus the linear relation between the pressure and the particle velocity $p/v = \rho_0 c_0$ is not valid in the second approximation. For the backward wave the signs in front of the first terms in expressions (IV.32) and (IV.33) must be reversed.

We obtained previously a relation between the pressure and the density in the second approximation, for example, in the form of Eq. (IV.14), which, substituting $s = \Delta\rho/\rho_0$, $K = \rho_0 c_0^2$, and $B/K = 2\epsilon_0 - 2$, can be written in the same form

$$p = c_0^2 \Delta\rho + (\epsilon_0 - 1)(c_0^2/\rho_0)(\Delta\rho)^2, \qquad \text{(IV.34)}$$

and, conversely,

$$\Delta\rho = p/c_0^2 - (\epsilon_0 - 1)p^2/(\rho_0 c_0^2). \qquad \text{(IV.35)}$$

It should be emphasized that the acoustic parameters p, $\Delta\rho$, and v, entering into Eqs. (IV.30)–(IV.35) linearly, must be evaluated in the second-order approximation; in the quadratic terms, however, the variables can be substituted for one another using the equations of linear acoustics, i.e., according to Table 5, since the addition of quadratic terms to these variables will lead to third and fourth-order infinitesimals, which can be neglected in the second-order approximation.

§ 5. Distortion of the form of a finite-amplitude wave during propagation

On substituting the expression (IV.17) the equation for the forward finite-amplitude plane wave (IV.12) assumes the form

$$v(x, t) = v_{max} \sin \omega [t - x/(c_0 + \epsilon_0 v)] . \qquad (IV.36)$$

Figure 14 shows the profile of the wave described by expression (IV.36), i.e., the instantaneous distribution of the particle velocity v along the x-axis. According to Eq. (IV.36), each point in this profile, characterized by the particle velocity v, moves along the x-axis with a different local phase velocity $c' = c_0 + \epsilon_0 v$, depending on v. For example, point 1, which corresponds to the maximum positive oscillatory velocity v_{max}, i.e., maximum condensation (the crest of the wave), moves with the velocity $c'_{max} = c_0 + \epsilon_0 v_{max}$; point 3, i.e., the

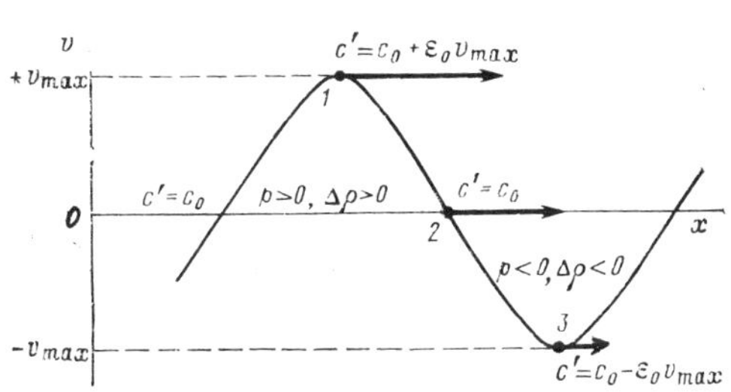

Fig. 14.

phase of maximum rarefaction (the trough of the wave), moves with minimum local velocity $c'_{min} = c_0 - \epsilon_0 v_{max}$; point 2, which corresponds to the phase with zero compression, moves with velocity $c' = c_0$, i.e., with the velocity of infinitesimal-amplitude sound. The remaining "points" propagate with velocity $c' > c_0$ in the compression half-wave ($v > 0$, $\rho > 0$, $\Delta\rho > 0$) and with velocity $c' < c_0$ in the rarefaction half-wave ($v < 0$, $\rho < 0$, $\Delta\rho < 0$). Thus, in a system of coordinates moving along the x-axis with velocity c_0, all points on the wave profile will move with velocity c_0 relative to the "zero" points, which remain stationary in this system of coordinates. As a result, a wave that is sinusoidal at the source* ($x = = 0$, $t = 0$) will become distorted during

*Strictly speaking, for finite-amplitude harmonic oscillations of the source the wave formed by the source will no longer be sinusoidal. This difference is, however, negligible compared to the subsequent distortion of the wave form during propagation.

propagation in the manner illustrated in Fig. 15, acquiring first at some distance a nearly sawtooth (shock-wave) form and then later the form of a "breaking" wave. This form of the wave profile, illustrated in Fig. 15 by the dashed line, is, however, physically unrealistic, because it

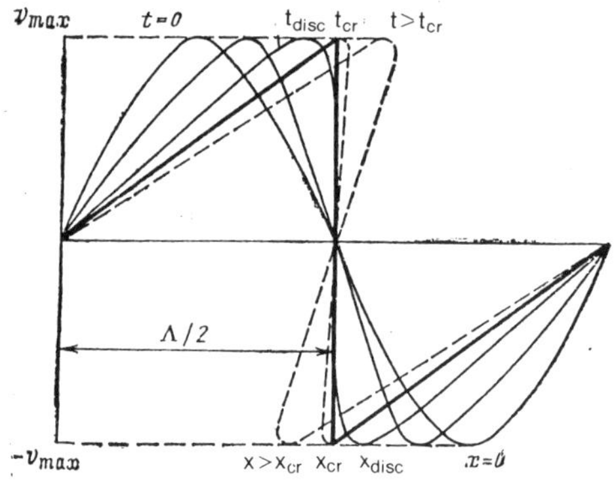

Fig. 15.

corresponds to three values of the particle velocity v at the same point x. Therefore, Riemann's solution, leading to this result, is physically meaningful only as long as the function (IV.36) is single-valued. This function, however, becomes multivalued at some distance from the source x_{disc}, which can be determined from the following condition: at this distance the slope of the leading edge becomes infinite, i.e., $(\partial v/\partial x)_{x=x_{disc}} = \infty$. In addition, the point $x = x_{disc}$ is the point of inflection of the function $v(x,t)$, and therefore $(\partial^2 v/\partial x^2)_{x=x_{disc}} = \infty$. From these conditions, differentiating (IV.36) with respect to x, we find:

$$x_{disc} = c_0^2/(\epsilon_0 \omega v_{max}) = c_0 \Lambda/(2\pi \epsilon_0 v_{max}). \qquad \text{(IV.37)}$$

Thus the starting equations of hydrodynamics can adequately describe a finite-amplitude wave only up to distances $x < x_{disc}$ and for $x > x_{disc}$ they are no longer valid. The reason for this is that the term accounting for the internal friction $\eta(\partial/\partial t)(\partial v/\partial x)$ (see § III.4), which in a real low-viscosity medium can indeed be neglected in studying the propagation of sinusoidal disturbances, is omitted in the equation of motion (IV.2). However, when there are distortions in the form of the wave resulting from nonlinear effects, the velocity gradient $\partial v/\partial x$ at the leading edge of the wave increases, and friction forces increase together with it. Near $x = x_{disc}$ the gradient $\partial v/\partial x \to \infty$ and the sharply increasing viscous losses prevent further distortion of the form of the wave, which begins to be absorbed intensely even in a medium with very low viscosity.

If, as before, we remain within the framework of the model of a hypothetical ideal medium in which there is no internal friction, then the multivaluedness of Riemann's solution for $x > x_{disc}$ for such a medium means that a plane discontinuity, leading to reflection of the wave, forms in the medium. Although in a real medium, with realistic amplitudes of ultrasonic waves, the situation as a rule never reaches a discontinuity*, the term "discontinuity distance" is sometimes used in nonlinear acoustics to denote the distance at which the discontinuity appears.

If we suppose that dissipative processes become significant only in the limit as $\partial v/\partial x \to \infty$, thereby preventing a discontinuity, then the distortions of the wave form will also accumulate beyond the point $x = x_{disc}$, leading to the formation of a sawtooth shock-type wave.** The distance x_{cr} at which the sawtooth wave forms can be found from the condition that at this distance the crest of the wave catches up with the neighboring trough. Here, the crest, moving relative to the trough with relative velocity $2\epsilon_0 v_{max}$, traverses an additional distance equal to one-half the wavelength over a time $t_{cr} = \Lambda/(4\epsilon_0 v_{max})$. Over the same time the entire wave, propagating with average velocity c_0, recedes as a whole from the source to a distance of

$$x_{cr} = c_0 t = c_0 \Lambda/(4\epsilon_0 v_{max}), \qquad \text{(IV.38)}$$

at which a sawtooth wave forms with $v_{max} = const$ and which we shall call the *critical* distance. The quantity x_{cr}, like the previously obtained value of x_{disc}, depends on the wavelength, i.e., on the frequency, of

*Such discontinuities are sometimes observed in the form of layers in plastic metals with low absorption, for example, in aluminum single crystals, subjected to intense ultrasound.

**With regard to sawtooth ultrasonic waves, we can talk about weak periodic shock waves, distinguishing them from aperiodic strong shock waves produced, for example, by intense explosions.

ultrasound. We can therefore introduce a dimensionless critical distance (in units of the wavelength or the number of periods of the oscillations):

$$N_{cr} = \frac{x_{cr}}{\Lambda} = \frac{t_{cr}}{T} = \frac{c_0}{4\epsilon_0 v_{max}}. \tag{IV.39}$$

If we take the limiting distortion in an ideal sawtooth wave to be one, then the inverse of N_{cr} will characterize the degree of distortion over a wavelength or a single period:

$$\Delta = 1/N_{cr} = 4\epsilon_0 v_{max}/c_0 = 4\epsilon_0 Ma. \tag{IV.40}$$

This quantity, naturally, depends on the acoustic Mach number and on the nonlinear properties of the medium. The values of Δ are presented in Table 8 for several values of the ultrasonic intensity in two liquids with the same characteristic impedances but considerably different nonlinear properties and in air under normal conditions. The amplitudes of the displacement velocities v_{max}, the corresponding Mach numbers, the velocity of sound c_0, and the density of the medium ρ_0 are also shown in the table; the critical distances for two frequencies ν ($x_{cr} = c_0/(\nu\Delta)$) are shown in the last column. According to this table, nonlinear distortions in gases at the intensities indicated can reach a considerable magnitude in the immediate vicinity of the source. However, in addition to the low efficiency of emission of ultrasound in a gas, noted above, absorption of ultrasonic waves is also very high in gases. In liquids, on the other hand, nonlinear distortions over a distance of a wavelength do not exceed 1%, even at the highest Mach numbers $\approx 10^{-3}$. For average intensities of the order of 1 W/cm^2, however, the distortion at a distance of one wavelength is a negligibly small quantity, which can always be neglected, assuming that the wave has a perfectly sinusoidal form at the source and that the wave form varies little over a distance of several wavelengths. However, the relatively weak ultrasonic absorption in many low-viscosity fluids makes it possible for this distortion to accumulate during propagation, which leads to the formation of nearly sawtooth waves at comparatively short distances from the source of ultrasound; this is confirmed directly by experiments. As an example, Fig. 16 shows oscillograms

Table 8

Characteristics of the Distortion of Ultrasonic Waves of Different Intensity for Several Media at T = 20°C

Medium	ϵ_0	ρ_0, g/cm^3	c_0, m/s	P_0, atm	I, W/cm^2	v_{max}, m/s
Carbon tetra-chloride	6	1.6	940	1300	0.1 5 100	0.035 0.25 1.1
Water	4	1.0	1490	3200	0.1 5 100	0.036 0.25 1.1
Air	1.2	$1.3 \cdot 10^{-3}$	330	1.0	0.1 5	2.0 14

Continuation of Table 8

Medium	p_{max}, atm	Ma	Δ	x_{cr}, cm	
				0.5 *MHz*	10 *MHz*
Carbon tetra-chloride	0.54 3.8 17.3	$4 \cdot 10^{-5}$ $3 \cdot 10^{-4}$ $1.2 \cdot 10^{-3}$	10^{-3} $7 \cdot 10^{-3}$ $3 \cdot 10^{-2}$	190 27 6.3	9.0 1.4 0.3
Water	0.54 3.8 17.3	$2.4 \cdot 10^{-5}$ $1.7 \cdot 10^{-4}$ $7 \cdot 10^{-4}$	$4 \cdot 10^{-4}$ $3 \cdot 10^{-3}$ $1.1 \cdot 10^{-2}$	750 100 28	37 5 1.4
Air	0.01 0.07	$6 \cdot 10^{-3}$ $4 \cdot 10^{-2}$	0.03 0.2	2.0 0.3	0.1

of the pressure in a plane wave obtained in water at different distances from the source at a frequency of 1 *MHz* with ultrasonic intensities of $I \simeq 50$ W/cm^2.[24] It is evident from the figure that at a distance $x = x_{cr}$ the wave acquires a nearly sawtooth form (Fig. 16c) without a significant decrease in amplitude. At the same time, the absorption of the wave increases sharply and further distortions are delayed due to the decrease in amplitude. Later, dissipative processes smooth out the pressure gradient and decrease the distortion.

The distortion of the form of the wave during propagation in a fluid is also clearly observed in the diffraction of light by large-amplitude ultrasonic waves. This phenomenon is based on the fact that the change in the optical refractive index of the medium, accompanying a change in the density of the medium in the ultrasonic wave (see §II.2), is equivalent to the formation of a phase diffraction grating at whose output the initially plane front of the light beam acquires a "fluted" form repeating the form of the ultrasonic wave profile. Due to phase modulation of the light beam by the ultrasound, a diffraction spot is observed at the focal point of the usual spectrograph, in which the light intensity distribution over the orders of diffraction is uniquely related to the slope of the ultrasonic wavefronts. If the ultrasonic waves have a sinusoidal form, then the diffraction pattern created by them is symmetric. If, on the other hand, the form of the wave is distorted, then a characteristic asymmetry appears in the diffraction orders, which enables the form of the profile of the ultrasonic wave to be reconstructed.[25] Figure 17 shows two diffraction patterns (a, c) and the corresponding ultrasonic wave profile (b, d) in water at two distances from the source (5 *cm* and 30 *cm*), obtained with an initial ultrasonic intensity of 15 W/cm^2 at a frequency of ≃ 570 *KHz*.[26] Under these conditions the maximum distortion of the wave form (about 90%) was observed at a distance of ≃ $1.5\,x_{cr}$ from the source.

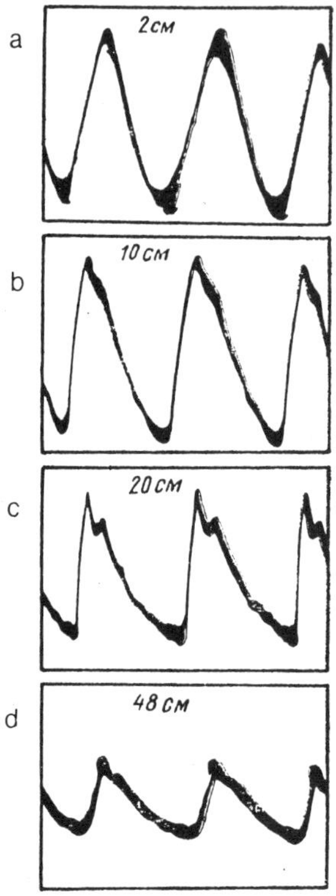

Fig. 16.

Thus, due to dissipative processes, whose intensity increases in a low-viscosity medium with the distortion of the waveform, the actual distance from the source at which the distortion reaches the limiting

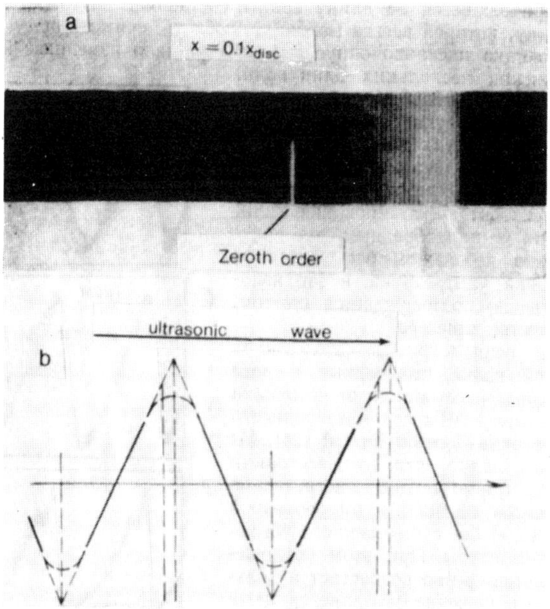

Fig. 17.

magnitude exceeds the value of x_{cr} determined by Eq. (IV.38) and the depth of the leading edge at maximum distortion always remains finite. Nevertheless, the values of x_{cr} and x_{disc} calculated for an ideal medium remain convenient spatial parameters in calculations of different nonlinear effects. These effects, which result from the distortions of the waveform, include a change in the spectral composition of a finite-amplitude wave during propagation, singularities in the absorption characteristics of such waves, the nonlinear interaction of ultrasonic beams, etc.

§ 6. Spectral analysis of a finite-amplitude wave

The distortion of the wave form during propagation is equivalent to the appearance in the wave of higher-order harmonics of the frequency of the fundamental tone ω, fixed by the source. It is not difficult to show that in addition to the fundamental frequency ω the wave (IV.36) also contains its harmonics. To this end, we write Eq. (IV.36) in the form

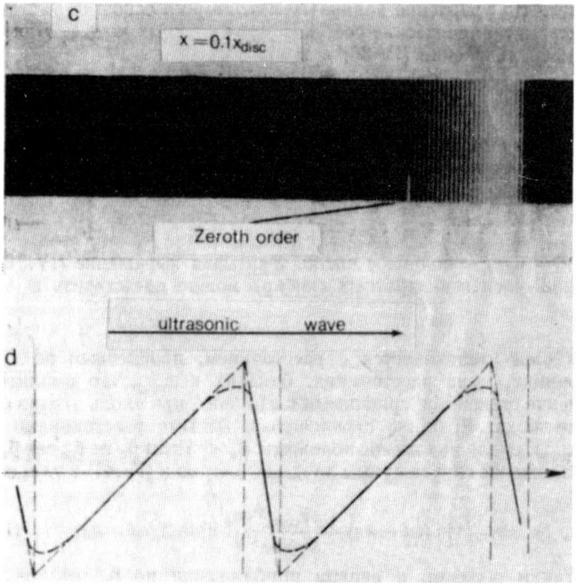

Continuation of Fig. 17.

$$v(x,\ t) = v_{max} \sin \omega \left[t - \frac{x}{c_0} \left(1 + \frac{\epsilon_0 v}{c_0} \right)^{-1} \right].$$

We expand the expression in parentheses in a series in powers of $\epsilon_0 v/c_0$ and, based on the smallness of the Mach numbers v_{max}/c_0, we retain only the first two terms of this expansion, which gives

$$v(x,\ t) \simeq v_{max} \sin [\omega t - kx) + \omega(\epsilon_0 v/c_0^2)x], \qquad \text{(IV.41)}$$

where $k = \omega/c_0$ is the wave number. Using expression (IV.37), the last term in square brackets can be represented in the form

$$(\omega \epsilon_0 v/c_0^2)x \equiv \beta_0 = (v/v_{max})\,(x/x_{disc})\,. \qquad \text{(IV.42)}$$

This term increases with the distance traversed by the wave from the source, and at distances close to x_{disc} its amplitude assumes a value of the order of one for arbitrarily small Mach numbers. If, however, we restrict ourselves to small distances ($x \ll x_{disc}$), then for these distances we can set $\beta_0 \ll 1$, $\sin\beta_0 \simeq \beta_0$, $\cos\beta_0 \simeq 1$, and the expansion of the sine of the sum of two arguments in (IV.41), in this case, yields

$$v(x,t) \simeq v_{max}\sin(\omega t - kx) + \left(\frac{v_{max}}{c_0}\right)^2 \frac{\omega\epsilon_0 x}{2}\sin 2(\omega t - kx) \qquad \text{(IV.43)}$$

Thus we obtain to a first approximation with respect to β_0, aside from the wave of the fundamental tone, a second harmonic (first overtone) with the amplitude $v_{max\,2} = (v^2{}_{max}/c_0^2)(\omega\epsilon_0/2)x$. In this approximation, the amplitude of the second harmonic, which is proportional to the square of the Mach number and the frequency of the fundamental tone, increases linearly with distance from the source. In the next approximation with respect to β_0 we would obtain the third harmonic, fourth harmonic, etc., in accordance with the distortion of the wave which accumulates during propagation.

When the wave acquires a sawtooth form, its spectrum is determined by a Fourier series for a sawtooth function, i.e.,

$$v(x,\ t) = \frac{2}{\pi}v'_{max}\sum_{n=1}^{\infty}\frac{1}{n}\sin n(\omega t - kx), \qquad \text{(IV.44)}$$

where v'_{max} is the amplitude of the sawtooth wave, i.e., the peak value of the particle velocity, which in the ideal situation illustrated in Fig. 15 corresponds to the value of the amplitude of the sinusoidal wave v_{max} at the source. According to Eq. (IV.44), the relation $v_{max\,1} = (2/\pi)v'_{max}$ between the amplitude of the sawtooth wave v'_{max} and the amplitude of its first harmonic ($n = 1$) is satisfied, from which it follows that even if the wave amplitude is constant the amplitude of the first harmonic must decrease as the wave becomes distorted in order to maintain the energy balance, which the approximate equation (IV.43) does not reflect.

To obtain a more detailed picture of the changes in the spectral composition of the wave without the restriction to small values of x/x_{disc}, we shall represent the wave (IV.41) in the form of a Fourier series

$$v(x,\ t) = v_{max}\sum_{n=1}^{\infty}B_n\sin n(\omega t - kx), \quad x < x_{disc}, \qquad \text{(IV.45)}$$

with coefficients

$$B_n = \frac{1}{\pi} \int_0^{2\pi} \sin \psi \sin n(\omega t - kx)\, d(\omega t - kx), \qquad \text{(IV.46)}$$

where $k = \omega/c_0$ is the wave number and

$$\psi \equiv \omega t - kx + \beta_0. \qquad \text{(IV.47)}$$

From Eqs. (IV.47) and (IV.42) we have

$$\omega t - kx = \psi - \beta_0 = \psi - \frac{x}{x_{disc}} \frac{v}{v_{max}} = \psi - \frac{x}{x_{disc}} \sin \psi .$$

Substituting this result into expression (IV.46) we obtain:

$$B_n = \frac{1}{\pi} \int_0^{2\pi} \sin \psi \sin (n\psi - n \frac{x}{x_{disc}} \sin \psi)\,(1 - \frac{x}{x_{disc}} \cos \psi)\, d\psi .$$

Integration of this expression gives

$$B_n = \frac{2x_{disc}}{nx} J_n (n \frac{x}{x_{disc}}), \qquad \text{(IV.48)}$$

where $n = 1, 2, 3, \ldots,$ and J_n is an n-th order Bessel function of the first kind. Substituting expression (IV.47) into Eq. (IV.45), we obtain finally

$$v(x, t) = 2v_{max} \sum_{n=1}^{\infty} \frac{J_n(nx/x_{disc})}{nx/x_{disc}} \sin n(\omega t - kx) . \qquad \text{(IV.49)}$$

This result, which is called the *Bessel–Fubini solution*, is a different form of the general solution of the system of nonlinear equations of hydrodynamics (IV.2) and (IV.3). Expression (IV.49) represents the spectral composition of a finite-amplitude wave as a function of the distance traversed by the wave from the source in the range $0 < x < x_{disc}$. The Bessel–Fubini solution, as also the approximate solution (IV.43), shows that a finite-amplitude wave becomes increasingly nonmonochromatic as it propagates. Increasingly higher harmonics, which are amplified with distance, appear in the spectrum of the wave. In addition, unlike the approximate result (IV.43), the more exact solution (IV.49) includes the reduction of the amplitude of the wave at the fundamental tone due to the transfer of its energy to the higher-order harmonics.

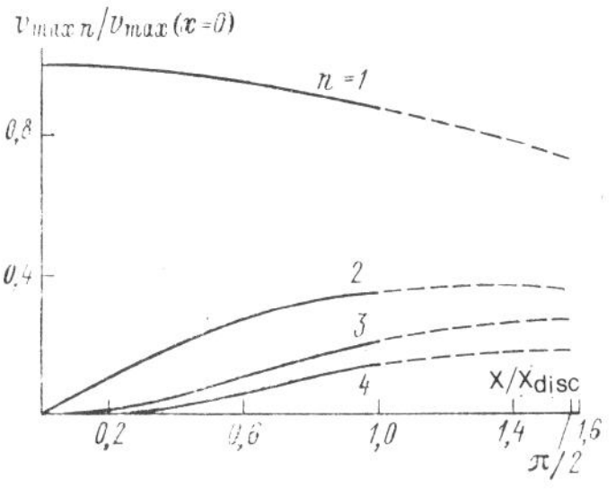

Fig. 18.

The quantitative picture of the change in the spectral composition of the wave in a nondissipative medium, corresponding to the solution (IV.49), is shown in Fig. 18 in the form of the dependence of the relative magnitude of the amplitudes of the first four harmonics v_{maxn}/v_{max} on the dimensionless distance x/x_{disc}, where $x_{disc} = c_0^2/(\omega\epsilon_0 \times v_{max})$ and v_{max1} is the amplitude of the first harmonic at $x = 0$, which equals the amplitude of the wave at the fundamental tone at the source. As is well known from the theory of Bessel functions, the series (IV.40) converges at $x = x_{disc}$ and diverges for $x/x_{disc} > 1$. Substituting $x = x_{disc}$ into Eq. (IV.49) we obtain

$$v(x, t) = v_{max}[0.88 \sin(\omega t - kx) + 0.35 \sin 2(\omega t - kx) + + 0.2 \sin 3(\omega t - kx) + 0.14 \sin 4(\omega t - kx) + \ldots].$$

Thus the spectrum of the wave at a distance $x = x_{disc}$ still differs from the spectrum of a sawtooth wave, in which, according to Eq. (IV.44), the ratio of the amplitudes of the harmonics has the form

$$v_{max\,n}/v_{max\,1} = 1/n \qquad \text{(IV.50)}$$

and which, by definition, forms at a distance

$$x = x_{cr} = (\pi/2)x_{disc}. \qquad \text{(IV.51)}$$

from the source. Extrapolating the curves in Fig. 18 to the values of the amplitudes of harmonics corresponding to (IV.50) at $x = x_{cr}$, we obtain the dashed curves illustrating the change in the spectral composition of a finite-amplitude wave in the segment $x_{disc} \leqslant x \leqslant (\pi/2)x_{disc}$, i.e., up

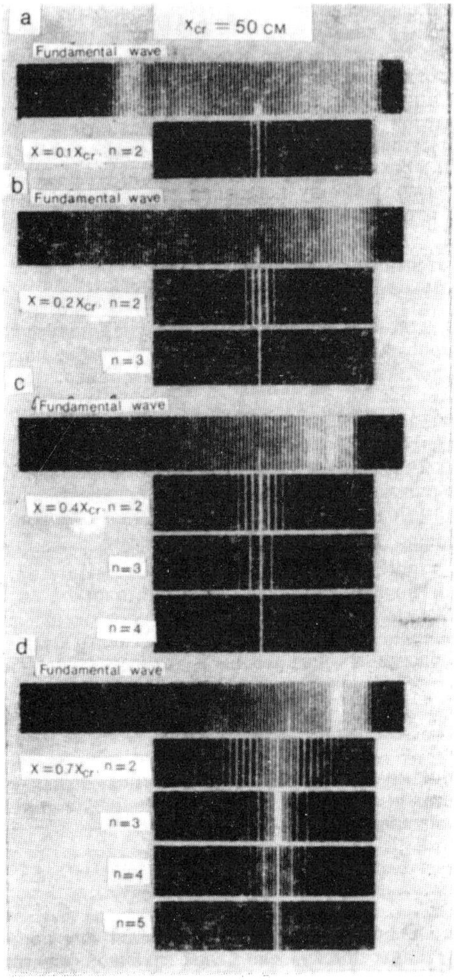
a
$x_{cr} = 50$ cm
Fundamental wave
$x = 0.1 x_{cr}$, $n = 2$
b
Fundamental wave
$x = 0.2 x_{cr}$, $n = 2$
$n = 3$
c
Fundamental wave
$x = 0.4 x_{cr}$, $n = 2$
$n = 3$
$n = 4$
d
Fundamental wave
$x = 0.7 x_{cr}$, $n = 2$
$n = 3$
$n = 4$
$n = 5$

Fig. 19.

to the formation of a sawtooth wave.

The appearance and growth of harmonics during propagation of a large-amplitude ultrasonic wave in a liquid can be clearly observed, for example, in the diffraction of light by placing a filter consisting of a plane-parallel plate (see below) which transmits the chosen harmonic and reflects waves with other frequencies in the path of the ultrasonic beam.[27] In this case, because the wavelength (playing the role of the period of the "ultrasonic grating") of the n-th harmonic singled out by the filter is shorter by a factor of n than the wavelength of the fundamental tone the distance of the diffraction lines ("orders") in the diffraction pattern from the harmonics will be a factor of n greater than in diffraction pattern from a standing wave; however, the number of observed orders depends on the amplitude of the harmonic. Thus, by moving the filter tuned to the given harmonic along the ultrasonic beam it is possible to follow the evolution of harmonics in the large-amplitude wave as it propagates from the source to the filter. Figure 19 shows the diffraction patterns formed by light diffracted by an ultrasonic wave with an intensity of 15 W/cm^2 and a frequency of 573 kHz propagating in distilled water and by its harmonics singled out with the help of glass filters at four distances x from the source.[27] The figure clearly shows how the distortion of the waveform, manifested as an asymmetry of the fundamental diffraction pattern, enriches its spectrum. (The "background" formed by the diffraction bands from the fundamental wave is due to the reflection of the fundamental wave from the inadequately deadened walls of the tank.) The quantitative results agree well with the curves presented in Fig. 18 only at low frequencies for the low-order harmonics and at comparatively short distances from the source.[28] On the whole, however, dissipative processes in real media at ultrasonic frequencies strongly distort the behavior of harmonics at large distances from the source.

§ 7. Intensity of distorted finite-amplitude ultrasonic waves

The equations obtained in § III.3 for the ultrasonic intensity of monochromatic waves, relating the intensity of a sinusoidal wave to its amplitude, for example,

$$I = v_{max}^2 \rho_0 c_0 / 2, \tag{IV.52}$$

where v_{max} is the amplitude of the particle velocity, are no longer valid when the form of the wave is distorted during propagation, and they must be modified. The intensity of a nonmonochromatic wave can be represented as a sum of the intensities of its harmonics: $I = \sum_{n=1}^{\infty} v_{max\,n}^2 \rho_0 c_0/2$, where $v_{max\,n}$ are the amplitudes of the particle velocity in the harmonics, defined as the coefficients of the Fourier expansion of the equation of a finite-amplitude plane wave (IV.45). In particular, when a sawtooth wave with a leading edge of width δ is formed, its Fourier spectrum for the particle velocity at a fixed point is expressed as

$$v(t) = \frac{2v'_{max}}{\pi(1-\delta)} \sum_{n=1}^{\infty} \frac{1}{n} \frac{\sin n\pi\delta}{n\pi\delta} \sin n\omega t,$$

where v'_{max} is the amplitude of the velocity in the sawtooth wave, ω is its fundamental frequency, and $n = 1, 2, 3, \ldots$. Correspondingly, we have for the intensity of the sawtooth wave:

$$I = \frac{2v'^2_{max}}{\pi^2(1-\delta)^2} \sum_{n=1}^{\infty} \frac{1}{n^2} \frac{\sin^2 n\pi\delta}{(n\pi\delta)^2}. \tag{IV.53}$$

This relation assumes its simplest form for a sawtooth wave with a front of infinitesimal depth. Setting $\delta = 0$ in expression (IV.53), we obtain:

$$I = \frac{2v'^2_{max}}{\pi^2(1-\delta)^2} \sum_{n=1}^{\infty} \frac{1}{n^2} \frac{\sin^2 n\pi\delta}{(n\pi\delta)^2}.$$

Comparing this result with Eq. (IV.52), we see that the intensity of the limiting distorted sawtooth wave is 3/2 times less than the intensity of a sinusoidal wave **with the same amplitude**. This in its turn means that with the formation of a sawtooth wave the conservation of energy flux leads to a 1.5-fold decrease in the amplitude of an initially sinusoidal wave, even in a nondissipative medium.

§ 8. Absorption of finite-amplitude plane waves

Qualitative analysis and an assessment of the role of dissipative effects. When a finite-amplitude wave propagates in a real medium, the increase in the gradient of the particle velocity at the leading edge of the wave accompanying nonlinear distortions of the wave form must be accompanied by an increase in dissipative losses due to viscosity and heat conduction in the medium. As a result, the amplitude of the wave will progressively decrease and, therefore, the distortion process will be slowed. At some distance from the source dissipative processes must completely compensate nonlinear effects; further distortion of the wave form then ceases — a process called *waveform stabilization*. Stabilization in the full sense of the word does not actually occur, because with continued propagation the amplitude of the wave continues to decrease, nonlinear effects become weaker, and the wave profile at large distances becomes smoother and the sinusoidal form is restored. Waveform stabilization should thus be understood as the maximum distortion of the wave, and the stabilization distance (x_{stab}) should be understood as the distance from the source at which this distortion is attained. It is true that the term "stable waveform" is justified to a certain extent by the fact that the profile of such a wave changes more slowly than the profile of any other wave with the same amplitude and frequency.

By virtue of the above circumstances the absorption coefficient a finite-amplitude wave (α) is not a constant: it increases with the distance of the wave from the source and with the distortion of the wave form, attaining a maximum value near the region of stabilization, and decreasing thereafter. For this reason, with respect to a finite-amplitude wave we must talk about a *differential absorption coefficient*, which in the region of stabilization can greatly exceed the absorption coefficient of an infinitesimal-amplitude wave α_0 defined by relations (III.37) or (III.40). At fixed distances from the source the differential absorption coefficient likewise depends on the amplitude of the wave at the source.

From the viewpoint of the spectral composition of a finite-amplitude wave the distortion of the wave in the course of propagation is equivalent to the appearance and amplification of harmonics. In addition, the amplitude of the wave at the fundamental tone will progressively decrease not only as a result of direct absorption, but also because the wave's energy is transferred to higher-order harmonics which, since the absorption increases, according to Eq. (III.37), as the square of the frequency, i.e., the square of the number of the harmonic, are absorbed more intensely. Therefore, at some distance from the source the growth of the harmonics ceases and after their relative stabilization the amplitude begins to decrease, i.e., the spectrum of the wave is depleted.

In this connection, it should be noted that the dissipation of the total energy of a finite-amplitude wave differs from the damping of its fundamental harmonic, i.e., the relation (III.47) between the amplitude and intensity coefficients of attenuation for finite-amplitude waves no longer holds. This is clearly evident for an ideal nondissipative medium: there are no energy losses in the medium (i.e., the intensity absorption coefficient equals zero), and the amplitude of the wave at the fundamental tone is damped (see Fig. 18) according to the Bessel–Fubini law (IV.49), i.e.,

$$v_{\max 1}(x) = 2v_{\max 0}(x_{\mathrm{disc}}/x)J_1(x/x_{\mathrm{disc}}), \qquad \text{(IV.54)}$$

where $v_{\max 0}$ is the amplitude of the wave at the source, and J_1 is a Bessel function of the first kind. This attenuation can be assigned a nonvanishing absorption coefficient α_1, which is easily calculated from (IV.54) using the general definition (III.41). For higher-order harmonics, however, the "differential" absorption coefficient in the region of stabilization is generally negative.

The maximum possible distortion of a finite-amplitude wave at the stabilization distance will evidently depend on the ratio of nonlinear and dissipative effects. Nonlinear effects, in their turn, depend on the nonlinearity parameter of the medium and on the amplitude of the wave, while dissipative effects depend on the (shear and bulk) viscosity, the thermal conductivity of the medium, and the ultrasonic frequency. Thus the higher the amplitude of the wave at the source and the lower its absorption, the greater the maximum distortion of the waveform in a given medium will be. In particular, in the case of an ideal nondissipative medium examined above, the limiting, stable, distorted form of the wave with arbitrarily small amplitude and with arbitrary frequency will be a sawtooth wave with infinitely small depth at the leading edge, formed at the distance x_{cr} defined in § 5. In any real medium, as noted there, the depth of the leading edge remains finite, but for very large amplitudes and low absorption the limiting distorted form of the wave can be nearly sawtooth, even in a real medium, while the distance at which this distortion is attained can be close to x_{cr}. In another limiting case, when the amplitude of the wave is small and the ultrasonic absorption in the given medium at a given frequency is large, the stable wave form will be nearly sinusoidal.

Since the differential absorption coefficient α of a finite-amplitude wave attains a maximum magnitude in the region of stabilization of the wave form, i.e., at maximum distortion, and the maximum distortion itself is larger for a given wave amplitude at the source the lower the small-amplitude absorption coefficient α_0, the following important qualitative result follows: the lower the viscosity of the medium and the higher the ultrasonic frequency, the more the absorption of the wave at a given finite amplitude in the region of maximum distortion will exceed the absorption of a monochromatic wave with the same frequency in the same medium. In the limiting case of a nondissipative medium, when $\alpha_0 \to 0$, the ratio $\alpha/\alpha_0 \to \infty$, which is the criterion for the formation of an almost sawtooth wave without a discontinuity. Experiments show that in real low-viscosity media, at megahertz frequencies, the absorption coefficient of large-amplitude waves can be **several orders of magnitude** higher than α_0.

This completes the qualitative picture of propagation of finite-amplitude waves in a dissipative medium. For a quantitative analysis, terms accounting for dissipative losses must be included in the equation of motion (IV.2) and the equation must be solved simultaneously with the nonlinear equation of continuity (IV.3) and the adiabatic equation of state (IV.4). Generally speaking, when the form of the wave is distorted, the process occurring in the wave is no longer adiabatic and in order to describe accurately the propagation of finite-amplitude waves the nonlinear equation of heat conduction must be added to the equations indicated above. As the theory of shock waves shows, however, the deviation from adiabaticity remains small even when the shock-wave front, in which the change in entropy occurs primarily as a result of heat conduction, is crossed. This makes it possible to linearize the heat-conduction equation, retaining the Navier–Stokes equation, to which the nonlinear hydrodynamic term must be added, for the analysis of the propagation of finite-amplitude waves. In this case, the one-dimensional nonlinear Navier–Stokes equation assumes the form

$$\rho \frac{\partial v}{\partial t} + \rho v \frac{\partial v}{\partial x} = -\frac{\partial p}{\partial x} + b \frac{\partial^2 v}{\partial x^2}, \tag{IV.55}$$

where (see Eq. (III.40) and (III.25)) the coefficient

$$b \equiv (4/3)\eta_s + \eta_o + \lambda_0(1/c_v + 1/c_p)$$

includes the coefficients of shear viscosity η_s, bulk viscosity η_b, and thermal conductivity. The role of heat conduction in most media remains, as before, insignificant, so that the quantity b can be taken to be the coefficient of total viscosity η.

Equation (IV.55) differs from the equation of motion (IV.2) only by the addition of the "dissipative" term $b(\partial^2 v/\partial x^2)$. Nevertheless, the rigorous solution of the system of nonlinear equations (IV.55), (IV.3), and (IV.4) for a dissipative medium encounters great difficulties and, for this reason, the system has been solved by different approximate methods. The best results have been achieved with the method of successive approximations, which is based on a preliminary estimate of the ratio of the magnitude of the dissipative term $b(\partial^2 v/\partial x^2)$ to that of the nonlinear term $\rho v(\partial v/\partial x)$ in Eq. (IV.55). Making this estimate for sinusoidal disturbances, as done in § 1, we find that the maximum ratio of the dissipative term to the nonlinear term equals $\rho_0 c_0 v_{max}/(b\omega)$, which is known in hydrodynamics as *Reynolds number*, *Re*. From the linear relation between the amplitude of the particle velocity v_{max} and the pressure amplitude p_{max} in the plane wave (III.10) we find that $Re = p_{max}/(b\omega)$.

Thus the Reynolds number is precisely the quantity that, in the qualitative analysis performed above, determines the maximum degree of distortion of the waveform at the stabilization distance of the wave in a given medium. On the mathematical level, however, it determines the relative magnitude of the nonlinear and dissipative terms in Eq.

(IV.55). If Reynolds number is large, then the dissipative term can be neglected to a first approximation and the system of equations (IV.2)–(IV.4) can be solved in this approximation for a nondissipative medium, including dissipative effects in the next approximation. For low Reynolds numbers, however, the nonlinear term in Eq. (IV.55) can be neglected in the first approximation; we then obtain the linear Navier–Stokes equation (III.27), and the nonlinear effects must be included in the second aproximation.

We shall examine separately the basic consequences of these approximations.

Large Reynolds numbers. The case $Re \gg 1$ concerns high ultrasonic intensity at the source in a medium with low viscosity and comparatively low ultrasonic frequency. We note that the condition $Re \gg 1$ is not incompatible with the smallness of Mach numbers in a high-intensity ultrasonic wave and is realized in the ultrasonic frequency band. Indeed, for ultrasonic intensities of, for example, 100 W/cm^2 in water ($c_0 = 1.5 \cdot 10^3\ m/s$, $\alpha_0/\nu^2 = 25 \cdot 10^{-17}\ s^2/cm$, $p_{max} \simeq 20\ atm$, $v_{max} \simeq 1.5\ m/s$), Mach's number is only $\approx 10^{-3}$, while Reynolds number equals approximately 100 at a frequency of 1 $MHz = 10^6\ s^{-1}$ and approximately 200 at 500 kHz.

The analysis of the equations of hydrodynamics at high Reynolds numbers directly adjoins the case of a nondissipative medium, for which $Re \to \infty$, studied in the preceding section. The approximation which admits the realizable condition $Re \gg 1$ essentially means that absorption can be neglected up to distances $x_{disc} = \Lambda c_0/(2\pi \epsilon_0 v_{max})$ from the source, after which dissipative processes that prevent a discontinuity appear. The distortion of the waveform in this case continues to increase up to the distance $x_{cr} = \pi x_{disc}/2$ until a stable sawtooth wave, whose amplitude decreases with further propagation owing to intense energy dissipation at its leading edge, forms. The law of attenuation of the stabilized sawtooth wave can be determined quite simply if it is assumed that the interval where the finite jump in the particle velocity in the plane of the "discontinuity" occurs is very small, which is justified by the smallness of Mach's numbers.

Using the equation of a finite-amplitude wave in the form (IV.41) and the equality (IV.42), i.e.,

$$v = v_{max} \sin\left(\omega t - kx + \frac{v}{v_{max}} \frac{x}{x_{disc}}\right), \qquad \text{(IV.56)}$$

we shall follow some fixed value of the particle velocity v up to the moment that it reaches the plane of the discontinuity, where a jump in the particle velocity occurs (Fig. 20). The plane of the discontinuity moves in space with the velocity c_0 of the zero phase and up to time t traverses a distance $x = c_0 t$ from the source. Therefore, the position of

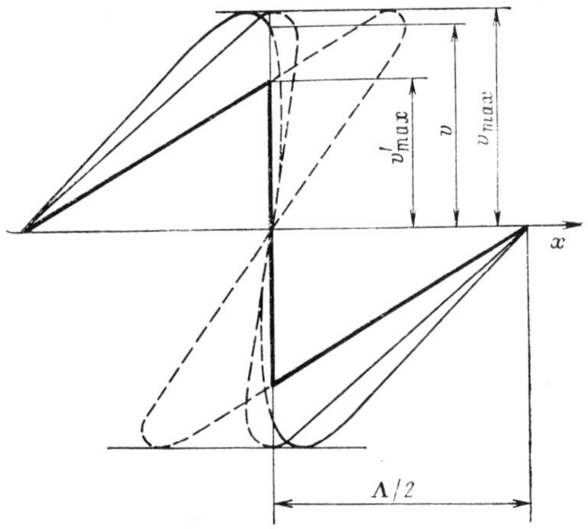

Fig. 20.

the plane of discontinuity is determined by the equation $\omega t - kx = 0$ and by the condition that the jump in the particle velocity at the plane of the discontinuity equals v, according to (IV.56), and will be

$$v = v_{max} \sin\left[(v/v_{max})(x/x_{disc})\right], \qquad \text{(IV.57)}$$

where $x \geqslant x_{disc}$; in addition, the coordinate $x = x_{disc}$ corresponds to the coordinate at which the point on the wave profile with the particle velocity $v = 0$ appears on the plane of discontinuity, which agrees with the previous definition of the distance x_{disc} as the distance up to the discontinuity. As the distance to the plane of discontinuity increases further, increasingly higher values of the particle velocity v appear on the surface, i.e., the velocity jump at the leading edge increases up to a distance from the source $x = x_{cr} = \pi x_{disc}/2$. At this distance $v = v_{max} = v'_{max}$, which corresponds to the formation of a sawtooth wave whose amplitude v'_{max} then decreases (see Fig. 20) according to the same law (IV.57), where v'_{max} must be substituted for v, i.e.,

$$v'_{max} = v_{max} \sin [(v'_{max}/v_{max})(x/x_{disc})]. \qquad \text{(IV.58)}$$

This law is especially simple for large values of x/x_{disc}, when the quantity $\sin[v'_{max}\, x/(v_{max}\, x_{disc})]$ can be expanded about the point $[v'_{max}\, x/(v_{max}\, x_{disc})] = \pi$. Indeed, setting $[v'_{max}\, x/(v_{max}\, x_{disc})] = \pi - \delta$, where $\delta \ll \pi$, we have

$$\sin \frac{x}{x_{disc}} \frac{v'_{max}}{v_{max}} = \sin(\pi - \delta) = \sin\delta \simeq \delta = \pi - \frac{v'_{max}}{v_{max}} \frac{x}{x_{disc}}.$$

Substituting this result into (IV.58) we obtain an equation describing the decrease in the amplitude of the sawtooth wave by not less than a factor of 2–3 at distances exceeding x_{cr}:

$$v'_{max} = v_{max}\pi/(1 + x/x_{disc}), \qquad \text{(IV.59)}$$

where, by definition, v_{max} is the amplitude of the wave at the source and x is the distance from the source. In deriving this equation, which is based on relation (IV.58) in which $v_{max} = const$, the decrease of the amplitude on the section where the sawtooth wave forms from x_{disc} to x_{cr} was neglected. Thermodynamic calculations, however, lead to the same expression for v_{max} with an insignificant expected difference in the numerical coefficient, which can be ascribed to the indicated error, interpreting v_{max} in Eq. (IV.58) and (IV.59) as the actual initial amplitude of the sawtooth wave.

A very important property of the propagation of intense ultrasound over large distances from the source follows immediately from relation (IV.59). For $x \gg x_{disc}$ this relation gives

$$v'_{max} = v_{max}\pi x_{disc}/x = c_0\Lambda/(2\epsilon_0 x), \qquad \text{(IV.60)}$$

i.e., **the amplitude of the sawtooth wave at a large distance from its**

source does not depend on the amplitude of the oscillations of the source. In this case, however, the amplitude at the source must attain a magnitude such that the sawtooth wave is formed at a distance $\simeq x_{disc}$, less than the given fixed distance x. As the ultrasonic intensity is further increased the amplitude of the sawtooth wave at this distance x will approach asymptotically the quantity defined by Eq. (IV.60); a subsequent increase of the wave amplitude at the source will be completely compensated by its decay in the region of formation and propagation of the sawtooth wave. Equation (IV.60) therefore determines the limiting magnitude of the amplitude of the particle velocity in the sawtooth wave $(v'_{max})_{max}$, which is reached at a fixed distance from the source in a given medium. Because sawtooth waves are formed at large Reynolds numbers, however, this equation determines the criterion for transmission of intense ultrasound in a low-viscosity medium over comparatively large distances from the source. This criterion turns out to be quite strict. Thus, for water ($c_0 = 1.5 \cdot 10^3$ m/s, $\epsilon_0 = 4$) at a frequency of 0.5 MHz for the maximum intensity that can be transmitted to a distance of 1 m away from the source ($x = 330\ \Lambda$) it gives $I_{max} = \rho_0 c_0 (v'_{max})^2_{max}/2 \simeq 30$ W/cm^2, and at 1.5 MHz ($x = 1000\Lambda$) the intensity is only $\simeq 3$ W/cm^2.

These results agree, at least qualitatively, with experiment (Fig. 21).[19] Part of the disagreement is attributable to the neglected factors, such as absorption and scattering of ultrasound prior to the formation of a shock wave, smearing of the wave front due to absorption beyond the region of formation, diffraction effects in a real ultrasonic beam, etc., which on the whole decrease the computed maximum values of the intensity. Equation (IV.60) thus gives the theoretical upper limit.

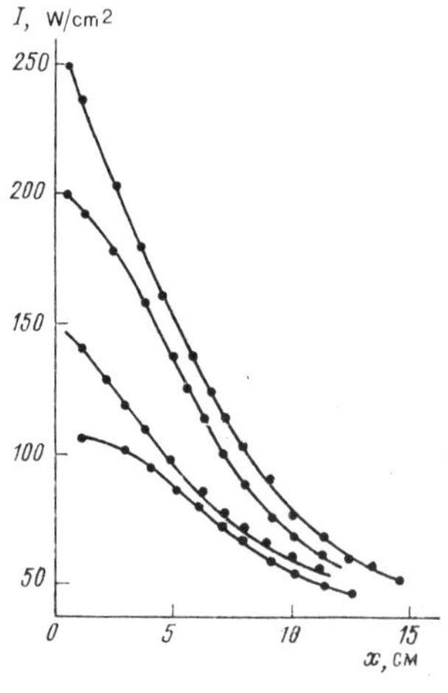

Fig. 21.

Starting from relation (IV.59), we find the amplitude absorption coefficient of the sawtooth wave

$$\alpha = -\frac{1}{v'_{max}}\frac{dv'_{max}}{dx} = \frac{1}{x + x_{disc}}. \qquad (IV.61)$$

Thus at large distances from the region where the sawtooth wave forms ($x >> x_{disc}$) the absorption coefficient of the wave decreases inversely as the reciprocal of the distance. Dividing Eq. (IV.61) by the expression for the absorption coefficient of an infinitesimal-amplitude wave $\alpha_0 = b\omega^2/2\rho_0 c_0^3$ we obtain

$$\frac{\alpha}{\alpha_0} = \frac{2\epsilon_0\rho_0 c_0 v_{max}}{(1 + x/x_{disc})\,b\omega} = \frac{2\epsilon_0 Re}{1 + x/x_{disc}}, \qquad (IV.62)$$

where $Re = \rho_0 c_0 v_{max}/b\omega$ is Reynolds number at the ultrasonic source. Using relation (IV.59) it is also possible to introduce an "intantaneous" Reynolds number, which is more convenient for measurements, in the sawtooth wave:

$$Re_x = \rho_0 c_0 v_0/(b\omega) = \pi Re/(1 + x/x_{disc}), \qquad (IV.63)$$

Expression (IV.62) then assumes the form

$$\alpha/\alpha_0 = (2/\pi)\,\epsilon_0 Re_x. \qquad (IV.64)$$

In ultrasonic technology the amplitude of the first harmonic (the wave of the fundamental tone) is usually measured. From the fact that, according to expansion (IV.44), the amplitude of the particle velocity in the first harmonic $v_{max\ 1}$ of the sawtooth wave is related to its peak value v'_{max} by the relation $v_{max} = (2/\pi)\,v'_{max}$ and introducing the instantaneous Reynolds number for the first harmonic $Re_{1x} = \rho_0 c_0 v_{max}/(b\omega) = p_{max\ 1}$, where $p_{max\ 1}$ is the pressure amplitude in the first harmonic, we obtain

$$\alpha/\alpha_0 = \epsilon_0 Re_{1x}, \qquad (IV.65)$$

where ϵ_0 is the nonlinear coefficient. The excess of the absorption coefficients of the sawtooth wave or its first harmonic above the absorption coefficient α_0 of a small-amplitude wave is maximum in the region of formation of the sawtooth wave where the instantaneous Reynolds number assumes its highest value; in addition, this excess is determined not only by the Reynolds number (which in low-viscosity fluids can reach values of several hundreds and even thousands at not very high frequencies) but also by the nonlinear coefficient ϵ_0, which for liquids falls in the range 4–6. Then, the absorption coefficients α/α_0 and α_1/α_0 decrease with distance according to the law (IV.61). In addition, the relations (IV.62), (IV.64), and (IV.65) remain in force as long as the sawtooth form of the wave exists. When the leading edge of the wave is smeared, the absorption coefficient of the wave decreases. Equation (IV.64) therefore determines the **maximum excess absorption coefficient** α/α_0 for fixed Reynolds number in a medium with a nonlinear coefficient ϵ_0, and this maximum value of α/α_0 is realized in the sawtooth wave. It is interesting to note that the expressions for the absorption coefficient α of the sawtooth wave (IV.61)– (IV.64) do not include the dissipative characteristics of the medium; the absorption coefficient depends only on the jump in the particle velocity (pressure, density, etc.) at the leading edge of the wave. Actually, dissipative processes were implicitly included in the derivation of Eq. (IV.65) with the help of the relations which determine the magnitude of this jump.

A more detailed analysis of the structure of the leading edge of the shock wave for a given value of the Reynolds number gives the following expression for the dimensionless depth of the front[19]:

$$\delta = (1 + x/x_{disc})/(\pi \epsilon_0 Re), \tag{IV.66}$$

where Re is the Reynolds number at $x = 0$ and the depth of the front δ equals the ratio of its thickness to the wavelength Λ, so that δ is related to the degree of distortion of the wave Δ (see (IV.40)) by the relation $\delta = (1 - \Delta)/2$.

According to expression (IV.66), the depth of the shock-wave front is smallest in the region where the discontinuity appears ($x \simeq x_{disc}$) and thereafter it increases in proportion to x. But, in addition, as we have already noted, the excess coefficient of the wave also decreases. Furthermore, as $Re \to \infty$, the depth of the front $\delta \to 0$; however, the actual value of Reynolds number in a given medium is limited by the dissipative properties of the medium. Therefore, dissipative processes also limit the depth of the shock-wave front and thereby its absorption, which thus indeed depends only on the pressure (velocity) jump, which also determines the magnitude of Re. We can obtain a quantitative relation between the excess absorption of the sawtooth wave and the depth of its front δ by comparing expressions (IV.66) and (IV.62), which gives

$$\alpha/\alpha_0 = 2/(\pi\delta), \tag{IV.67}$$

and for the first harmonic measured in the experiment, according to (IV.65), we have $\alpha_1/\alpha_0 = 1/\delta$. In addition, these relations are valid for strong distortions, i.e., for $\delta \ll 1/2$. Thus the depth of the front of the distorted wave δ or the degree of its distortion Δ can be determined by measuring the excess absorption coefficient of the fundamental harmonic.

An approximate expression for the spectral composition of a finite-amplitude wave for large Reynolds numbers can be obtained by combining the Bessel–Fubini solution (IV.49) for $0 \leqslant x \leqslant x_{disc}$ and expanding the sawtooth wave (IV.44) in a Fourier series for $x \gg x_{disc}$ taking into account its attenuation in accordance with (IV.59), which gives

$$v(x, t) = v_{max} \sum_{n=1}^{\infty} B_n \sin n(\omega t - kx), \tag{IV.68}$$

where v_{max} is the amplitude of the wave at the source, k is its wave number, and

$$B_n = \begin{cases} \dfrac{2x_{disc}}{nx} J_n\left(n\dfrac{x}{x_{disc}}\right), & 0 \leqslant x \leqslant x_{disc}, \\[2ex] \dfrac{2}{n(1 + x/x_{disc})}, & x \gg x_{disc}, \end{cases}$$

where J_n is the n-th order Bessel function of the first kind. Joining these solutions at the point $x = x_{cr}$ (the point x/x_{disc} in Fig. 18), we obtain the picture of the change in the spectral composition of the wave for large Reynolds numbers illustrated in Fig. 22, where the dashed sections for the amplitudes of the harmonics correspond to regions where the wave amplitude varies according to (IV.58).

Thus the amplitudes of the higher-order harmonics reach maximum magnitudes in the region of formation of the sawtooth wave, where according to (IV.65) the absorption of the fundamental (first) harmonic is highest. At distances somewhat greater than x_{cr}, the amplitudes of all harmonics (including the fundamental harmonic) decay according the same law $v_{max\,n} = 2v_{max}/[n(1 + x/x_{disc})]$ with the same absorption coefficient $\alpha_n = 1/(x + x_{cr}) = \alpha$, which is characteristic for a sawtooth wave and corresponds to conservation (stabilization) of its form, i.e., conservation of high instantaneous Reynolds numbers Re_x, for which the depth of the shock front is much smaller than the wavelength Λ. As the amplitude decreases, however, the shock front of the wave is smeared, i.e., its depth increases. If the finite depth of the front δ determined by relation (IV.66) and its relation to the attenuation in the form (IV.67) are included in the calculation, then the Fourier expansion will have the form

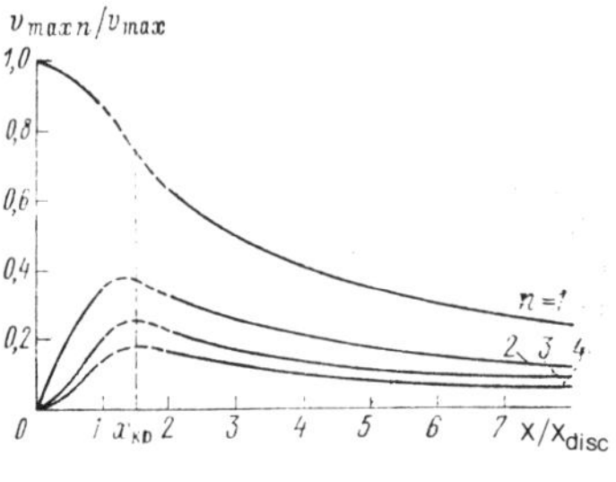

Fig. 22.

$$v = \frac{v_{max}}{\epsilon_0 Re} \sum_{n=1}^{\infty} \frac{\sin n(\omega t - kx)}{\mathrm{sh}[n(1 + x/x_{cr})/(2\epsilon_0 Re)]} \qquad \text{(IV.69)}$$

where Re is the initial Reynolds number and $x > x_{cr}$. For large values of Re the expression (IV.69) is the same as (IV.68) for $x = x_{cr}$. For strong smearing of the shock front at large distances from the source, when the wave again approaches a sinusoidal form, relation (IV.69) gives approximately

$$v = \frac{v_{max}}{\epsilon_0 Re}[e^{-\alpha_0 x}\sin(\omega t - kx) + e^{-2\alpha_0 x}\sin 2(\omega t - kx) + \ldots]. \text{(IV.70)}$$

Thus the absorption coefficients of the harmonics differ at large distances from the region of formation of the shock wave: the high-order harmonics are absorbed more strongly than the low-order harmonics. This difference, however, is still smaller than expected from the quadratic frequency dependence of the absorption coefficient: in Eq. (IV.7) the absorption coefficient of the second harmonic is only two, and not four, times greater than that of the first harmonic. This is attributable to the continuous transfer of energy from the low-order harmonics to the high-order harmonics along the entire propagation path of the wave up to the point at which its initial sinusoidal form is restored at distances satisfying the condition $\alpha_0 x \gg 1$, for which relation (IV.70) gives

$$v \simeq [v_{max}/(\epsilon_0 Re)] \exp [(-\alpha_0 x) \sin(\omega t - kx)] .$$

If, however, we include the fact that for $\alpha_0 x \gg 1$, $\exp(-\alpha_0 x) \simeq 1/(\alpha_0 x)$, where $\alpha_0 = b\omega^2/(2\rho_0 c_0^3$ and $Re = \rho_0 c_0 v_{max}/(b\omega)$, then based on (IV.70) we again arrive at the conclusion that at large distances from the region of the discontinuity the wave amplitude does not depend on the intensity of the radiation from the source, and for the limiting amplitude at a fixed distance $x \gg x_{cr}$ we obtain the expression $v_{max} = c_0\Lambda/(\pi\epsilon_0 x)$, which to within $\pi/2$ agrees with the previous result (IV.60).

Small Reynolds numbers ($Re \ll 1$). The small degree of distortion of the wave form at its stabilization distance is of greatest interest in the analysis of the conditions under which the propagation of real waves with finite but small amplitude obeys the laws of linear acoustics. This case is almost always realized, for example, in measurements of the velocity and absorption of ultrasound: the intensities used in this case match the losses due to absorption and the Reynolds numbers $Re = p_{max}/(b\omega)$, as a rule, remain small.

For small Reynolds numbers the problem of the propagation of finite (but small) amplitude waves in a viscous medium can be solved by the method of successive approximations, in which the solution for the acoustic parameters v, $\Delta\rho$, etc. is sought in terms of series expansions:

$$v(x, t) = v_1 + v_2 + \ldots, \quad \Delta\rho = \rho - \rho_0 = \rho_1 + \rho_2 + \ldots, \tag{IV.71}$$

where $v_2 \ll v_1$ and $\rho_2 \ll \rho_1$. Substitution of these series into the exact equations of hydrodynamics (IV.3) and (IV.5) enables the linear equations of the first approximation, including viscous losses, to be separated from the equations of the second approximation, which include the second-order nonlinear terms. Solving these equations and adding the results obtained, we find, according to (IV.71), the complete solution in the second approximation including weak nonlinear effects, in which the third-order terms can be neglected.

The system of linear equations of hydrodynamics including viscous losses was solved in § III.4. For harmonic oscillations of a flat source and low absorption over a wavelength ($\alpha_0\Lambda \ll 1$), neglecting heat conduction in the medium, this solution (i.e., the first approximation) describes a damped plane wave with frequency ω:

$$v_1(x, t) = v_{\max 1}\exp[(-\alpha_0 x)\sin(\omega t - kx)], \qquad \text{(IV.72)}$$

where $\alpha_0 = b\omega^2/(2\rho_0 c_0^3)$ and $b = \eta = (4/3)\eta_s + \eta_b$.

In the second approximation, the system of nonlinear equations, including the equation of state in the second approximation (IV.14), assumes the following form:

$$\frac{\partial \rho_2}{\partial t} + \rho_0\frac{\partial v_2}{\partial x} + \frac{\partial}{\partial x}(\rho_1 v_1) = 0;$$

$$\text{(IV.73)}$$

$$\rho_0\frac{\partial v_2}{\partial t} + \rho_1\frac{\partial v_1}{\partial t} + \rho_0 v_1\frac{\partial v_1}{\partial x} + c_0^2\frac{\partial \rho_2}{\partial x} - b\frac{\partial^2 v}{\partial x^2} + B\frac{\partial}{\partial x}\left(\frac{\rho_1}{\rho_0}\right)^2 = 0,$$

where $B = \rho_0^2(\partial c^2/\partial \rho)_{\rho=\rho_0}$ and $c^2 = \partial p/\partial \rho$ is the coefficient in front of the quadratic term in the equation of state. The solution of the equations in the second-order approximation (IV.73) gives *), [29,30]

$$v_2(x, t) = \frac{\rho_0 c_0 \epsilon_0 v_{\max 1}^2}{2b\omega}(e^{-2\alpha_0 x} - e^{-4\alpha_0 x})\sin 2(\omega t - k) \qquad \text{(IV.74)}$$

and, in addition, $p_2 = \rho_0 c_0 v_2$, $\Delta\rho_2 = v_2\rho_0/c_0$, i.e., it determines the relation between the acoustic parameters in the second-order approximation, analogous to the relations of the first approximation. Equation (IV.74) describes a damped plane wave with frequency 2ω, i.e., the second harmonic of the fundamental wave (IV.72). In the next approximation we would obtain the third harmonic, etc. However, the condition of applicability of the approximation used $v_2/v_1 \ll 1$ gives

*A calculation of the attenuation of the wave due to the diffraction of a finite beam of quasiplanar finite-amplitude waves leads to an analogous result.[31]

$$\left(\frac{v_2}{v_1}\right)_{max} \simeq \frac{\rho_0 c_0 \epsilon_0 v_{max\,1}}{2b\omega} = \frac{\epsilon_0 p_{max\,1}}{2b\omega} = \frac{3}{2} Re \ll 1. \quad \text{(IV.75)}$$

In this case the third harmonic can be neglected and the complete solution for $v(x, t)$ can be represented in the form

$$v(x, t) = v_1 + v_2 = v_{max\,1} e^{-\alpha_0 x} \sin(\omega t - kx) + + v_{max\,2}(e^{-2\alpha_0 x} - e^{-4\alpha_0 x}) \sin 2(\omega t - kx), \quad \text{(IV.76)}$$

where, according to Eq. (IV.74),

$$v_{max\,2} = \frac{\rho_0 c_0 \epsilon_0 v^2_{max\,1}}{2b\omega} = \frac{v_{max\,1} \epsilon_0 Re}{2} = \frac{\epsilon_0 \omega v^2_{max\,1}}{4c_0^2 \alpha_0}.$$

The solution obtained in this manner has the drawback that it includes damping of the fundamental wave only as a result of dissipative processes, but, as also the approximate solution for a nondissipative medium, it does not reflect the transfer of energy in the fundamental wave into the second harmonic. This can be taken into account by subtracting the known energy of the second harmonic from the energy of the wave at the fundamental tone. This, however, gives a negligible correction, because in accordance with condition (IV.75) the amplitude of the second harmonic is small. For the second harmonic, however, the fact that it is correlated with the wave at the fundamental tone is already included together with the direct absorption of the medium. According to Eq. (IV.74), the amplitude of the second harmonic equals zero at the origin and for small x increases with distance approximately linearly: for $\alpha_0 x \ll 1$, $\exp(-2\alpha_0 x) - \exp(-4\alpha_0 x) \simeq 2\alpha_0 x$. Substituting this expression into Eq. (IV.74) we obtain $v_{max\,2} = \epsilon_0 \omega v^2_{max\,1} x/(2c_0^2)$, which is identical to the result (IV.43) obtained previously for a nondissipative medium. With further propagation, however, at a distance $x_{stab1} = \ln 2/(2\alpha_0)$ the growth of the second harmonic slows down, and its amplitude reaches a maximum and then decreases. In this case, the second harmonic decays more rapidly for $x > x_{stab\,1}$ than the wave at the fundamental tone, but more slowly than the simple absorption of the wave at the doubled frequency. The ratio of the amplitude of the second harmonic to the amplitude of the fundamental wave

$$\frac{v_{max\,2}}{v_{max\,1}} = \frac{\rho_0 c_0 \epsilon_0 v_{max\,1}}{2b\omega}(e^{-\alpha_0 x} - e^{-3\alpha_0 x}) \qquad (IV.77)$$

has a maximum value at the point

$$x_{stab} = \ln 3/(2\alpha_0), \qquad (IV.78)$$

i.e., at a somewhat larger distance from the source. At this distance the wave form undergoes maximum distortion. At $x = x_{stab}$ expression (IV.77) gives

$$\left(\frac{v_{max\,2}}{v_{max 1}}\right)_{max} = \frac{\rho_0 c_0 \epsilon_0 v_{max\,1}}{3\sqrt{3}b\omega} = \frac{\epsilon_0 p_{max\,1}}{3\sqrt{3}b\omega} = \frac{\epsilon_0 Re}{3\sqrt{3}}. \qquad (IV.79)$$

In the limiting case of a sawtooth wave (degree of distortion $\Delta = 1$) this ratio, as we know, equals 1/2. Since our solution was obtained for Re << 1, according to Eq. (IV.79), $(v_{max\,2}/v_{max\,1})_{max} << 1/2$, i.e., the degree of maximum distortion of the wave in this case is much less than one; the stable wave form is therefore far from a sawtooth form and its profile is only a slightly distorted sinusoid.

If the distribution of the particle velocities in the wave (IV.76) is known, it is possible to find the amount of energy lost by the wave due to dissipation and to calculate the differential absorption coefficient. The following expression is then obtained for the amplitude absorption coefficient of the **entire wave** [30]:

$$\alpha = \alpha_0 + 3\alpha_0\left(\frac{\epsilon_0 \rho_0 c_0 v_{max\,1}}{2b\omega}\right)^2 (e^{-\alpha_0 x} - e^{-3\alpha_0 x})^2 =$$

$$= \alpha_0[1 + \frac{3\epsilon_0^2 Re^2}{4}(e^{-\alpha_0 x} - e^{-3\alpha_0 x})^2]. \qquad (IV.80)$$

Dividing this by α_0 and introducing the notation $\Phi(\alpha_0 x) \equiv \exp(-\alpha_0 x) - \exp(-3\alpha_0 x)$ we obtain

$$\alpha/\alpha_0 = 1 + (3\epsilon_0^2 Re^2/4)\Phi(\alpha_0 x). \qquad (IV.81)$$

Therefore, the relative coefficient of absorption of a finite-amplitude wave for small Re numbers changes with distance from the source as the function $\Phi(\alpha_0 x)$, whose graph is presented in Fig. 23. At $x = 0$ and $x = \infty$, $\Phi(\alpha_0 x) = 0$ and $\alpha = \alpha_0$. The excess absorption reaches a maximum value at the point x_{max}, where the distortion of the wave form is also a maximum. Substituting into (IV.81) the value $x_{max} = x_{stab} = \ln 3/(2\alpha_0)$ we obtain

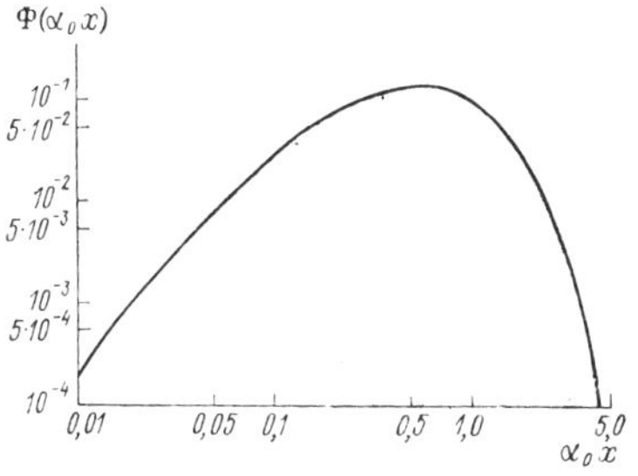

Fig. 23.

$$(\alpha/\alpha_0)_{max} = 1 + 0.12\epsilon_0^2 Re^2. \tag{IV.82}$$

Thus, as in the case $Re \gg 1$, the excess absorption of a finite-amplitude wave in the region of maximum distortion is determined only by the nonlinear coefficient of the medium ϵ_0 and by the Reynolds number, but for small Reynolds numbers the relative absorption coefficient increases in proportion to Re^2, in contrast to the linear dependence obtained previously (see relation (IV.62)).

The theoretical results obtained for the relative absorption coefficient for $Re < 1$ and $Re \gg 1$ are in very good agreement with the experimental data. As an illustration, Fig. 24 summarizes the measurements of the absorption coefficient of ultrasonic waves with different amplitude in water at the waveform stabilization distance, i.e., in the region of maximum absorption. The local Reynolds numbers, determined from the pressure amplitude at the point where the measurement is made, are plotted along the abscissa axis. The continuous curve is the theoretical curve constructed for $\epsilon_0 = 4$. The points refer to measurements performed by different methods at different ultrasonic frequencies in the range 1–10 *MHz*.

In conclusion, we note that Eq. (IV.82) gives the criterion for determining the maximum Reynolds number at which the contribution of nonlinear effects to the measured ultrasonic absorption can be neglected. Indeed, if the error in the absorption measurements, performed by a given method, is $\Delta\alpha/\alpha_0$, then according to (IV.82)

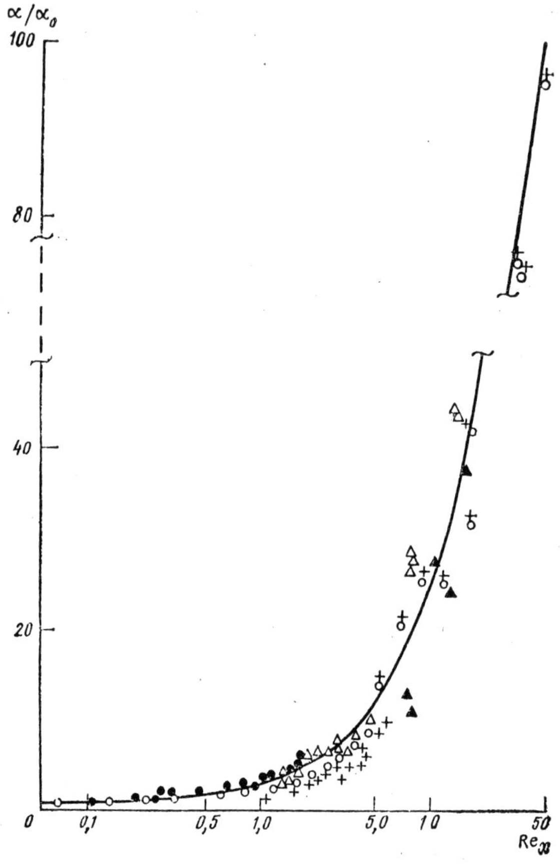

Fig. 24.

nonlinear effects will exceed the random errors in the experiment if

$$\left|\frac{\Delta\alpha}{\alpha_0}\right| > \frac{\alpha_{max} - \alpha_0}{\alpha_0} = 0.12\epsilon_0^2 Re^2. \qquad \text{(IV.83)}$$

Thus, for an accuracy of absorption measurements $\simeq$ 10%, based on condition (IV.83) we obtain: $Re_{max} \leqslant 0.7$ for gases ($\epsilon_0 \simeq 1.2$) and $Re_{max} \leqslant 0.2$ for liquids ($\epsilon_0 \simeq 5$). Taking into account the fact that Reynolds number is related to the pressure amplitude and the ultrasonic intensity by the relations $p_{max} = 2\alpha_0\rho_0 c_0^3\, Re/\omega$ and $I = 2\alpha_0^2\rho_0 c_0^5 Re/\omega^2$, it is not difficult to calculate the limiting magnitudes of the acoustic pressures and intensities at which the contribution of nonlinear effects to the measured ultrasonic absorption can be neglected even for measurements in the region of waveform stabilization, where this contribution is highest. The results of such a calculation for several media in which the absorption coefficient of a low-amplitude wave depends quadratically on frequency are presented in Table 9, where the waveform stabilization distances ($x_{stab} \simeq 0.5/\alpha_0$) and the Mach numbers Ma, which determine the absolute magnitude of the nonlinear effects, are also indicated. The data presented in Table 9 are for room temperature.

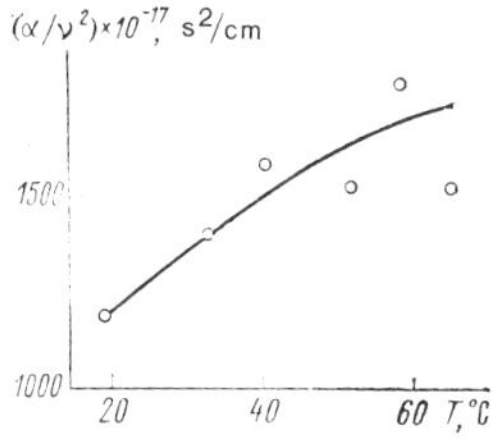

Fig. 25.

As the temperature increases, the viscosity of liquids decreases and the Reynolds number increases. If the absorption coefficient of a monochromatic wave α_0 depends on the temperature more strongly than does the viscosity, then the absorption of a finite-amplitude wave can increase with temperature due to the increase in nonlinear effects. This is illustrated by the experimental curve presented in Fig. 25, showing the anomalous temperature dependence of ultrasonic absorption in transformer oil at a frequency of 1.5 MHz and for an intensity of 9 W/cm^2 at the source.[32]

As is evident from Table 9, nonlinear effects can appear in strongly absorbing liquids only for very high ultrasonic intensities. In low-viscosity liquids, such as water, alcohol, and others, nonlinear effects

Table 9

Pressure Amplitudes, Ultrasonic Intensity, and

Medium	$\alpha_0\gamma^2\cdot 10^{17}$, s^2/cm	ρ_0, g/cm^3	$c_0\cdot 10^{-5}$, cm/s	$\gamma\cdot 10^{-6}$, s^{-1}
Water	25	1.0	1.5	1.0 10
Methyl alcohol	34	0.8	1.12	1.0 10
Mercury	6.0	13.6	1.45	1.0 10
Benzene	900	0.88	1.3	1.0 10
Glycerin	2500	1.26	1.9	1.0 10
Air	$2\cdot 10^4$	$1.3\cdot 10^{-3}$	0.33	1.0

are already appreciable in ultrasonic absorption at 1 *MHz* at intensities of the order of 10^{-4} *W/cm*, which correspond to acoustic Mach numbers $\simeq 10^{-6}$. In this case, the nonlinear terms proportional to Ma^2 in the equations of hydrodynamics are only $\simeq 10^{-12}$. Ultrasonic absorption is thus very sensitive to nonlinear effects and it gives rise to the strongest criterion for the smallness of the amplitude of a real wave in a specific medium. This criterion, as we can see, depends on the ultrasonic frequency and on the accuracy of the measurements of the absorption coefficient, as well as on the region in which the measurements are performed.

Condition (IV.83) and the corresponding limiting amplitude values presented in Table 9 concern regions of waveform stabilization in which the contribution of nonlinear effects to absorption is maximum. In its turn, however, the distance of stabilization x_{stab}, by definition

Continuation of Table 9

Mach Numbers for $\Delta\alpha/\alpha_0 \simeq 10\%$ ($T = 20°C$)

p_{max}, atm	I, W/cm^2	Ma	x_{stab}, cm
0.05	$8 \cdot 10^{-4}$	$2 \cdot 10^{-6}$	$2 \cdot 10^3$
0.5	0.08	$2 \cdot 10^{-5}$	20
0.02	$2 \cdot 10^{-4}$	10^{-6}	$1.6 \cdot 10^3$
0.2	0.02	10^{-5}	16
0.13	$4 \cdot 10^{-4}$	$5 \cdot 10^{-7}$	10^4
0.3	0.04	$5 \cdot 10^{-6}$	100
1.0	0.4	$6 \cdot 10^{-5}$	60
10	40	$6 \cdot 10^{-4}$	0.6
15	125	$3 \cdot 10^{-3}$	22
150	12500	$3 \cdot 10^{-2}$	0.2
$2 \cdot 10^{-3}$	$1.5 \cdot 10^{-3}$	$1.4 \cdot 10^{-3}$	2.7

(IV.78), depends on the degree of attenuation of the wave (i.e., on the value of the absorption coefficient α_0 of the low-amplitude wave) and for small α_0, as is evident from Table 9, it can greatly exceed the distances from the source at which ultrasonic measurements are usually performed. In this case, the condition (IV.83) is weakened in accordance with the form of the function $\Phi(\alpha_0 x)$, illustrated in Fig. 23. It should be kept in mind, however, that relations the (IV.80) and (IV.83) define the absorption coefficient of the **entire** wave, while in practice the amplitude of its first harmonic is measured. In the region of waveform stabilization, where the second harmonic remains almost unchanged, the absorption coefficient of the entire wave α must be practically equal to the absorption coefficient of its first harmonic α_1 and, therefore, the relations (IV.82) and (IV.83), which determine the maximum absorption of a finite-amplitude wave at the point x_{stab}, are equally applicable to the absorption coefficient of the first harmonic α_1 at this point. At distances from the source less than the stabilization distance, however, the absorption coefficient of the first harmonic α_1 can greatly exceed the absorption coefficient of the entire wave, since part of the energy of the first harmonic is transferred to the growing second harmonic (as well as the higher overtones, which we have neglected here). Even so, the quantity α_1 cannot exceed the value α_{max} defined by Eq. (IV.82), which thus gives the **theoretical upper limit of the excess absorption** of a finite-amplitude wave at small Reynolds numbers. Thus a practical criterion for the wave amplitude to be small from the viewpoint of the influence of nonlinear effects on the absorption is that the absorption coefficient, measured by a given method, should be independent of the distance to the source and independent of the ultrasonic intensity.

V. Steady Forces Arising in an Ultrasonic Field

§ 1. Radiation pressure

Radiation pressure, or the *pressure of ultrasonic radiation*, which, in particular, appears in the form of steady ponderomotive forces acting on obstacles in the propagation path of an ultrasonic wave, can in a certain sense be regarded as a nonlinear effect. In a free ultrasonic field, ultrasonic radiation pressure also exists in the form of a constant component of the pressure. Radiation pressure is characteristic of any wave process, independent of its nature; it is related to the change in the magnitude of the momentum carried by the wave at an obstacle. The ponderomotive forces arising in this manner are weak. It is well known, for example, that very sensitive instruments are required to record the pressure of light. Ultrasonic radiation pressure is also a small quantity compared to the amplitude of the variable pressure in the ultrasonic wave. Nevertheless, the radiation effect follows directly from the linear equations of electrodynamics and the linearized equations of hydrodynamics. The nonlinearity of the exact equations of hydrodynamics adds corrections to the ultrasonic radiation pressure which are comparable in magnitude to the effect calculated in the first approximation, in contrast to the nonlinear corrections to other acoustic parameters, for example, the velocity of sound, the energy density,* etc., in which they appear as second and higher order quantities. These comparatively large corrections to the ultrasonic radiation pressure constitute the nonlinear effect. The difference between

*Nonlinear corrections to the ultrasonic energy density and ultrasonic intensity, with the highest ultrasonic wave amplitudes, remain considerably lower than the existing accuracy of absolute measurements of energy quantities. For this reason, we do not present their detailed calculation in the second-order approximation, and instead refer the interested reader to the speciallized literature on nonlinear acoustics.[19,20]

acoustic and electromagnetic radiation forces is also related to the fact that under the action of an ultrasonic wave the surface of an obstacle undergoes oscillations, which change the ultrasonic field. All this leads to the fact that calculations of the ultrasonic radiation pressure yield different results under different conditions: an infinite wave front, a bounded ultrasonic beam, an unbounded unperturbed medium, a closed ultrasonic field when the mass of the medium supporting the oscillations occur remains constant, the case of a "free" ultrasonic field, or the case when the radiation forces acting on an obstacle are calculated.

Since an obstacle distorts the ultrasonic field, radiation forces in this case are determined not only by the change in the momentum flux of the wave incident on the obstacle, but also by the momentum flux of the scattered wave. The problem of calculating the radiation forces acting on an obstacle thus includes the problem of diffraction of an acoustic wave by an obstacle. Radiation forces also depend on the reflective properties of an obstacle. For this reason, we shall calculate radiation forces in detail when we describe specific radiometric systems, in particular, systems used to measure the ultrasonic intensity. In this section we shall derive the general equations required for such calculations, and we shall study the case of a free ultrasonic field.

So, the radiation force acting on some fixed surface of a volume singled out in a medium is determined by the momentum flux through this surface. The expression for the components of this force was presented in §II.2. It has the form

$$F_i = -\oint_S \Pi_{ik} n_k dS, \tag{V.1}$$

where n_k are the vector components of the unit outer normal to the surface S. We are interested in the constant component of this force, which can be found by averaging expression (V.1) with respect to time, i.e.

$$F_i = -\oint_S \overline{\Pi}_{ik} n_k dS. \tag{V.2}$$

The radiation pressure is therefore determined by the time-averaged value of the "strain tensor" Π_{ik} (II.15) introduced in the same section: $\overline{\Pi}_{ik} = \bar{p}\delta_{ik} + \overline{\rho v_i v_k}$.

Thus, unlike the scalar hydrostatic or acoustic pressure, according to Eq. (V.2) the radiation pressure is a vector quantity: it depends on the orientation of the area dS relative to the direction of propagation of the ultrasonic wave. Because, however, the term "radiation pressure" is widely used in the literature, we shall retain it, though to avoid misunderstandings it would be better to speak about radiation stress or strain.

We shall study, as we have done above, the case of plane waves propagating along the x-axis. The time-averaged momentum flux density tensor in this case has the following form:

$$\overline{\Pi}_{xk} = \begin{vmatrix} \bar{p} + \overline{\rho v_x^2} & 0 & 0 \\ 0 & \bar{p} & 0 \\ 0 & 0 & \end{vmatrix}.$$

In the first approximation the acoustic pressure averaged over a period equals zero: $\bar{p} = p_{max}\int_0^T \cos(\omega t - kx)\,dt = 0$ and $\rho = \rho_0$. Only one component of the radiation pressure acting on an area perpendicular to the x-axis will differ from zero:

$$\overline{\Pi}_{xx} = \Pi = \rho_0 \bar{v}_x^2 = \rho_0 v^2 = 2\overline{w}_{kin}, \qquad \text{(V.3)}$$

where $\overline{w}_{kin} = \rho_0 \bar{v}^2/2$ is the average kinetic energy density in the ultrasonic wave. As shown in §I.7, in an unbounded ultrasonic field the average kinetic energy density $\overline{w}_{kin}$ equals the average potential energy density $\overline{w}_{pot}$. In this case, for which the constancy of the amount of liquid in the ultrasonic field is important,

$$2\overline{w}_{kin} = \overline{w}_{kin} + \overline{w}_{pot} = \overline{w}, \qquad \text{(V.4)}$$

where $\overline{w}$ is the average total energy density in the ultrasonic wave. Expression (V.3) then assumes the following form: $\Pi = \overline{w}$. If the conditions for propagation of ultrasound are such that the amount of liquid in the ultrasonic field can vary, then the equality $\overline{w}_{kin} = \overline{w}_{pot}$ is at the least not obvious, and only Eq. (V.3) for the radiation pressure remains indisputable.

We shall now calculate the average value of the tensor Π_{xk} in the second-order approximation, i.e., up to quadratic terms. We obtained the expressions for the acoustic pressure in the second approximation in the preceding chapter. For the forward plane wave, they have the same form as expressions the (IV.32) and (IV.34). Quadratic terms must also be included in the variables v and $\Delta\rho = \rho - \rho_0$, which appear in these expressions linearly. In the second approximation, these variables have the form (IV.30) and (IV.33). The relations (IV.30) and (IV.32)–(IV.34) show that additional conditions must be imposed in order to calculate the average pressure p in an ultrasonic field. Retaining the requirement that the total amount of liquid be constant, i.e., setting $\Delta\rho = 0$, we obtain from expressions (IV.32) and (IV.33):

$$\bar{p} = \rho_0 c_0 \bar{v} + \frac{\epsilon_0}{2}\rho_0\bar{v}^2 = \frac{\epsilon_0 - 2}{2}\rho_0\bar{v}^2 + \frac{\epsilon_0}{2}\rho_0\bar{v}^2 = (\epsilon_0 - 1)\rho_0\bar{v}^2. \qquad \text{(V.5)}$$

We note that the average velocity v in the second approximation does not vanish in this case; a constant, nonvanishing component of the velocity appears in the acoustic field:

$$\bar{v} = v_0 = \frac{\epsilon_0 - 2}{2}\frac{\bar{v}^2}{c_0} = \frac{\epsilon_0 - 2}{4}\frac{v_{max}^2}{c_0},$$

where v_{max} is the amplitude of the particle velocity in the ultrasonic wave.

The expression for the time-averaged momentum flux density tensor now assumes the form:

$$\overline{\Pi}_{xk} = \begin{vmatrix} (\epsilon_0 - 1)\rho_0\bar{v}^2 + \rho_0\bar{v}^2 & 0 & 0 \\ 0 & (\epsilon_0 - 1)\rho_0\bar{v}^2 & 0 \\ 0 & 0 & (\epsilon_0 - 1)\rho_0\bar{v}^2 \end{vmatrix}. \qquad \text{(V.6)}$$

The quantity $\rho_0\bar{v}^2$ with $\overline{\Delta\rho} = 0$ equals the average total energy density $\bar{w}$. Thus, as a result of the nonlinearity of the hydrodynamic equations, a pressure additional to the hydrostatic scalar pressure $\bar{p} = (\epsilon_0 - 1)\bar{w}$ appears in an ultrasonic field. Since for liquids $\epsilon_0 = 4 - 6$ (see Table 7), this nonlinear increment even exceeds the magnitude of the radiation pressure calculated in the first approximation.* According to expression (V.6), the radiation pressure acting on a plane perpendicular to the x-axis in the case of an unbounded (along the y- and x-axes) ultrasonic field or an ultrasonic beam bounded by rigid walls preventing inflow of liquid into the beam (the so-called *Rayleigh radiation pressure*) equals $\Pi = \overline{\Pi}_{xx} = \epsilon_0 \rho_0 \overline{v^2} = \epsilon_0 \overline{w}$. If, on the other hand, the beam of ultrasonic waves is of finite size and has a boundary with the unperturbed liquid, as most often occurs in practice, then the presence of the positive scalar pressure (V.5) in it must "constrict" the beam, as a result of which the static pressure in the beam equals the hydrostatic pressure in the unperturbed medium. The directed radiation pressure acting on a surface area parallel to the ultrasonic wavefront in this case will equal

$$\Pi = \rho_0 \overline{v^2} = \rho_0 v_{max}^2/2 = 2\overline{w}_{kin}. \qquad \text{(V.7)}$$

This radiation pressure, measured in a finite beam surrounded by unperturbed liquid, is sometimes called the *Langevin pressure*.

As already noted, the radiation pressure is a very small quantity: it is many orders of magnitude smaller than the amplitude of the variable pressure in the ultrasonic wave. Indeed, expressing the amplitude of the particle velocity v_{max} in relation (V.7) in terms of the amplitude of the pressure $p_{max} = \rho_0 c_0 v_{max}$, we obtain: $\Pi = p_{max}^2/(2\rho_0 c_0^2) = (p_{max}/2)$, where Ma is the acoustic Mach number. Thus the ratio of the radiation pressure to the amplitude of the variable pressure equals one-half the Mach number. The latter, however, as is evident from Table 9, does not exceed the magnitude 10^{-3}–10^{-4} in liquids at the highest ultrasonic intensities. For example, for a pressure amplitude in the ultrasonic plane wave $p_{max} = 1$ atm, the ultrasonic intensity is $I \simeq 0.3$ W/cm^2, $Ma = 4 \cdot 10^{-5}$, and the radiation pressure at normal incidence of the wave on a plane obstacle $\Pi \simeq 20 \cdot 10^{-6}$ *atm*. Exact measurements of such pressures, which appear in the presence of a noisy background formed, for example, by "acoustic flows" in a viscous medium, reflected waves, etc., encounter enormous experimental difficulties. On the other hand, such measurements give a comparatively simple method for finding directly the absolute energy characteristics of an ultrasonic

---

*We note that this additional pressure would be absent in a medium in which $\epsilon_0 = 1$, i.e., $n = 1$.

field. In this case, the Langevin pressure (V.7) is often written in terms of the ultrasonic intensity (energy flux density) by means of the linear relation (III.21) between the intensity and the average energy density in the plane wave $\bar{w}$

$$I = \bar{w} c_0 \tag{V.8}$$

and the relation (V.4), i.e., it is assumed that the quantity $\rho \bar{v}^2$ in (V.6) equals $\bar{w}$. An accurate analysis, however, including the nonlinear effects,[33] leads to more complicated relations between the intensity and the energy density, which also depend on the initial and boundary conditions of the problem. In addition, as already noted, the equality (V.4) is valid only under certain initial and boundary conditions: in some particular cases, the potential and kinetic energy densities can differ considerably even in the linear approximation.[33,34] For this reason, universal application of the expressions (V.8) and (V.4) in the equations for the radiation pressure can only be justified by the fact that the low accuracy of absolute measurements of radiation pressure and the absence of more exact methods for making independent measurements of the acoustic energy density make it impossible, at the present time, to compare the results of such measurements under different experimental conditions. The nonlinear corrections to Eq. (V.8) can be neglected, especially in measurements of radiation pressure in a real viscous medium with low Reynolds numbers, since the nonlinear effects in this case are weak.

Accepting the relations (V.8) and (V.4) with these reservations, we obtain the following expression for the radiation pressure (V.7), acting on a flat surface oriented perpendicular to the direction of propagation of ultrasound, in the form

$$\Pi = \bar{w} = I/c_0. \tag{V.9}$$

We have calculated the radiation pressure for the case of an undistorted field of plane waves, i.e., for a "free" field. The Langevin pressure, on the other hand, is manifested in the form of radiation forces acting on some surface. If this surface is not to distort the field, it must completely absorb the energy in the sound wave. Thus the result obtained in the form (V.7) and (V.9) refers to the case of a completely absorbing obstacle.

§ 2. Radiation pressure forces on an obstacle

If the cross-sectional area of an ultrasonic beam incident normally on a plane obstacle equals S, then the radiation force

$$F = S\bar{w} = SI/c_0, \qquad \text{(V.10a)}$$

which can serve as a direct measure of the ultrasonic intensity, will act on a perfectly absorbing obstacle in the direction of propagation of the ultrasonic wave. Equation (V.10a) holds when the dimensions of the obstacle greatly exceed the transverse dimensions of the beam, while the size of the beam is much greater than the ultrasonic wavelength, so that diffraction effects can be neglected. If a beam of plane waves is incident on a flat perfectly absorbing obstacle at an angle θ to its normal, then due to the vector nature of the radiation pressure, expressed by the relation (V.2), normal and tangential components of the radiation force will act on the obstacle: $F_n = S\bar{w}\cos\theta$ and $F_T = S\bar{w}\sin\theta$. The resulting force in the direction of propagation of the ultrasonic wave in this case equals $F_x = \sqrt{(F_n^2 + F_T^2)} = S\bar{w}$, i.e., it does not depend on the orientation of the surface of the obstacle. If a plane obstacle completely reflects the plane wave incident normally on it, then the momentum flux density at such an obstacle changes by a factor of two and the radiation pressure

$$\Pi = 2\bar{w} = 2I/c_0. \qquad \text{(V.10b)}$$

will act on the obstacle.

At oblique incidence on a totally reflecting obstacle, the normal and tangential components of the radiation force equal $F_n = 2S\bar{w}\cos\theta$ and $F_T = 0$, respectively, and the force

$$F_x = 2S\bar{w}\cos^2\theta. \qquad \text{(V.11)}$$

will act in the direction of propagation of the ultrasound.

Using the laws of geometric reflection without corrections for diffraction, it is not difficult to find the radiation force for simple geometric figures, used in practice to make absolute measurements of the ultrasonic intensity. Such figures include a sphere and a cone, which do not create backwards-reflected waves. Since we are talking about geometric reflection, it is assumed that the dimensions of these figures greatly exceed the wavelength of the ultrasonic wave.

For a cone which is oriented so that its tip points into the ultrasonic beam and at whose surface the ultrasound is totally reflected the radiation pressure force will apparently be determined by the same equation (V.11) as for an inclined flat obstacle, because in both cases the angle θ is the same at all points on the surface. Thus, for a cone

$$F_x = 2S\bar{w}\cos^2\theta = 2S\bar{w}\sin^2\varphi = (2SI/c_0)\sin^2\varphi, \qquad \text{(V.12)}$$

where the ultrasonic wave propagates along the x axis, which coincides with the axis of the cone; φ is one-half the cone angle; and, S is the area of the base (if it is less than the cross-sectional area of the ultrasonic beam).

For a reflecting sphere the expression (V.12) must be integrated over the angle φ in the range from $\varphi = \pi/2$ (for the normal ray) to $\varphi = 0$ (for the tangential ray). This gives

$$F = \frac{2I}{c_0} S \int_{\pi/2}^{0} \sin^2\varphi d\varphi = \frac{I}{c_0} S.$$

Next, we shall study the case when a plane ultrasonic wave incident normally on a plane boundary between two media partially penetrates into the second medium and is partially reflected from the boundary. Let the relative fraction of the reflected energy be ρ_I and that of the transmitted energy be d_I, so that $\rho_I + d_I = 1$. The radiation pressure on the boundary will be determined by the energy density of the incident and reflected waves as well as the transmited wave:

$$\Pi = \bar{w}_1 + \bar{w}_2 - \bar{w}_3, \qquad \text{(V.13)}$$

where the indices 1, 2, and 3 refer to the energy density in the incident, reflected, and transmitted waves, respectively. If the intensity of the incident wave is I, then based on the relation $w = I/c_0$ and (V.13) we have

$$\Pi = \frac{I}{c_{01}} + \rho_I \frac{I}{c_{01}} - d_I \frac{I}{c_{02}} = I\left(\frac{1+\rho_I}{c_{01}} - \frac{d_I}{c_{02}}\right), \qquad \text{(V.14)}$$

where c_{01} and c_{02} are the velocities of sound in the first and second media. Since $I = \bar{w}c_{01}$ and $d_I = 1 - \rho_I$, we obtain from (V.14)

$$\Pi = \bar{w}[(1+\rho_I) - (1-\rho_I)c_{01}/c_{02}]. \qquad (V.15)$$

For $\rho_I = 1$ (total reflection), Eq. (V.15) reduces to (V.10b), and for $\rho_I = 0$ and $c_{01} = c_{02}$, $\Pi = 0$.

In the case of two liquid media with different velocities of sound the ultrasonic reflection coefficient at the boundary between the liquids can vanish. For this case, Eq. (V.15) gives

$$\Pi = \bar{w}(1 - c_{01}/c_{02}). \qquad (V.16)$$

Since the energy density is a scalar quantity, it follows from the equality (V.16) that the radiation pressure will be oriented into the liquid with the higher velocity of sound, irrespective of the direction of propagation of the ultrasonic waves. This effect is illustrated in Fig. 26, which shows the radiation effect of ultrasonic beams on the boundary between two immiscible liquids: water (top layer) and aniline (bottom layer).[12] The primary ultrasonic beam is incident from top to bottom; then, after being reflected from the plane reflector, it propagates from bottom to top. It is evident that irrespective of the direction of propagation of ultrasound (indicated in the figure by the arrows), the inflection of the interface under the action of the radiation pressure is oriented toward the aniline, in which the velocity of sound is greater than in water.

If the ultrasonic beam is incident on the free surface of a liquid, then radiation forces acting on the surface cause a stationary "upwelling" of the surface, which transforms into a characteristic "ultrasonic fountain" when the radiation pressure exceeds the surface tension of the liquid. For ultrasonic waves with low amplitude, the radiation-induced upwelling of the liquid surface produces an amplitude distribution pattern over the cross-sectional area of the ultrasonic beam. This phenomenon is sometimes used to visualize objects located in the propagation path of ultrasonic waves in an optically nontransparent liquid.

General case. The equations obtained above for the radiation pressure refer to particular cases which can be simply calculated. In

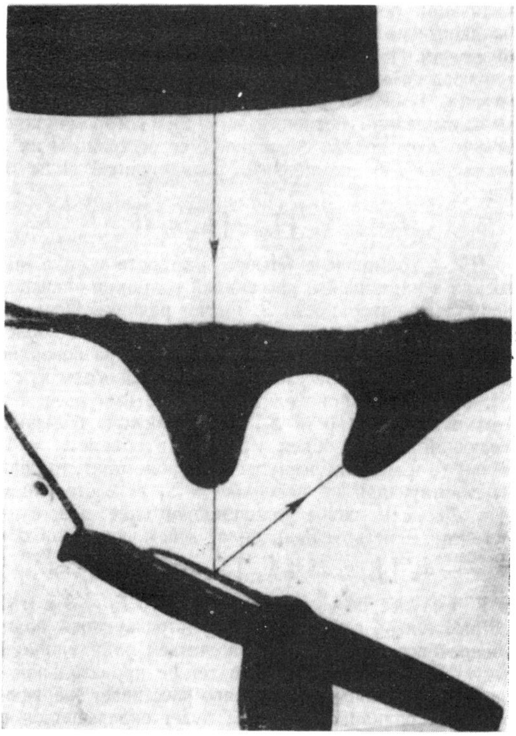

Fig. 26.

general, radiation forces can be calculated based on relation (V.2), in which the change in the momentum of the wave due to its scattering by the obstacle must be included. Then, for the i-th component of the radiation force we obtain the expression

$$\bar{F}_i = -\oint_S (\bar{\Pi}'_{ik} + \bar{\Pi}''_{ik}) n_k dS,$$

where Π'_{ik} and Π''_{ik} are the components of the momentum flux density tensor for the incident and scattered waves, respectively; $\mathbf{n}$ is the outer unit normal to the surface S. The calculation of the radiation forces acting on an obstacle with a complicated shape is greatly simplified by the fact that the radiation force does not depend on the choice of the surface S. To prove this we consider a volume V_1 with surface area S_1, in which the momentum change occurs. Let us draw an auxiliary surface S_2, enclosing a larger volume V_2, which includes the volume V_1. If there is no additional change in momentum in the volume $V' = V_2 - V_1$, then the radiation force acting on the surface S_1 due to the momngeentum change in the volume V_1 equals the radiation force acting on the surfacc S_2. Indced, the force acting on the volume V' is:

$$F_i = -\oint_{S_2} \Pi_{ik} n_k dS + \oint_{S_1} \Pi_{ik} n_k dS = F_{i2} - F_{i1}.$$

The momentum does not change in the volume V', so $F_i = 0$ and $F_{i2} = F_{i1}$.

We shall now examine a beam of plane ultrasonic waves with cross-sectional area S, and we shall calculate the radiation force acting along the beam on an obstacle with arbitrary shape, which partially absorbs and scatters the incident wave. This radiation force will consist of two components. The first component is determined by the difference between the energy flux densities before and after the obstacle. Therefore,

$$\bar{F}_1 = S(I_1/c_0 - I_2/c_0), \tag{V.17}$$

where I_1 is the intensity of the incident wave; I_2 is the intensity of the transmitted wave; and, c_0 is the speed of sound in the medium. In its turn, the change in the energy flux density (intensity) in the beam is determined by the to absorption and scattering of energy by the obstacle. For this reason, the law of conservation of energy flux (power) can be written in the form $SI_1 - SI_2 = D + D_{sc}$, where D is the power absorbed by the obstacle; D_{sc} is the power scattered by the obstacle. The latter quantity, based on the general definition of power (III.22b), can be represented in the form:

$$D_{sc} = \oint_{S_1} I(\theta, \psi)\, dS_1, \tag{V.18}$$

where $I(\theta, \psi)$ is the intensity of the waves scattered in a direction determined by the polar angles θ and ψ, and S_1 is the area of any closed surface enclosing the obstacle, for example, a sphere centered on the obstacle.

In addition to F_1, a radiation force related to the momentum flux in the scattered wave acts on the obstacle. According to the general definition (V.2), and using (V.18), the component of this force along the direction of propagation of the incident wave is

$$\bar{F}_{2\parallel} = -\oint_{S_1} \frac{I(\theta, \psi)}{c_0} \cos\theta dS, \tag{V.19}$$

where θ is the angle between the wave vectors of the incident and scattered waves. As will be shown in Chap. VII, the scattered wave assumes a simple ("asymptotic") form at distances which are much larger than the wavelength and the dimensions of the scattering object. Since the calculation of the radiation force acting on this object due to the change in the momentum flux on the object does not depend on the form and dimensions of the surface S_1 enclosing it, the surface can be assumed to be a sphere with very large radius R. In addition, we shall neglect the ultrasonic absorption in the medium at the distance R.

Adding expressions (V.17) and (V.19), we now obtain a general equation for the radiation force acting on an obstacle in a beam of plane ultrasonic waves in the direction of propagation:

$$\bar{F}_{\parallel} = \bar{F}_1 + \bar{F}_{2\parallel} = \frac{1}{c_0}\left(D + D_{sc} - \int_{S_1} I(\theta, \psi) \cos\theta dS_1\right),$$

where D_{sc} is determined by the relation (V.18). The component F_1 vanishes in a direction perpendicular to the beam (it acts only along the beam), and the radiation force

$$\bar{F}_{\perp} = \bar{F}_{2\perp} = -\frac{1}{c_0} \oint_{S_1} I(\theta, \psi) \sin\theta dS. \tag{V.20}$$

acts on the obstacle in this direction.

Thus the problem of calculating the radiation forces acting on an arbitrary obstacle, in general, almost completely reduces to the problem of diffraction and scattering of the ultrasonic wave by the obstacle. We shall study this problem in a special chapter. We shall now present without derivation the results of a calculation of the radiation forces acting on small suspended spherical particles, and we shall also study other types of steady forces acting on particles in an ultrasonic field.

§ 3. Steady forces acting on suspended particles in an ultrasonic field

Radiation pressure forces. The radiation forces acting on a small spherical particle, whose radius R is much less than the ultrasonic wavelength Λ, are calculated in Refs. 35–37. For an absolutely incompressible sphere in the field of a traveling plane wave with $kR \ll 1$, the calculation yields the following expression for the force in the direction of the wave vector **k** [35]:

$$\bar{F}=4\pi R^2\bar{w}\,(kR)^4[1+(1-a)^2/(2+a^2)], \tag{V.21}$$

where $a = \rho_0/\rho_p$ is the ratio of the density of the medium to the density of the particle. In a standing ultrasonic wave,

$$\bar{F}=\frac{4}{3}\pi R^2\bar{w}kR\,\frac{2.5-a}{2+a}\sin kx, \tag{V.22}$$

where x is the distance from the center of the sphere to a node in the particle velocity. When the compressibility of the sphere is included, the same equations assume a different form [36]:

$$\bar{F}=4\pi R^2\bar{w}\,(kR^4)\left[\left(\frac{1}{a}-\frac{2+a}{3b^2}\right)^2+\frac{2}{9}\left(\frac{a-1}{a}\right)^2\right], \tag{V.23}$$

for a traveling wave and

$$\bar{F}=4\pi R^2\bar{w}kR\left[\frac{1+2(1-a)/3}{2+a}-\frac{a}{3b^2}\right]\sin 2kx, \tag{V.24}$$

for a standing wave, where $b = c_p/c_0$ is the ratio of the velocity of sound in the particle material to that in the medium surrounding the particle. In both cases, other conditions being equal, the radiation pressure in a standing wave is much higher than in a traveling wave, because in the latter wave $\bar{F} \sim (kR)^4$ while in the standing wave $\bar{F} \sim kR$ $(kR \ll 1)$.

In the case of a standing ultrasonic wave, the factor sin kx or sin $2kx$, which indicates that the radiation pressure forces are periodic in space, appear in the corresponding equations (V.22) and (V.24). This periodicity (with a change in the sign of the force) leads to the fact that small particles in the field of a standing ultrasonic wave will move to some equilibrium positions, which can be nodes as well as antinodes of the standing wave, depending on the ratio of the density of the particle to the density of the medium.

If the density of the suspended particle is very low, when $a \simeq (kR)^{-2}$, the expressions for the radiation forces including compressibility assume the form[36]

$$\bar{F} = 4\pi R^2 \bar{w} (kR)^4 \frac{1}{(kR)^6 + [3b^2/a - (kR)^2]^2} \tag{V.25}$$

for a traveling wave and the form

$$\bar{F} = \frac{4\pi R^2}{kR} \bar{w} \frac{b^2[3b^2/a - (kR)^2]}{(kR)^6 + [(3b^2/a - (kR)^2]^2} \sin 2kx \tag{V.26}$$

for a standing wave. Equations (V.25) and (V.26) correspond to the case realized for gas bubbles in a liquid. It is evident from these equations that if the equality

$$(kR)^2 = 3b^2/a, \tag{V.27}$$

is satisfied, then the radiation force acting on the bubbles in a travelling wave $\bar{F} = 4\pi R^2 \bar{w}/(kR)^2 = \Lambda^2 \bar{w}/\pi$ becomes very large compared to the force acting on an incompressible sphere or a sphere whose compressibility is almost equal to that of the surrounding medium. In this case, the force acting on the bubbles in the standing wave vanishes.

It is not difficult to see that the condition (V.27) corresponds to the condition of resonance of the gas bubble and determines the frequency of the characteristic radial oscillations of the bubble:

$$\nu_{res} = [1/(2\pi R)] \sqrt{3\gamma P/\rho_0}, \tag{V.28}$$

where $\gamma = c_p/c_v$ is the ratio of the specific heat capacities of the gas filling the bubbles and P is the gas pressure in the bubble. The latter quantity is usually equal to the hydrostatic pressure plus the pressure due to surface tension: $P = P_0 + 2\sigma/2$. It follows from (V.26) that if $kR > b(3/a)^{1/2}$, i.e., the dimensions of the bubble exceed the resonant value for the given frequency ν, then the radiation force acting on the bubble will be oriented in such a way that the bubble will move toward the node of the standing wave. If the opposite inequality, when the dimensions of the bubble are less than the resonant value, holds, then the radiation force will displace the bubble toward the antinode. Bubbles whose dimensions correspond to their resonance at a given frequency will not feel the radiation pressure forces. All this is confirmed by appropriate experiments.

In addition to the radiation pressure, forces of a different origin can act on suspended particles in an ultrasonic field.

Bjerknes forces. It is known from hydrodynamics that when a body undergoes oscillatory motion in a liquid near the surface of a stationary obstacle, the particle velocity of the fluid particles will be higher on the side of the obstacle which is turned toward the oscillating body.[39] As a result, the pressure in the liquid between the obstacle and the body will be lower than on the opposite side, and a pressure force oriented toward the oscillating body will act on the obstacle. Forces of this type are called Bjerknes forces. They also arise in an ultrasonic field when suspended particles and bubbles oscillate in the field with different velocities and phases relative to one another. Analysis of Bjerknes forces shows that if two spherical particles with radii R_1 and R_2, whose centers are a distance L apart, undergo a pulsating oscillation with identical frequency but different velocities v_1 and v_2, then a steady force

$$\bar{F} = 4\pi\rho R_1^2 R_2^2 (v_1 v_2 / L^2) \cos\beta, \tag{V.29}$$

where β is the phase difference between the particle velocities of the spherical surfaces, which depends on the ratio of the distances between the spheres and the acoustic wavelength Λ, arises between them. Equation (V.29) is valid for $\Lambda \gg L$, i.e., for the case when the distance between the particles is much smaller than the wavelength. In the opposite case, Bjerknes force (V.29) will be very small, since it decreases inversely as the first power of the distance L. According to Eq. (V.29), Bjerknes force acting between pulsating spheres can be attractive as well as repulsive, depending on the phase difference β. If, however, $kL \ll 1$, the phases of forced oscillations of particles in an ultrasonic field differ by less than $\pi/2$ and the force acting on them will be positive, i.e., repulsive.

In addition to pulsating oscillations, suspended particles in an ultrasonic field can also undergo translational oscillatory motion in a direction that depends on their distribution relative to the front of the ultrasonic wave. If two spherical particles oscillate along a line connecting their centers, then the Bjerknes force acting on them is given by the equation[40]: $\bar{F} = 6\pi\rho R_1^3 R_2^3 [(v_1 - v_2)/L^4] \cos\beta$ if $L \gg R_1 + R_2$. If the direction of oscillation of the spherical particles is perpendicular to the line joining them, for example, the particles are distributed along the front of an ultrasonic wave and oscillate in phase with identical velocities v, the Bjerknes force is given by[40]

$$\bar{F} = -3\pi\rho\,(R_1^3 R_2^3/L^2)\,v^2.$$

It has been established experimentally that Bjerknes forces are primarily responsible for the agglomeration of gas bubbles accompanying the appearance of cavitation in the process of degasification of liquids in a low-frequency ultrasonic field; this process will be studied below.

Bernoulli forces. Analogous steady forces arise between solid particles, if, due to their inertia, they cannot follow the motion of the liquid and the liquid flows around them. It is well known from hydrodynamics that if two stationary spheres are positioned in a fluid flowing with velocity v perpendicular to the line connecting their centers, then due to the decrease in the pressure between the spheres an "attractive" force

$$\bar{F} = (3/2)\,\pi\rho\,(R_1^3 R_2^3/L^4)\,v^2, \qquad \text{(V.30)}$$

called *Bernoulli's force*, acts on them. As is evident from Eq. (V.30), Bernoulli's force is determined by the square of the flow velocity, i.e., it does not depend on the sign of the velocity, and it therefore arises during the oscillatory motion of the fluid in an acoustic wave. Bernoulli forces explain the coagulation of solid particles in a high-frequency ultrasonic field.[41]

Stokes forces. A friction force, determined by the well-known Stokes equation $F = 6\pi\eta_s Rv$, where η_s is the coefficient of shear viscosity, acts on a spherical body with radius R moving with velocity v in a viscous medium. If it is assumed that the force is constant, then for oscillatory harmonic motion of the particle in an acoustic field, the time-averaged Stokes force must vanish. The viscosity, however, depends on the temperature, which oscillates in an acoustic wave. The change in temperature is related to the velocity of the particles in the medium v by the relation (see §IV.3)

$$\Delta T = (c_0 \alpha_T T_0 / c_P) v, \qquad \text{(V.31)}$$

where T_0 is the average temperature of the medium; α_T is its coefficient of thermal expansion; c_p is the specific heat capacity at constant pressure; and, c_0 is the velocity of the wave. Taking into account the temperature dependence of the viscosity, the coefficient of viscosity $\eta_s(T)$ can be represented by the first two terms of its series expansion:

$$\eta_s(T) = \eta_{s0} + (d\eta_s/dT)\Delta T. \qquad \text{(V.32)}$$

Averaging Stokes formula with respect to time $F = 6\pi R\eta_s(T)v$, substituting here $\eta_s(T)$ in the form of Eq. (V.32), and using the relation (V.31) as well as the value of the particle velocity in an ultrasonic wave including second-order quantities (see §IV.4), we obtain a nonvanishing steady Stokes force:

$$F = 6\pi R\eta_s \frac{v_{max}^2}{2c_0}\left(-1 + 2\frac{d\eta_s}{dT}\frac{\alpha_T c_0^2 T_0}{c_P}\right).$$

Driven by this force the particle will move in one direction with constant velocity

$$v_0 = \frac{v_{max}^2}{2c_0}\left(-1 + 2\frac{d\eta_s}{dT}\frac{\alpha_T c_0^2 T_0}{c_P}\right).$$

A calculation shows that appreciable motion of particles under the action of Stokes forces can occur only in gases.

§ 4. Streaming

Aside from the periodic displacement of particles, different types of steady flows — differing in both their nature and their origin — arise in acoustic fields. In a real viscous medium such flows arise both in a free field and near obstacles. The latter flows are due to the interaction of a viscous liquid (or gas) with the solid walls of obstacles, as a result of which the velocity of the tangential displacements of the particles in the medium near the wall must vanish. The thickness of the layer in which this interaction appears is of the order of the penetration depth of a shear wave in the medium. As shown in §II.5, the absorption coefficient of a shear wave in a liquid $\alpha_s = [\omega\rho_0/(2\eta_s)]^{1/2}$; the penetration depth of a shear wave (i.e., the distance at which it is attenuated by a factor of e) $\Delta \simeq [2\eta_s/(\omega\rho_0)]^{1/2}$. Eddy currents, which are observed in practice only at low acoustic frequencies, arise in a layer with this thickness. At ultrasonic frequencies the thickness of the "acoustic boundary" layer Δ becomes negligibly small. In water, for example, at a frequency of $\nu = 1$ *MHz* it constitutes only 10^{-4} *cm*.

Flows arising in a free ultrasonic beam are characteristic for an ultrasonic field: they are observed only at the high intensities which are realized at ultrasonic frequencies. Such flows were first called "quartz streaming" because of the fact that intense ultrasonic beams were obtained with the help of piezoelectric quartz plates. Since many different methods for exciting intense ultrasound now exist, however, it is more natural to call this effect *ultrasonic streaming*. These flows are stimulated by the radiation pressure in an ultrasonic beam, which in a real dissipative medium is related to the absorption of the ultrasonic wave in the medium. Since the average energy density of the wave decreases due to absorption, a radiation force

$$\bar{F} = d\bar{w}/dx = \nabla\bar{w}, \tag{V.33}$$

which is what causes the stationary flow, will act on a unit volume of the absorbing medium along the direction of propagation x of the ultrasonic waves. The velocity of this flow v_0 can be approximately calculated starting from the hydrodynamic equation of motion for an ideal liquid (II.4), assuming that the pressure gradient in it $\partial p/\partial x =$

$\partial \bar{w}/\partial x$ and $v = v_0$, which gives

$$-\frac{\partial \bar{w}}{\partial x} = \rho\frac{\partial v_0}{\partial t} + \rho v_0\frac{\partial v_0}{\partial x}. \tag{V.34}$$

Since the radiation force $\partial \bar{w}/\partial x$ is small, with respect to it the liquid may be assumed to be incompressible, making the assumption that $\rho = \rho_0 = const$. Then, for a steady-state flow $(\partial v_0/\partial t = 0)$, based on (V.34) we have $-\partial \bar{w}/\partial x = \rho_0 v_0 \partial v_0/\partial x$. Integrating this equation from the location of the radiator at $x = 0$ up to the point of observation x and taking into account the fact that the ultrasonic stream velocity vanishes at $x = 0$, we obtain

$$\rho_0 v_0^2(x)/2 = \bar{w}(0) - \bar{w}(x). \tag{V.35}$$

The left side of this equation is the kinetic energy per unit volume of the medium at the point x, and the expression on the right side represents the difference between the average acoustic energies at the radiator and at a distance x from it. Thus Eq. (V.35) expresses the law of conservation of energy: the decrease in the acoustic energy due to absorption in the medium is compensated by the kinetic energy of the directed flow of the medium. Based on the fact that the average energy density in an ultrasonic beam determines the magnitude of the Langevin radiation pressure in a given section of the beam, Eq. (V.38) can be interpreted as an expression of the well-known Bernoulli theorem, which asserts that the dynamic pressure of the flow $(\rho_0 v_0^2/2)$ and the hydrostatic (in this case radiation) pressure are constant. According to (V.35), the drop in the radiation pressure along the beam due to absorption of energy in the ultrasonic wave is compensated by the dynamic pressure of the flow resulting from this absorption, so that the sum of the radiation and dynamic pressures on a plane obstacle located at any point of the cross section of the beam must remain constant. Here, however, we neglected the viscous resistance of the medium, which transforms part of the kinetic energy of the flow into heat. Nevertheless, the graphic result obtained is well confirmed by measurements of the radiation pressure and of the ultrasonic stream velocity in low-viscosity liquids at comparatively short distances from the ultrasonic source.

Actually, viscous forces will retard the increase in velocity, so that a stationary flow with a constant velocity depending on the shear viscosity of the medium η_s must be established in the beam. Therefore, in order to calculate the stationary ultrasonic streaming, the resistance forces of the viscous medium as well as the real velocity distribution over the cross section of the beam, determined, in particular, by the boundary conditions, must be included in the equation of motion (V.34).

The velocity of the stationary flow can easily be calculated approximately for the idealized case of a sharply collimated ultrasonic beam, which is uniform over its cross section and at whose boundaries the velocity of the flow vanishes. Such conditions are realized to a certain extent, for example, if the beam is bounded by the walls of a rigid tube, which, however, must have openings in order to establish hydrodynamic contact between the liquid in the ultrasonic field, i.e., inside the tube, and the unperturbed external liquid. Without such a contact the radiation pressure in the beam will only give rise to rarefaction of the medium and, naturally, a flow will not arise in it (if, of course, the beam is uniform over its cross section). The velocity of the stationary acoustic flow on the beam axis in this case can be found from the well-known Poiseuille equation:

$$v_0 = \Delta P R^2 / (4\eta_s x), \tag{V.36}$$

where R is the radius of the tube (beam); η_s is the shear viscosity of the medium; and, ΔP is the difference between the static pressures in two sections separated by a distance x. Since the flow in this case arises due to the radiation pressure gradient, under the assumption that this gradient equals the difference in the acoustic energy densities along the beam and taking into account the exponential attenuation of plane waves (III.48), we obtain

$$\Delta P = \bar{w}(0) - \bar{w}(x) = \bar{w}(0)\,[1 - \bar{w}(0)\, e^{-2\alpha_0 x}], \tag{V.37}$$

where α_0 is the ultrasonic amplitude absorption coefficient. For $\alpha_0 x \ll 1$, which is satisfied for most liquids in the megahertz frequency band at reasonable distances from the source (of the order of several centimeters), the expression (V.37) gives $\Delta P \simeq 2\overline{w}(0)\,\alpha_0 x$. Substituting this value of ΔP into Eq. (V.36) and taking into account the relation between the energy density and the ultrasonic intensity, we find:

$$v_0 = I_0 R^2 \alpha_0 / (2\eta_s c_0), \qquad \text{(V.38)}$$

where I_0 is the intensity of the ultrasonic beam in the initial section of the tube (from which the coordinate x is measured) or, in the case of a free beam, the intensity at the ultrasonic source. Thus the velocity of the stationary acoustic flow in a viscous medium is proportional to the ultrasonic intensity and can serve as its approximate measure. As far as the dependence of v_0 on α_0 and η_s is concerned, we note immediately that if the ultrasonic absorption is determined solely by the shear viscosity, so that the absorption coefficient can be calculated using Stokes equation (III.39)

$$\alpha_0 = \frac{8\pi^2 \eta_s \nu^2}{3\rho_0 c_0^3} = \frac{2}{3}\frac{\omega^2 \eta_s}{\rho_0 c_0^3},$$

then the expression (V.38) can be written as

$$v_0 = \frac{4}{3}\pi^2 \frac{I_0 R^2 \nu^2}{\rho_0 c_0^4} = \frac{I_0 R^2 \omega^2}{3\rho_0 c_0^4}, \qquad \text{(V.39)}$$

i.e., the velocity of the acoustic flow does not depend on the viscosity of the liquid and is proportional to the square of the ultrasonic frequency. This result is to be expected, because the coefficient of shear viscosity, on the one hand, determines the viscous forces retarding the flow, while on the other hand it increases the radiation pressure gradient. If, however, the bulk viscosity contributes to the ultrasonic absorption coefficient, as occurs in most liquids, then by virtue of the relations (III.38) and (III.25), Eq. (V.38) assumes the form

$$v_0 = \frac{I_0 R^2 \omega^2}{4\rho_0 c_4^0 \eta_s}\left(\eta_b + \frac{4}{3}\eta_s\right). \qquad \text{(V.40)}$$

Thus, for a fixed ultrasonic intensity and for the conditions under which Eq. (V.40) was derived, the velocity of the acoustic flow is a direct measure of the bulk viscosity of the medium. Actually, as noted in § III.4, ultrasonic waves can also be attenuated by diffraction effects, scattering by inhomogeneities in the medium, the effect of thermal conductivity, etc. All these factors contribute to the radiation pressure in the beam, which determines the velocity of the acoustic flow for a given kinematic viscosity of the medium. In addition, in a real beam the ultrasonic intensity is distributed nonuniformly over the cross section of the beam. As a result, acoustic flows arise even in a beam bounded by a closed tube. If, however, the beam of ultrasonic waves is bounded by an unperturbed liquid, then the conditions at its boundaries are more complicated than those assumed in the derivation of Eqs. (V.38)–(V.40), which, in this case, give only an approximation to the velocity of the acoustic flow.

For an ultrasonic intensity of $I_0 = 1\ W/cm^2$ in water ($\alpha_0 = 25 \cdot 10^{-17}\ cm^{-1}$, $c_0 = 1.5 \cdot 10^5\ cm/s$, $\eta_s = 0.01\ g/(cm \cdot s)$) at a frequency of 1 *MHz* and with a beam diameter of 2 *cm*, Eq. (V.38) gives $v_0 \simeq 1\ cm/s$. Under these conditions the flow in a free beam is laminar. In addition, the flow of liquid away from the ultrasonic radiator is accompanied by inflow of liquid from the unperturbed regions; when a solid obstacle is encountered, liquid flows out of the beam into the unperturbed sections of the medium. For this reason, in a closed vessel whose transverse dimensions exceed the diameter of the ultrasonic beam, stationary circulating currents, which can be visualized with the help of particles suspended in the liquid, are established. Figure 27 shows the pattern of circulating flows in a plane separating two immiscible liquids (glycerin and fine lubricating oil), between which a film of colored water of intermediate density is injected in order to visualize the flow. The interface at this film is oriented along the axis of the ultrasonic beam, whose source is located to the right of the figure. Figure 28 shows the ultrasonic streaming pattern in a vessel containing benzene in which aluminum powder is suspended.[44]

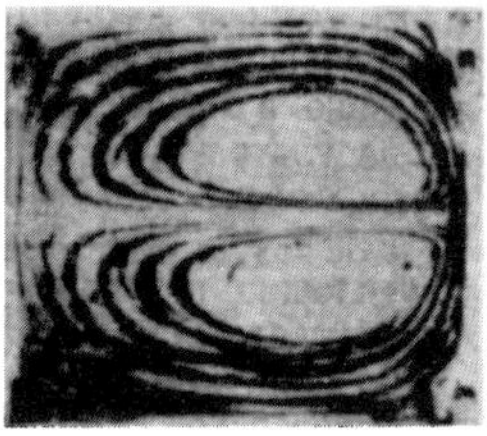

Fig. 27.

At high ultrasonic wave intensities acoustic flows become turbulent. In this case, an intense ultrasonic beam gives rise to intense mixing of the liquid, which can play an important role in a number of processes stimulated by ultrasound. In addition, as shown in the preceding chapter, at large Reynolds numbers the form of an ultrasonic wave propagating in a liquid can greatly differ from the sinusoidal form, and its absorption can increase sharply. This, in its turn, will intensify the flow, which can thus become turbulent at some distance from the ultrasonic source.

Fig. 28.

So far, we have discussed only acoustic flows induced by Langevin radiation pressure, arising due to the absorption of ultrasonic waves and the change in their momentum in a viscous medium. It follows, however, from the analysis presented in the preceding section that under certain conditions acoustic flows can also arise in a nondissipative medium. In particular, the time-averaged velocity of particles in a medium in the field of plane finite-amplitude waves may differ from zero. Of course, this does not always indicate the presence of a directed stationary flow of the medium. For example, such a flow in the field of waves with infinite fronts is precluded by the law of conservation of mass: the constant component of the displacement velocity in this case is compensated by the nonzero constant component of the acoustic pressure or density. In the case of a bounded ultrasonic beam in contact with an unperturbed liquid, however, Rayleigh radiation pressure in the beam can give rise to circulating currents of nonlinear origin. The existence of such "nonlinear" acoustic flows was confirmed experimentally by L. K. Zarembo.[42]

VI. Ultrasonic Cavitation

§ 1. Rupture strength of liquids

The term cavitation refers to the formation and collapse of cavities (holes) in a liquid. In general, cavitation can appear with any local rarefaction in liquids: in a hydrodynamic flow, in flow past solids, in a wake, etc. In an acoustic wave, which creates periodic rarefaction, cavitation is observed at very high wave intensities, which are realized at ultrasonic frequencies. It is thus a characteristic feature of ultrasound and is called *ultrasonic cavitation*. Because in cavitation the continuity of the medium is destroyed this phenomenon must also be regarded as a nonlinear effect.

In general terms, the elementary act of ultrasonic cavitation can be represented as follows. During the rarefaction phase of an ultrasonic wave, a hole filled with the saturated vapor of the liquid forms in the liquid. During the compression phase the vapor condenses and the cavity collapses under the high pressure, which is "helped" by the surface tension of the walls of the cavity, as if the cavity were empty. Some of the gas dissolved in the liquid, which during rapid collapse is subjected to strong adiabatic compression, diffuses into the cavity through its surface. At the instant of collapse the pressure and temperature of the gas reach considerable magnitudes, which leads to the formation of a secondary spherical shock wave, decaying rapidly in space, in the surrounding liquid.

Cavitation processes play an important role in practical applications of ultrasound; for this reason, a great deal of attention is devoted to the study of ultrasonic cavitation. Although a number of problems in cavitation remain unsolved, its physical nature as a whole has now been studied in detail. We shall briefly describe the basic results of these investigations, referring the interested reader to the literature for more detailed information.[19, 45–49]

One of the basic questions arising in the study of ultrasonic cavitation is the question of how the liquid ruptures in an ultrasonic wave at acoustic pressures far below the theoretical rupture strength of the liquid. Indeed, for a cavity with radius R to form in an ideal liquid a tensile stress P' equal to the Laplace pressure, generated by the surface tension σ of the liquid, must be applied, i.e.,

$$P' \simeq 2\sigma/R. \qquad \text{(VI.1)}$$

To rupture an ideal liquid, its constituent particles must be moved apart to a distance equal approximately to twice the intermolecular distance. For water, for example, this distance equals about 2 Å. Substituting this value of R into Eq. (VI.1), for water ($\sigma \simeq 80$ dynes/cm) we obtain $P' \simeq 10{,}000$ atm. It is true that in a real, clean liquid there can be a local decrease in strength as a result of the spontaneous formation of a gas bubble due to thermal fluctuations. A calculation of the probability of the formation of the vapor phase leads in this case to a lower rupture pressure for water $P' \simeq 1000$ atm,[50] which is still much higher than the experimental value.[47]

Cavitation nuclei. Almost all real liquids, and in particular water, contain different dissolved substances, including dissolved gas, which in liquids results in the existence of vapor-gas bubbles, which decrease the local strength of the liquid and form *cavitation nuclei*. There arises, however, the question of how such bubbles can exist for a long time, i.e., how they can be stable. After all, large bubbles must rise under the action of the buoyancy force, which decreases with decreasing bubble radius, while very small bubbles must be dissolved due to the high pressure created by surface tension forces, which increase with decreasing radius. There are two hypotheses along these lines, each of which is indirectly confirmed by experiments. According to the first one, the dissolved substances in liquids include surfactants, which, being absorbed on the surface of a bubble, create a monomolecular layer that decreases the surface tension. The second hypothesis explains the stability of very small gas bubbles by the adsorption of singly charged ions of dissolved salts on their surface. Interacting with oppositely charged ions located near the bubble, these ions prevent the closure of the bubble, thereby stabilizing its size. This mechanism of stabilization is formally equivalent to a decrease in the effective surface tension forces. For this reason, without considering the specific factor responsible for the stabilization of gas bubbles, whose existence has been reliably established experimentally, we can write down the condition for the equilibrium of a bubble with radius R_0 in the following form

$$P = P_s + P_g - 2\sigma/R_0, \tag{VI.2}$$

where P is the external (static) pressure; P_s is the saturation vapor pressure of the liquid; P_g is the pressure of the gas that has diffused into the bubble from the liquid; and, σ is the effective surface tension. If the external pressure P decreases, then the size of the bubble increases. We shall now examine the condition for a liquid to rupture, using the equation of state of a stable cavitation nucleus (VI.2).

§ 2. Cavitation strength of a liquid

Let the hydrostatic pressure P decrease below some equilibrium value P_0 corresponding to an initial radius of the vapor-gas bubble R_0. How will the bubble size change in this case? To calculate the dependence $P(R)$, to a first approximation we can neglect the diffusion of gas out of the liquid as well as the change in the saturation vapor pressure P_s, which is nearly independent of the curvature of the surface. Since the mass of the gas in the bubble is small and the heat capacity of the liquid surrounding the bubble is quite high, it can be assumed that the gas pressure P_g in the bubble changes in this case by means of an isothermal process, i.e., $P_g = (P_0 - P_s + 2\sigma/R_0)R_0^3/R^3$. Substituting this expression into Eq. (VI.2), we obtain

$$P = P_s + (P_0 - P_s + \frac{2\sigma}{R_0})\frac{R_0^3}{R^3} - \frac{2\sigma}{R}. \tag{VI.3}$$

Figure 29 shows graphs of $P(R)$ for water at room temperature, calculated on a computer using Eq. (VI.3), for four values of the initial radius R_0 with a starting hydrostatic pressure $P_0 = 1$ *atm*. The saturated vapor pressure in a bubble P_s is assumed to equal $2\cdot10^{-2}$ *atm* (dashed line). It is evident from the figure that at sufficiently high

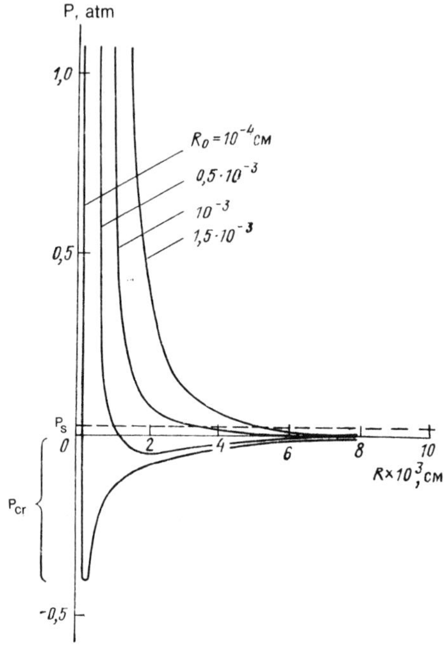

Fig. 29.

pressures, exceeding the saturated vapor pressure of the given liquid P_s, the equilibrium bubble radius R remains practically constant as the presssure increases. In this region the isothermal dependence $P(R)$ plays the basic role. According to this dependence the pressure in the bubble varies inversely as the cube of the radius for small changes in the bubble radius, while the pressure due to the surface tension forces changes inversely as the first power of the radius.

The situation is different at low pressures, close to or less than the vapor pressure P_s. It is evident from Fig. 29 that for small R_0 there are two equilibrium values of the radius for negative pressures. In this case, for the right side of the curves $P(R)$, corresponding to large values of R, the gas pressure in the bubble remains virtually constant as the bubble radius increases (the term with the factor R_0^3/R^3 can be neglected), while the surface-tension pressure decreases. In this case the saturated vapor pressure expands the bubble to infinity, i.e., the bubble becomes unstable. When the tensile stress is removed, surface tension forces reduce the size of the bubble to a stable size, corresponding to the left points of the curves $P(R)$ in Fig. 29.

It is also evident from Fig. 29 that for negative pressures, exceeding some critical value P_{cr} corresponding to the minima on the curves $P(R)$, an equilibrium bubble size cannot exist at all. Thus, when the external pressure drops below the starting value P_0, the bubble will remain stable up to a tensile stress P_{cr}, after which it begins to expand without limit. When the tension force is removed and replaced by compression, the bubble will spontaneously collapse under the action of surface tension forces and the positive external pressure. The negative quantity P_{cr} therefore determines the *cavitation strength of a liquid* for a given value of the radius R_0 of a cavitation nucleus. This quantity can evidently be found by equating to zero the derivative of Eq. (VI.3) with respect to R:

$$\frac{dP}{dR} = -3\left(P_0 - P_s + \frac{2\sigma}{R_0}\right)\frac{R_0^3}{R_{cr}^3} + \frac{2\sigma}{R_{cr}^2} = 0,$$

whence

$$R_{cr} = \left[\frac{3R_0^3}{2\sigma}\left(P_0 - P_s + \frac{2\sigma}{R_0}\right)\right]^{1/2}.$$

Substituting the value of R_{cr} into Eq. (VI.3), we obtain the magnitude of the negative critical pressure:

$$P_{cr} = P_s - \frac{2}{3\sqrt{3}}\sqrt{\frac{(2\sigma/R_0)^3}{P_0 - P_s + 2\sigma/R_0}}. \qquad \text{(VI.4)}$$

Figure 30 shows the dependence of P_{cr} on the radius of the cavitation nucleus R_0 in water, calculated according to Eq. (VI.4). It is evident from the figure that negative pressures from 1 to 10 *atm*, for which ultrasonic cavitation usually develops at moderate frequencies in untreated water, correspond to radii of stable cavitation nuclei of the order of $10^{-4} - 10^{-5}$ cm. For $R_0 \simeq 10^{-7}$ cm the critical pressure increases up to thousands of atmospheres, which corresponds to the theoretical rupture strength of a liquid with no gas bubbles.

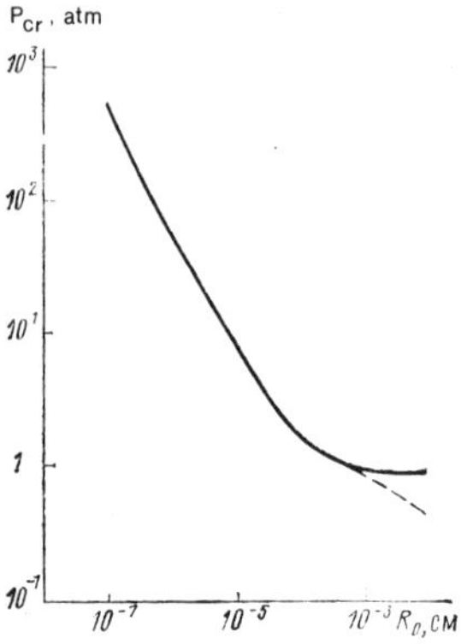

Fig. 30.

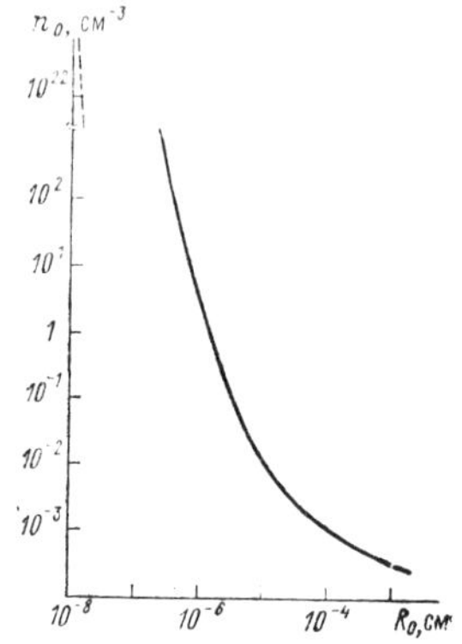

Fig. 31.

Actually, in real liquids there is a distribution of nuclei over the initial radius R_0. As an example, Fig. 31 shows the experimental curve of the distribution of nuclei in settled distilled water, taken from Ref. 48. This figure shows the density of nuclei n_0, which increases with decreasing R_0. For values of R_0 corresponding to intermolecular distances ($R_0 \simeq 10^{-7} - 10^{-8}$ cm), the extrapolated part of the curve $n_0(R_0)$ (the broken curve) approaches the theoretical density of "nuclei" created by thermal fluctuations. Thus, for all practical purposes, there is no upper threshold of cavitation, while the lower threshold is diffuse and lies near pressures corresponding to the critical pressure for bubbles with radius $\simeq$ of 10^{-3} cm, i.e., $|P_{cr}| \simeq 1$ atm (see Fig. 30).

In our calculation of the cavitation strength of a liquid we neglected the diffusion of gas, which, of course, occurs if the pressure changes slowly. This approximation is justified by the fact that for the rapid changes in pressure occurring at ultrasonic frequencies, diffusion processes indeed have little effect on the cavitation strength. On the other hand, for rapid changes in pressure the inertial properties of the liquid surrounding the bubble and the resonance properties of the bubbles can affect the cavitation strength. A more detailed calculation of P_{cr} taking into account the inertia of the liquid, however, gives appreciable corrections to the curve shown in Fig. 30 only in the region with the largest nuclei, whose number in the liquid, as evident from Fig. 31, is small. This correction is shown in Fig. 30 by the broken curve. As far as resonance phenomena are concerned, they can have a large effect on the cavitation process if the ultrasonic frequency equals the frequency of characteristic oscillations of the gas bubbles ν_{res}. For small oscillations this frequency is determined by the expression (V.28) presented above, i.e.,

$$\nu_{res} = \frac{1}{2\pi R_0} \sqrt{3\gamma (P_0 + 2\sigma/R_0)/\rho}, \qquad \text{(VI.5)}$$

where $\gamma = c_p/c_v$ is the ratio of the specific heat capacities of the gas filling the bubble and ρ is the density of the liquid. The resonant frequencies of air bubbles with different radius calculated using Eq. (VI.5) are shown for water in Fig. 32. It is evident from this figure that a bubble radius $R_0 = 10^{-3}$ cm corresponds to a characteristic frequency of radial oscillations $\nu_{res} \simeq 500$ kHz. Therefore, the entire computed curve of critical pressures in Fig. 31 refers to frequencies of pressure variations below several hundred kilohertz. If the ultrasonic frequency exceeds the resonant frequency of the bubble ν_{res}, then there will not be enough time for the size of the bubble to change in response to the rapid changes in pressure and such bubbles will not participate in the cavitation process. Thus, as the ultrasonic frequency increases, the cavitation threshold must increase to critical pressures corresponding to increasingly smaller cavitation nuclei, i.e., cavitation will begin at increasingly higher ultrasonic wave amplitudes.

All these results are confirmed by the corresponding experiments. It is well known that the cavitation threshold increases sharply as the ultrasonic frequency increases, and at frequencies above 10 MHz cavitation can be excited only at the focal spot of focusing radiators, where the pressure amplitude can reach hundreds and thousands of atmospheres.[51] The cavitation threshold increases as the static pressure P_0 in the liquid increases. This is explained by the fact the sizes of the nuclei present in the liquid decrease and the density of the gas in the nuclei increases. On the other hand, a drop in the static pressure lowers the cavitation threshold, as does a rise in the temperature of the liquid. The cavitation strength of a liquid is also increased by degasification. Experimental data indicate that an electric field, which affects the conditions of adsorption of "hydrophobic" ions on the surface of a bubble, changes the cavitation strength and that the cavitation threshold in water is lowered when salts that give rise to negative hydration are dissolved in the water.[52]

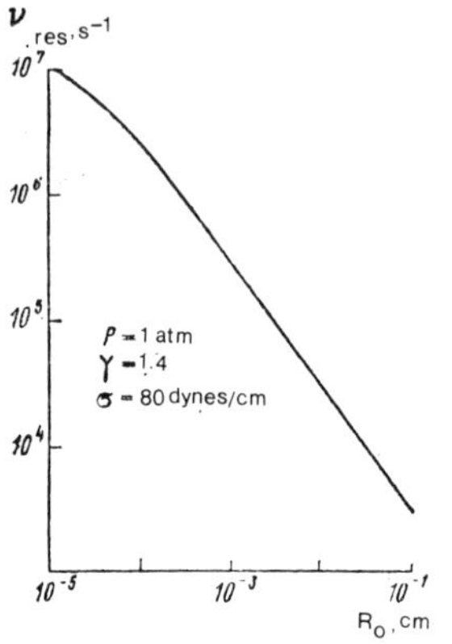

Fig. 32.

In all of these experiments different criteria for the onset of cavitation are used. One such criterion is the expansion of the cavitating liquid due to the formation of large gas bubbles in it.[53] In many experiments, "cavitation noise," produced by the collapse of cavitation holes, was used as the criterion for the onset of cavitation.[54] *Sonoluminescence* (luminescence of a liquid accompanying active cavitation), cavitation erosion of solids, and other phenomena accompanying ultrasonic cavitation can also be used as a criterion for the onset of cavitation.[48] These phenomena, however, arise or attain appreciable development at different stages of the cavitation process and, for this reason, quantitative data on cavitation thresholds determined by different methods differ considerably from one another; differences arise also because of the fact that the liquids under study are in different states. Nevertheless, experiments qualitatively confirm the basic results concerning the cavitation strength and the factors affecting it as well as the basic features which follow from the analysis presented here.

These results, however, were obtained without taking into account the dynamics of the pressure variation. A more detailed analysis of ultrasonic cavitation requires an investigation of the behavior of a cavitation nucleus in an ultrasonic wave. We shall return to this question later, after we study the process of collapse of a cavitation hole in greater detail.

§ 3. Collapse of a cavitation cavity

To find the kinematic characteristics of a collapsing cavitation bubble, we shall examine the simplest problem of the collapse of a spherical cavity in an incompressible liquid under the action of a constant pressure P, assuming at first that the cavity is filled only with the saturated vapor and that there is enough time for the vapor to condense within the time it takes for the bubble to close up, so that its pressure P_s can be neglected.

Let R be the instantaneous radius of the collapsing sphere and U the velocity of the walls of the sphere. We single out a spherical layer of thickness dr at a distance r from the center of the cavity (Fig. 33).

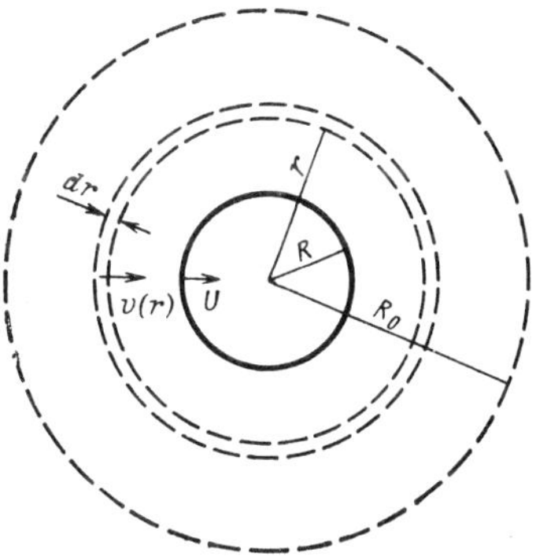

Fig. 33.

The velocity of the liquid $v(r)$ in this layer is determined by the condition of continuity, according to which the ratio of the velocities in two sections of the "stream tube" is inversely proportional to the ratio of the areas of these sections or, in this case, to the ratio of the areas of two spherical surfaces with radii r and R, i.e. $v(r)/U = R^2/r^2$, whence

$$v(r) = UR^2/r^2. \tag{VI.6}$$

The kinetic energy of a layer with volume $4\pi r^2 dr$ is $dW_{kin} = (\rho/2) v^2(r) \cdot 4\pi r^2 dr$, where ρ is the density of the liquid. The total kinetic energy of the mass of liquid filling the cavity is $W_{kin} = 2\pi\rho \int_R^\infty v^2(r) r^2 dr$. Substituting here the value of $v(r)$ from expression (VI.6) and performing the integration, we find:

$$W_{kin} = 2\pi\rho U^2 R^4 \int_R^\infty \frac{dr}{r^2} = 2\pi\rho U^2 R^3. \qquad \text{(VI.7)}$$

This kinetic energy equals the work performed by the pressure P in decreasing the volume of the cavity from the initial value $(4/3)\pi l R_0^2$ to the final value $(4/3)\pi R^3$, i.e.,

$$A = (4/3)\pi P (R_0^3 - R^3). \qquad \text{(VI.8)}$$

Equating Eq. (VI.7) and (VI.8), we obtain an expression for the velocity of the walls of the collapsing cavity:

$$U = \sqrt{(2/3)(P/\rho)(R_0^3/R^3 - 1)}. \qquad \text{(VI.9)}$$

This expression was obtained by Rayleigh in 1917. Using this expression it is easy to calculate the total time Δt for an empty cavity with initial radius R_0 to close up. Since $U = dR/dt$, and making the substitution $R = R_0 x^{1/3}$, we obtain from Eq. (VI.9) $dx/dt = [6P/(\rho R_0^2)]x^{1/3}(1 - x)$, whence

$$\Delta t = R_0 \sqrt{\frac{\rho}{6P}} \int_0^1 x^{-1/6}(1 - x)^{-1/2} dx =$$

$$= R_0 \sqrt{\frac{\rho}{6P}} \frac{\Gamma(5/6)\Gamma(1/2)}{\Gamma(4/3)} = 0.915 R_0 \sqrt{\frac{\rho}{P}}. \qquad \text{(VI.10)}$$

Thus the total time for an empty cavity in a liquid with density ρ to close up has a simple functional dependence on the pressure P and the initial radius R_0.

This model, however, neglects a number of factors that exist in the real world. These include the following: the surface tension, which creates an additional compressive pressure; the variable nature of the pressure in the acoustic wave; the compressibility of a real liquid; and, finally, the presence of some gas in the nucleus, which will retard the process of collapse. As far as the surface tension forces are concerned, a simple calculation shows that they are manifested in the "effective pressure" P only at the final stage of collapse, when the radius of the cavity becomes very small. The effective pressure in ultrasonic cavitation must be understood to be the hydrostatic pressure P_0 plus the pressure in the acoustic wave. For the wave, it is natural to take the amplitude value p_{max}. A more detailed analysis of the dynamics of a cavitation cavity in an acoustic field shows that the process of collapse sometimes begins at the intermediate stage of the compression phase, and a comparison of the theoretical results with the standard data gives the best agreement if the pressure averaged over a half-period $(2/\pi)p_{max}$ is used. We can therefore set $P = P_0 + (2/\pi)p_{max}$ in Eqs. (VI.8)–(VI.10).

In reality, a cavitation bubble contains some gas, whose mass remains practically constant with a rapid collapse. It is also necessary to take into account the fact that for rapid collapse there is not enough time for the saturated vapor filling the cavitation cavity to condense on the surface of the cavity. For this reason, the collapsing cavitation cavity can be assumed to be filled with a vapor–gas mixture, whose pressure under rapid compression obeys the adiabatic law:

$$P_g/P_{g0} = (R_0/R)^{3\gamma}, \qquad \text{(VI.11)}$$

where P_{g0} is the initial pressure of the gas in a nucleus with radius R_0 and γ is the ratio of the specific heat capacities of the mixture. Due to the presence of the adiabatically compressed vapor–gas mixture the velocity of the walls of the closing cavity will not approach infinity, as predicted by Eq. (VI.9), and the radius of the cavity will not decrease to zero but to some finite minimum value R_{min}, which is determined by the starting pressure of the vapor–gas mixture P_{g0}. This minimum radius can be easily found by calculating the work of adiabatic compression of a vapor–gas mixture, which for an arbitrary radius R equals:

$$A_g = \int_{R_0}^{R} P_{g0}(R_0/R)^{3\gamma}4\pi R^2 dR. \qquad \text{(VI.12)}$$

For a vapor—air mixture the adiabatic index γ can be set equal to 4/3. In this case, the integral (VI.12) yields

$$A_g = -4\pi P_{g0} R_0^3 (R_0 - 1), \tag{VI.13}$$

where the minus sign is determined by the direction in which the forces act. With complete compression of the cavity to the minimum radius all of the energy of the collapsing liquid, defined by Eq. (VI.8), goes into the work required to compress the vapor—gas mixture (VI.13) and, therefore, in this case the following equality holds: $(4/3)\pi P(R_0^3 - R_{min}^3) = 4\pi P_{g0} R_0^3 (R_0/R_{min} - 1)$. Assuming that $R_0/R_{min} \gg 1$, we obtain

$$R_{min} \simeq 3R_0 P_{g0}/P. \tag{VI.14}$$

The ratio $P_{g0}/P_0 \equiv q$ (this ratio is customarily called the gas-content parameter) for real cavitation bubbles falls in the range 0.02–0.03. Therefore, the radius of the bubble decreases by a factor of 10 when the bubble collapses, which justifies the assumption made above. In addition, at the time the bubble collapses a very high pressure develops in the vapor—gas mixture. Indeed, based on Eq. (VI.11), setting $\gamma = 4/3$ and $R = R_{min}$ and substituting the expression (V.14), we obtain $P_{gmax} = (P/3^4)q^{-3}$. For $P = P_0 = 1$ *atm* and $q = 0.02$, this gives $P_{gmax} \simeq$ 40,000 *atm*. This result is probably somewhat too high, because we neglected, for example, the compressibility and viscosity of a real liquid, which undoubtedly decrease the pressure arising as the cavity collapses. Furthermore, the equations of thermodynamics used are inaccurate at high pressures. In addition, experiments show that the assumed sphericity of the bubbles is destroyed when they collapse suddenly: they become formless and even break up into small fragments. Nevertheless, the collapse of cavitation bubbles in a real liquid is indeed accompanied by a large increase in the pressure, giving rise to spherical shock waves which decay rapidly in space and are a characteristic consequence of cavitation.

We shall now calculate the temperature of the vapor—gas mixture, arising under adiabatic compression of the mixture in a collapsing bubble. For an adiabatic process $T/T_0 = (R_0/R)^{3(\gamma-1)}$, where T is the instantaneous gas temperature during compression and T_0 is the initial temperature. At the moment of maximum compression, with (VI.14) and $\gamma = 4/3$, we have $T_{max} = (T_0/3)(P/P_{g0})$. Using again the minimum values $P = P_0 = 1\ atm$ and $P_{g0}/P_0 = q = 0.02$, at $T_0 = 300\ K$ we obtain $T_{max} \simeq 6000\ K$. This temperature could be the reason (or one of the reasons) for the characteristic luminescence observed to accompany ultrasonic cavitation (sonoluminescence).

To calculate the rate and time of collapse of a gas-filled bubble, we shall use the previous technique, taking into account the work peformed in compressing the gas as given by expression (VI.13). Combining Eqs. (VI.7), (VI.8), VI.12), and (VI.13), we now obtain:

$$U = \left\{\frac{2}{3}\frac{P}{\rho}\left[\frac{R_0^3}{R^3} - 1 - 3q\frac{R_0^3}{R^3}\left(\frac{R_0}{R} - 1\right)\right]\right\}^{1/2}. \qquad \text{(VI.15)}$$

For $q = P_{g0}/P_0 = 0$, this equation reduces to Rayleigh's equation (VI.9), and for a realistic gas-content of $q = 0.02$ and $P = P_0 = 1\ atm$ it gives a maximum velocity for water of about 500 m/s. Using again the fact that $U = dR/dt$ and making the substitution $R/R_0 = y$ in Eq. (VI.15), based on this equation we find the total collapse time:

$$\Delta t = R_0\sqrt{\frac{3}{2}\frac{\rho}{P}}\int_{R_{min}/R_0}^{1}\frac{y^2 dy}{[(-y^4+y)-3q(1-y)]^{1/2}}. \qquad \text{(VI.16)}$$

According to Ref.55, the integral on the right side of expression (VI.16) is close to 1 for $q = 0.02$–0.03. Therefore, the presence of gas in the cavitation bubble has virtually no effect on the collapse time given by Rayleigh's equation (VI.10) for an empty cavity. In any case, the total collapse time thus depends only on the initial radius of the bubble and the effective pressure. This dependence for water is shown in Fig. 34 by curves 1 and 2, which refer to different pressures. The broken curve in the same figure illustrates the periods $T_{res} = \nu_{res}^{-1}$ of the characteristic oscillations of spherical bubbles with radius R_0, calculated using Eq. (VI.5). If we recall that cavitation can occur at ultrasonic frequencies lying below the resonance frequency of the nuclei, then based on this figure it is evident that for most cases the collapse time Δt is shorter than the period of an ultrasonic wave capable of creating cavitation. This difference, however, is not large enough for the pressure P acting on the bubbles to be quasistatic for all bubbles and the behavior of the bubbles in the rapidly varying ultrasonic field turns out to be more complicated. Direct observations show, for example, that a cavitating bubble can undergo several oscillations before collapsing: small bubbles grow slowly and after reaching some critical radius they either collapse or they continue to grow, leaving the cavitation process when their size exceeds the resonant size for the given ultrasonic frequency. The last circumstance, incidentally, explains an anomalous result following from the analysis presented above: according to Eq. (VI.14), for example, the pressure inside a collapsing bubble P_{gmax} will increase as its initial radius R_0 increases. It is clear, however, that this increase cannot be infinite. In reality, for large values of R_0, the collapse time of a bubble becomes so large that for a given ultrasonic frequency it exceeds the period of the ultrasonic oscillations and even before the bubble completely collapses the acoustic pressure changes sign and becomes negative. The critical dimensions of bubbles, in their turn, depend on the amplitude of the pressure in the ultrasonic wave and prolonged pulsations of bubbles prior to their collapse occur only for quite high amplitudes.[56] For amplitudes that are too high, the cavitation efficiency can even decrease somewhat.

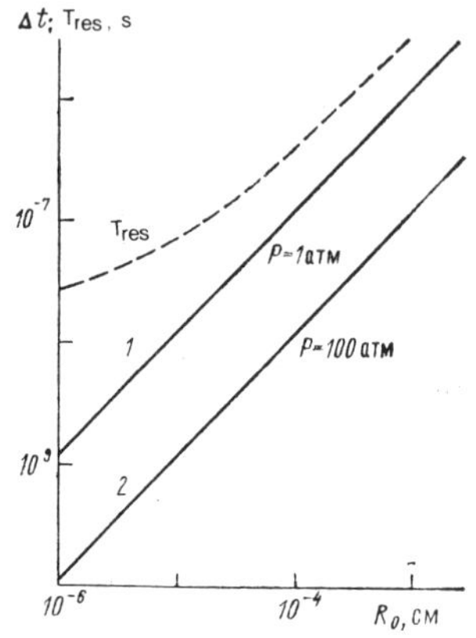

Fig. 34.

Since the variable nature of the effective pressure was neglected, the simplified approach used above does not, of course, permit determining all the quantitative characteristics of ultrasonic cavitation. For this reason, a more detailed analysis of the cavitation process requires an investigation of the dynamic behavior of a cavitation bubble in an ultrasonic field, i.e., it is necessary to solve the equation of motion of a cavitation cavity under the action of a variable pressure.

§ 4. Dynamics of a cavitation cavity in an ultrasonic wave

We shall study a spherical cavity with an instantaneous radius $R(t)$ and we shall write down the equation of motion for the velocity of the fluid particles $v(r)$ for $r \gg R$ in polar coordinates, whose origin coincides with the center of the cavity. Due to the spherical symmetry of the problem, the equation of motion (II.4) will be one-dimensional with one polar coordinate r:

$$\frac{\partial v}{\partial t} + v\frac{\partial v}{\partial r} = -\frac{1}{\rho}\frac{\partial p}{\partial r}. \qquad \text{(VI.17)}$$

We retain the nonlinear term $v(dv/dr)$ in this equation, because the velocities of the fluid particles in this case can be high. We shall again assume that the liquid is incompressible and has a constant density $\rho = \rho_0$. The equation of continuity of the incompressible liquid can be written in the convenient form (VI.6), which relates the velocity of the cavity walls U and radius of the cavity R to the velocity of the spherical layer with the radial coordinate r (see Fig. 33). The entire mass of the liquid participates in the motion, so that Eq. (VI.17) must be integrated from r to ∞. This is most easily done by introducing the velocity potential as defined in (II.7) (we assume that the motion is irrotational):

$$v = -\partial\varphi/\partial r. \qquad \text{(VI.18)}$$

Then, since in the limit $r \to \infty$, $v = 0$, $\varphi = 0$, and $p(r) = P(\infty)$, the integration gives

$$\frac{\partial\varphi}{\partial t} - \frac{v^2}{2} + \frac{1}{\rho_0}[P(\infty) - p(r)] = 0, \qquad \text{(VI.19)}$$

According to expressions (VI.6) and (VI.18), $\varphi = UR^2/r$. Differentiating this expression with respect to t and substituting into Eq. (VI.19), we obtain

$$\frac{1}{r}(R^2\frac{dU}{dt} + 2UR\frac{dR}{dt}) - \frac{1}{2}U^2\frac{R^4}{r^4} + \frac{1}{\rho_0}[P(\infty) - p(r)] = 0.$$

We are interested in the motion of the cavity wall. Setting $r = R$ and using the fact that $U = dR/dt$, we obtain:

$$R\frac{d^2R}{dt^2} + \frac{3}{2}(\frac{dR}{dt})^2 + \frac{1}{\rho_0}[P(\infty) - P(R)] = 0. \quad \text{(VI.20)}$$

This equation describes the pulsations of a spherical cavitation cavity with a pressure $P(R)$ on its walls and a pressure $P(\infty)$ far away from the cavity. We shall assume that the cavitation cavity is filled with gas with a partial pressure P_g and vapor, whose pressure we shall consider to be unchanged as before, making the assumption that there is enough time for condensation and vaporization to follow the change in the volume of the cavity. We shall assume that the gas pressure in general obeys the polytropic law: $P_g = P_{g0}(R_0/R)^{3n}$ with $1 \leqslant n \leqslant c_p/c_V$. The initial equilibrium gas pressure P_{g0} in a stable bubble with radius R_0 is $P_{g0} = P_0 - P_s + 2\sigma/R_0$, where P_0 is the hydrostatic pressure. Thus

$$P_g = (P_0 - P_s + 2\sigma/R_0)(R_0/R)^{3n}. \quad \text{(VI.21)}$$

Taking into account the curvature-dependent surface-tension forces, we find

$$P(R) = (P_0 - P_s + \frac{2\sigma}{R_0})(\frac{R_0}{R})^{3n} + P_s - \frac{2\sigma}{R}. \quad \text{(VI.22)}$$

The pressure $P(\infty)$ consists of the hydrostatic pressure P_0 and the acoustic pressure p, which varies sinusoidally with time with frequency ω:

$$P(\infty) = P_0 - p_{max}\sin\omega t. \quad \text{(VI.23)}$$

Substituting expressions (VI.22) and (VI.23) into Eq. (VI.20), we finally obtain:

$$R\frac{d^2R}{dt^2} + \frac{3}{2}\left(\frac{dR}{dt}\right)^2 + \frac{1}{\rho_0}\left[P_0 - P_s - p_{max}\sin\omega t + \frac{2\sigma}{R} - \right.$$

$$\left. - \left(P_0 - P_s + \frac{2\sigma}{R_0}\right)\left(\frac{R_0}{R}\right)^{3n}\right] = 0. \qquad \text{(VI.24)}$$

This nonlinear differential equation is known in the theory of cavitation as the *Nolting–Neppiras equation*.[57] It describes quite well the change in the radius of a cavitation cavity in the field of an ultrasonic wave with arbitrary frequency. Equation (VI.24) is inadequate only at the last stage of collapse of a cavitation bubble, when the velocity of the bubble walls is comparable to the velocity of sound in the liquid and the compressibility of the liquid must be included.

An equation describing the pulsations of a spherical cavitation cavity, which takes into account the compressibility of the liquid, was obtained by Herring and Flynn.[49] Omitting the derivation of this equation, we write out its final form:

$$R\left(1 - 2\frac{U}{c_0}\right)\frac{d^2R}{dt^2} + \frac{3}{2}\left(1 - \frac{4}{3}\frac{U}{c_0}\right)\left(\frac{dR}{dt}\right)^2 + \frac{1}{\rho_0}\left[P_0 - P_s - p_{max}\sin\omega t + \right.$$

$$\left. + \frac{2\sigma}{R} + \frac{4\eta U}{R} - \left(P_0 - P_s + \frac{2\sigma}{R_0}\right)\left(\frac{R_0}{R_0}\right)^{3n}\right] + \frac{R}{\rho_0}\frac{U}{c_0}\left(1 - \frac{U}{c_0}\right)\frac{dP(R)}{dR} = 0. \qquad \text{(VI.25)}$$

where η is the coefficient of viscosity and c_0 is the velocity of sound in the liquid in the linear approximation. It is not difficult to see that for $U/c_0 \ll 1$ the Herring–Flynn equation transforms into the Nolting–Neppiras equation (VI.24). In its turn, Eq. (VI.25) holds only for values of U that do not exceed c_0. A more accurate solution, describing pulsations with arbitrary velocities U, was obtained by Kirkwood and Bethe[58] in the form

$$R\left(1 - \frac{U}{c}\right)\frac{d^2R}{dt^2} + \frac{3}{2}\left(1 - \frac{U}{3c}\right)\left(\frac{dR}{dt}\right)^2 -$$

$$- \left(1 + \frac{U}{c}\right)H - \frac{U}{c}\left(1 - \frac{U}{c}\right)R\frac{dH}{dR} = 0. \qquad \text{(VI.26)}$$

where $H = \int_{P(\infty)}^{P(R)} dp/\rho$, ρ is the instantaneous value of the density of the liquid, and c is the local velocity of sound defined by relation (IV.27).

Unfortunately, the Nolting–Neppiras equation (VI.24), and especially Eqs. (VI.25) and (VI.26), cannot be solved in general form; they can only be solved numerically for specific values of the frequencies, amplitudes of ultrasonic oscillations, initial radii of the nuclei, etc. An analysis of these solutions, performed with a computer, shows that the dimensions of the cavitation bubbles increase rapidly during the first half-period of rarefaction, after which they can complete several pulsations prior to collapsing.[47] The maximum radius of a bubble R_{max}, the number of pulsations, and the collapse time increase with increasing amplitude of the ultrasonic wave. For low amplitudes p_{max}, less than some threshold value p'_{max}, the bubbles pulsate nonlinearly without collapsing. In this case bubbles whose size is less than or equal to the resonant size pulsate approximately at the ultrasonic frequency, while bubbles with dimensions exceeding the resonant value pulsate with a period close to the period of the characteristic oscillations determined by Eq. (VI.5). For pulsations of large bubbles, when $p_{max} > p'_{max}$, the lifetime of the bubbles before collapse must be greater than the period of the characteristic oscillations. For this reason, large bubbles pulsate at all values of p_{max} (collapsing when $p_{max} > p'_{max}$), while small bubbles undergo pulsations prior to collapsing only for quite high amplitudes. If the amplitude of the ultrasonic pressure is small (but larger than p'_{max}), then such bubbles expand during the stretching half-period and collapse during the first compression half-period, and they begin to collapse almost exactly at the maximum value of the positive pressure, as assumed in the preceding section.

All these results are clearly illustrated by the series of curves presented in Fig. 35a, which were calculated on a computer using Eq. (VI.24) for a specific initial bubble radius $R_0 = 10^{-4}$ cm.[47] The calculations were performed assuming adiabatic pulsations ($n = \gamma = 4/3$) in water with hydrostatic pressure $P_0 = 1$ atm for an ultrasonic frequency $\nu = 500$ kHz and different pressure amplitudes p_{max}, indicated on the curves in atmospheres. Figure 35b shows the variation of the pressure p in the ultrasonic wave as a function of time. The shaded region in Fig.

35a corresponds to a "structural instability" of Eq. (VI.24). It is evident that in these regions a small variation in p_{max} leads to a qualitative change in the dependence of R/R_0 on ωt.

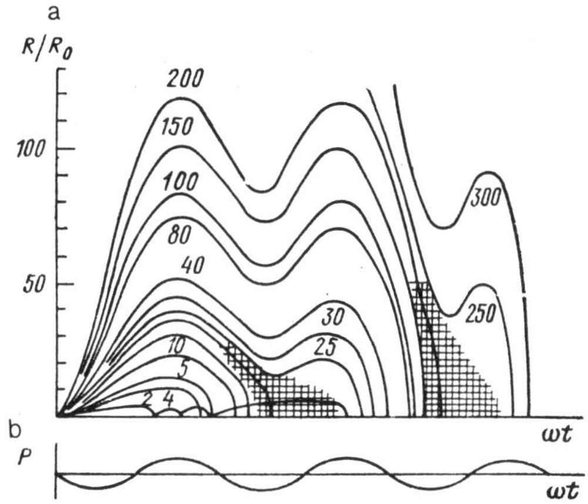

Fig. 35.

Thus a more detailed study of the dynamics of a cavitation cavity yields important information about its behavior prior to collapse and explains a number of experimental dependences, including the dependence on the amplitude of the ultrasonic wave. As far as the basic aspects of cavitation are concerned — the existence of stable nuclei and the collapse of cavitation cavities, their description remains at the level of the preceding sections. In particular, the collapse time following from the Nolting–Neppiras equation (VI.24) agrees well with Rayleigh's equation (VI.9) for well-developed cavitation if we set $R_0 = R_{max}$ and $P = P_0 + p_{max}$: $\Delta t = 0.915 R_{max} \sqrt{[\rho_0/(P_0 + p_{max})]}$. It is this pressure that determines the potential energy of a bubble which expands to a radius R_{max}, while the potential energy, in its turn, determines not only the time but also the rate of collapse. For this reason, the expression for the velocity U likewise remains the same with $R_0 = R_{max}$ and $P = P_0 + p_{max}$:

$$U = \sqrt{(2/3)(P_0 + p_{max})(R^3_{max}/R^3 - 1)\rho_0},$$

which holds for velocities $U \ll c_0$, when the compressibility can be neglected. Taking into account the compressibility using Eq. (VI.25), we obtain the more accurate equation

$$U = \sqrt{\frac{2}{3}\,\frac{P_0 + p_{max}}{\rho_0}\left[\frac{R^3_{max}}{R^3} - 1\right]\left[1 - \frac{4}{3}\,\frac{U}{c_0}\right]^{-1}}\,. \qquad \text{(VI.27)}$$

All these results are in good agreement with experiment. In addition, at pressures p_{max} slightly above the threshold value p'_{max} the best agreement, as already noted, is obtained when p_{max} is replaced by the average value of the pressure over a period of the wave.

§ 5. Acoustic properties of a cavitating liquid

When ultrasonic cavitation appears in a liquid, the acoustic properties of the liquid change considerably. First of all, the cavitation bubbles scatter ultrasound; we shall study scattering by bubbles below. As a result, the energy of the ultrasonic wave will decrease rapidly as a function of distance. Scattering is not, however, the only reason for the decrease in energy accompanying cavitation: a large part of the energy is expended on the creation of cavitation bubbles, i.e., on expanding them to maximum radius R_{max}. After the cavitation cavity collapses, this energy is partially transformed into the energy of cavitation-induced shock waves, but it is completely removed from the primary ultrasonic wave.

The work performed on the expansion of N cavitation bubbles up to a radius R_{max} from the initial radius R_0 can be calculated using the expression (VI.8):

$$A_{cav} = \frac{4}{3}\pi P\,(R^3_{max} - R^3_0)\,N, \qquad \text{(VI.28)}$$

where P is the sum of the hydrostatic and acoustic pressures. Keeping in mind that $R_0 \ll R_{max}$ expression (VI.28) can be written in the form $A_{cav} \simeq (4/3)\pi PNR^3_{max}$ or $A_{cav} \simeq P\Delta V$, where ΔV is the total volume of the N cavitation bubbles with maximum radius or the maximum cavitation-induced increase in the volume of the liquid. Since in reality the bubbles have different initial radii (see Fig. 31) and maximum sizes, the total increase in their volume is difficult to calculate. It can, however, be easily measured using the simple dilatometric method mentioned previously,[48,53] in which the liquid under study is poured into a hermetically sealed vessel, fitted with acoustically transparent windows and a capillary for determining ΔV. It is interesting that although the volume increase ΔV measured in this manner provides a rough measure of the cavitation energy, it is well correlated with the relative strength of its different manifestations: the brightness of sonoluminescence, the magnitude of the cavitation erosion of solids in the cavitation zone and subjected to cavitation failure, the intensity of cavitation-induced shock waves, etc.[48]

Measurements and estimates both show that a considerable amount of energy is expended on the excitation of cavitation. Together with scattering by cavitation bubbles, this leads to rapid attenuation of the ultrasonic wave, whose amplitude in the region of cavitation decreases according to the standard exponential law (for plane waves) $p_{max} = p_{max0}\exp(-\alpha_{cav}x)$, but with an attenuation factor α_{cav} greatly exceeding the absorption coefficient α_0 of a noncavitating liquid. As a result of this attenuation the amplitude of the pressure in the ultrasonic wave at some distance x_{cav} from the source decreases to a value below the threshold value p_{th} for the appearance of cavitation, and cavitation ceases. Thus, in the field of plane or diverging ultrasonic waves, the cavitation zone has a quite distinct boundary (in a focused beam it is located in the region of the focal spot).

It should also be noted that the sharp ultrasonic attenuation in the cavitation zone causes strong acoustic flows to develop (see § V.4). In addition, directed radiation pressure forces act on the cavitation bubbles. As a result, in a finite beam intense fluid motion occurs in the cavitation zone.

Aside from the additional ultrasonic attenuation, cavitation also "loosens up" the liquid, as a result of which its density, compressibility, and, therefore, its characteristic impedance change. To describe these changes quantitatively we shall single out in the cavitation zone a volume V_0 whose linear dimensions are much smaller than the ultrasonic wavelength, so that the acoustic pressure in it can be assumed to be constant and in-phase, and at the same time much larger than the cavitation bubbles. In other words, the bubbles must be much smaller than the ultrasonic wavelength; this condition is usually satisfied at the comparatively low frequencies at which cavitation is observed. Then, the change in the given volume ΔV_0 due to cavitation can serve as a measure of the density ρ_{cav} and compressibility κ_{cav} of the cavitating liquid. Introducing the relative change in volume — *the cavitation index* — $D = \Delta V_0/V_0$, for the new values of the cavity and compressibility averaged over a period we can write: $\rho_{cav} = \rho_{liq}(1-\bar{D}) + \rho_g\bar{D}$, $\kappa_{cav} = \kappa_{liq}(1-\bar{D}) + \kappa_g\bar{D}$, where ρ_{liq} and κ_{liq} are the density and compressibility of the continuous liquid; ρ_g and κ_g are the density and compressibility of the vapor–gas mixture in the cavitation bubbles; and, $\bar{D}$ is the average cavitation index, corresponding to the average radius of a cavitation bubble about which the bubble oscillates. It is natural to set this average radius equal to one-half the maximum radius R_{max}; then, $\bar{D} \simeq 0.1\,D$. The time-averaged characteristic impedance of the cavitating medium can now be written in the form

$$\overline{\rho_{cav}c_{cav}} = \rho_{liq}c_{liq}\left[\frac{(1-\bar{D}) + (\rho_g/\rho_{liq})\bar{D}}{(1-\bar{D}) + (\kappa_g/\kappa_{liq})\bar{D}}\right]^{1/2}, \qquad \text{(VI.29)}$$

where c_{liq} and c_{cav} are the velocities of sound in the medium with and without cavitation. Since usually $\rho_g/\rho_{liq} \ll 1$, $\kappa_g \gg \kappa_{liq}$, and $\bar{D} \ll 1$, expression (VI.29) can be simplified as follows

$$\overline{\rho_{cav}c_{cav}} = \bar{z}_{cav} \simeq \rho_{liq}c_{liq}\kappa_{liq}/(\bar{D}\kappa_g). \qquad \text{(VI.30)}$$

However, for very small D the number 1 in the denominator of Eq. (VI.29) can be neglected, and in the limit $D \to 0$ the expression

$$\bar{z}_{cav} = \rho_{liq}c_{liq}(1 + \bar{D}\kappa_g/\kappa_{liq})^{-1/2}. \qquad \text{(VI.31)}$$

must be used.

The dependence of the relative change in the characteristic impedance $\bar{z}_{cav}/z_0$ on the cavitation index D (which can be measured experimentally using the dilatometric method) for water with $\kappa_g/\kappa_{liq} = 10^4$, calculated using Eqs. (VI.30) and (VI.31), is shown in Fig. 36.[59] The initial part of this dependence, corresponding to very low values of $\bar{D}$, is shown on an enlarged scale in Fig. 36b. It is evident that the acoustic impedance of the cavitating medium varies considerably, even for small

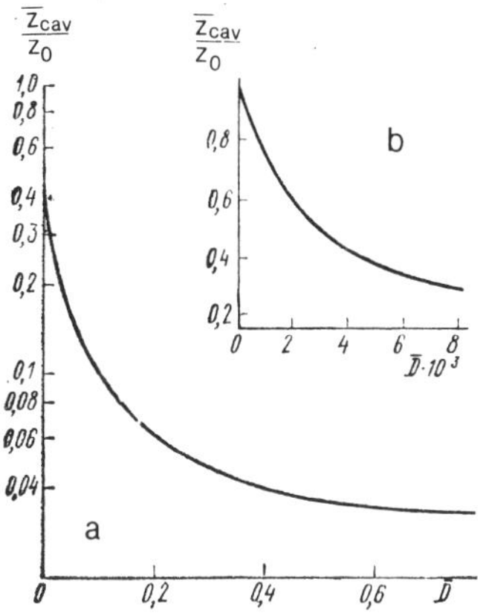

Fig. 36.

$\bar{D}$. For example, for $\bar{D} = 10^{-3}$ it decreases by 30%, while for $\bar{D} = 0.003$ it already drops by a factor of two. It is clear that this phenomenon cannot be neglected when performing the corresponding calculations.

VII. Reflection, Refraction, and Scattering of Ultrasonic Waves

§ 1. Transmission and reflection of plane waves at normal incidence on the boundary between two media

Up to now, we have been studying the propagation of ultrasonic waves in an unbounded medium. At the boundary separating two media a wave is partially reflected, interfering with the incident wave, and partially transmitted into the second medium. In this chapter, we shall determine the criteria for the reflection and transmission of plane waves under different conditions of oblique and normal incidence on the boundary between two media, and we shall study the structure of the interference field formed when the reflected wave is superposed on the incident wave. For the time being, we shall confine our attention to media which support only longitudinal waves, i.e., liquids and gases, since the results obtained also hold for other types of waves. At boundaries between solid media, in addition to being reflected and refracted, waves are also transformed from one form into another (see below), but the general energy balance and the laws of reflection and refraction remain the same for each wave. We shall confine our attention below to monochromatic plane waves with infinitesimal amplitude, taking into account the role of nonmonochromaticity, nonlinear effects, and attenuation in the adjacent media. The results which we shall obtain for these waves are generally also valid for waves with other configurations (spherical, cylindrical, etc.) with respect to their rays, i.e., normals to the wavefront. For this reason, we shall not study specially the transmission of spherical and cylindrical waves and waves of other configurations through a boundary, but rather we shall take into account any corrections attributable to differences in the angles of incidence. We shall start the analysis of the transmission of plane waves through the boundary between two media with the simplest cases, generalizing the results later to more complicated situations.

Let a plane monochromatic wave, propagating along the x-axis, be incident normally on the boundary between two media 1 and 2 with densities ρ_{01} and ρ_{02} and velocities of sound c_{01} and c_{02} (Fig. 37). Since we are analyzing waves with infinitesimal amplitudes, to simplify the notation we shall drop the zero indices referring to the linear approximation, noting specifically cases when the quantities ρ,

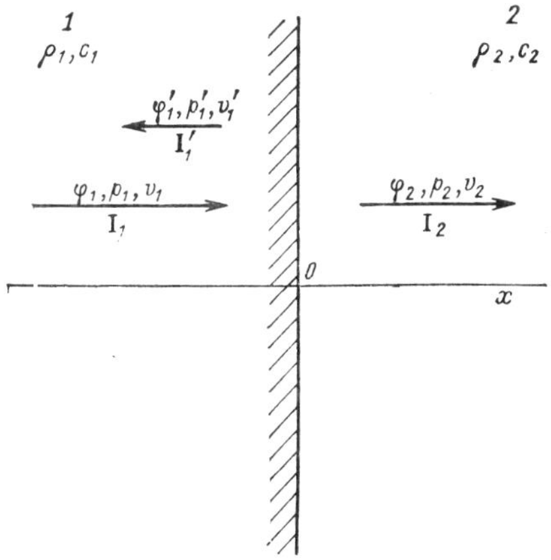

Fig. 37.

c, etc. refer to their total values. Under the action of the incident wave the boundary will undergo harmonic oscillations, creating in the adjacent media forward and backward waves propagating in the positive and negative directions along the x-axis, respectively. Thus two waves appear: a wave passing through the boundary and a wave reflected by the boundary, which add to the incident wave. Under the condition that the media are continuous, which also holds at the boundary, the resulting particle velocity (or particle displacement) in the incident and reflected waves must equal the particle velocity (displacement) in the transmitted wave, otherwise a discontinuity would appear at the boundary. In addition, at the boundary, as in any other section of the continuous medium, the equality of action and reaction implies that stresses must equal their normal components, i.e., pressures in this case. This leads to boundary conditions which determine the quantitative values of the acoustic parameters in the transmitted and reflected waves.

We shall denote the potential, pressure, and particle velocity by φ_1, p_1, and v_1 in the incident wave, by φ_1', p_1', and v_1' in the reflected wave, and by φ_2, p_2 and v_2 in the wave transmitted into medium 2. The equations for the velocity potentials of the incident, reflected, and transmitted waves, respectively, will have the following complex form:

$$\begin{aligned} \varphi_1 &= \varphi_{1\max} \exp[i(\omega t - k_1 x)], \\ \varphi_1' &= \varphi_{1\max}' \exp[i(\omega t + k_1 x)], \\ \varphi_2 &= \varphi_{2\max} \exp[i(\omega t - k_2 x)], \end{aligned} \tag{VII.1}$$

where $\varphi_{i\max}$ is the amplitude of the potential and $k_1 = \omega/c_1$ and $k_2 = \omega/c_2$ are the wave numbers in media *1* and *2*, respectively. On the basis of the above discussion, the following conditions hold at the boundary, to which we shall assign the coordinate $x = 0$:

$$\left.\begin{aligned} v_1 + v_1' &= v_2 \\ p_1 + p_1' &= p_2 \end{aligned}\right|_{x=0} \tag{VII.2}$$

All quantities here are variables: the boundary conditions (VII.2) must be satisfied at all times. Using the relation between the potential φ and the pressure and velocity in a plane wave

$$p = \rho\, \partial\varphi/\partial t, \quad v = -\,\partial\varphi/\partial x, \tag{VII.3}$$

performing the corresponding differentiation of Eqs. (VII.1), and substituting the results into the boundary conditions (VII.2), we obtain:

$$\left.\begin{aligned} \frac{\varphi_1}{c_1} - \frac{\varphi_1'}{c_1} &= \frac{\varphi_2}{c_2} \\ \rho_1\varphi_1 + \rho_1\varphi_1' &= \rho_2\varphi_2 \end{aligned}\right|_{x=0} \tag{VII.4}$$

where all values of the variable potentials refer to the boundary, which, in this case, is located at $x = 0$. The two equations (VII.4) enable finding the potentials φ_1' and φ_2' and, with the help of relations (VII.3), calculating in terms of these potentials the pressure and particle velocity in the reflected and transmitted waves for fixed parameters of the incident wave. In this case, however, the calculation of the potentials can be avoided and these quantities can be determined indirectly from the boundary conditions (VII.2), using the previously obtained relations (III.10) and (III.11) between the pressure and velocity in the forward and backward waves. Based on these considerations, we obtain the following relations for the incident (forward), transmitted (forward), and reflected (backward) waves, respectively, propagating along the negative x-axis,*

$$p_1 = \rho_1 c_1 v_1, \quad p_2 = \rho_2 c_2 v_2, \quad p_1' = -\rho_1 c_1 v_1'. \qquad \text{(VII.5)}$$

Substituting these values of the pressure into the second equation in (VII.2), we obtain:

$$\left. \begin{aligned} & v_1 + v_1' = v_2 \\ & z_1(v_1 - v_1') = z_2 v_2 \end{aligned} \right|_{x=0}, \qquad \text{(VII.6)}$$

where $z_1 = \rho_1 c_1$ and $z_2 = \rho_2 c_2$ are the specific acoustic impedances of the adjoining media. Expressing the particle velocity in relations (VII.5) in terms of the corresponding pressure and substituting their values into the first of equations (VII.2), we obtain analogously

$$\left. \begin{aligned} & p_1 + p_1' = p_2 \\ & (p_1 - p_1')/z_1 = p_2/z_2 \end{aligned} \right|_{x=0} \qquad \text{(VII.7)}$$

We now introduce the *pressure* and *velocity reflection* $\rho_{p,v}$ and *transmission* $d_{p,v}$ *coefficients*, defining them as the ratio of the pressure and velocity, respectively, in the reflected and transmitted waves to the pressure and velocity in the incident wave. Solving Eq. (VII.6) and (VII.7) for these ratios, we obtain

---

*The choice of the positive orientation of the x-axis is, of course, of no significance here, the only significant factor being that the reflected wave propagates in a direction opposite to the incident wave.

$$\rho_p = \frac{p_1'}{p_1} = \frac{z_2 - z_1}{z_2 + z_1}; \quad \rho_v = \frac{v_1'}{v_1} = \frac{z_1 - z_2}{z_1 + z_2}; \qquad \text{(VII.8)}$$

$$d_p = \frac{p_2}{p_1} = \frac{2z_2}{z_1 + z_2}; \quad d_v = \frac{v_2}{v_1} = \frac{2z_1}{z_1 + z_2}. \qquad \text{(VII.9)}$$

Thus the magnitudes of the parameters of the reflected and transmitted waves depend substantially on the ratios of the specific acoustic impedances of the adjoining media, and this dependence is of a different character for the reflection and transmission coefficients. We first note that the reflection coefficients ρ_p and ρ_v for any ratio of z_1 and z_2 have different signs. This means that the sign of the pressure or velocity is reversed on reflection. Since the time-varying values of p and v appear in Eqs. (VII.8) and (VII.9), the change in their sign on reflection corresponds to a change in phase by 180°. In the transmitted wave, on the other hand, according to the relations (VII.9), the pressure and velocity are always in phase with the pressure and velocity in the incident wave, i.e., with one another also. Neglecting for the time being the phase relations and taking into account the fact that Eqs. (VII.8) hold for any instantaneous values of the pressure and velocity, as well as for their amplitudes, we can introduce the generalized "amplitude" reflection coefficient:

$$\rho_\varphi \equiv \left| \frac{\varphi_{1\max}'}{\varphi_{1\max}} \right| = \left| \frac{p_{1\max}'}{p_{1\max}} \right| = \left| \frac{v_{1\max}'}{v_{1\max}} \right| = \left| \frac{z_1 - z_2}{z_1 + z_2} \right| \qquad \text{(VII.10)}$$

which will determine the absolute values of the parameters of the reflected wave relative to the incident wave. The form of the pressure and velocity amplitude transmission coefficients (VII.9) remains the same in this case. As we can see, the amplitude reflection coefficient ρ_φ is close to 1, i.e., the amplitude of the reflected wave is almost equal to the amplitude of the incident wave, if $z_1 \ll z_2$ or $z_1 \gg z_2$, i.e., at the boundary between two media with strongly differing acoustic impedances, independent of the direction of propagation.

The pressure and velocity transmission coefficients, which determine the amplitudes of the pressure and velocity in the transmitted wave, as follows from Eq. (VII.9), depend substantially on which medium the wave enters. If the incident wave propagates in an acoustically stiff medium and penetrates through the boundary into an acoustically flexible medium, i.e., if $z_1 \gg z_2$, then the amplitude of the pressure in the transmitted wave will be insigificant, while the amplitude of the particle velocity will be almost twice that in the incident wave. On the other hand, for $z_1 \ll z_2$, i.e., for a wave propagating in a flexible medium and incident on the boundary with a more rigid medium, for example, from a gas into a liquid, the pressure amplitude of the transmitted wave will be double that of the incident wave and the velocity amplitude will decrease correspondingly. The last circumstance should be specialy noted, because ultrasonic receivers generally record the **pressure** (for example, a piezoelectric crystal) and for this reason such a receiver, which, for example, is submerged in a liquid, records almost twice the amplitude (pressure) of the ultrasonic wave incident on the same liquid from the adjoining gaseous medium.

At the same time, of course, the energy balance must be maintained at the boundary between the media, i.e., the sum of the absolute values of the intensities of the transmitted and reflected waves must equal the intensity of the incident wave, i.e.,

$$I_1 = I_1' + I_2, \tag{VII.11}$$

which is easily checked by expressing all intensities with the help of Eq. (III.21) in terms of the corresponding amplitudes of the pressure or particle velocities, using the reflection and transmission coefficients. Dividing both parts of Eq. (VII.11) by the intensity of the incident wave and introducing the *energy reflection and transmission coefficients*

$$\rho_I = I_1'/I_1, \quad d_I = I_2/I_1, \tag{VII.12}$$

we obtain the equation of conservation of energy in the form

$$\rho_I + d_I = 1. \tag{VII.13}$$

Since the energy of the wave is proportional to the square of its amplitude and the reflected and incident waves propagate in the same medium, the energy reflection coefficient is expressed as

$$\rho_l = \rho_\varphi^2 = \left[\frac{z_1 - z_2}{z_1 + z_2}\right]^2 \qquad \text{(VII.14)}$$

Subtracting this quantity from unity, we obtain from Eq. (VII.13) an expression for the *energy transmission coefficient*:

$$d_l = 1 - \rho_l = 4z_1z_2/(z_1 + z_2)^2, \qquad \text{(VII.15)}$$

in which, as expected, the specific acoustic impedances of both adjacent media appear symmetrically.

Thus the reflective properties of the boundary between two media are completely determined by the **difference** of their specific acoustic impedances. If $z_1 = z_2$, then the reflection coefficient equals zero, there is no reflected wave, and the boundary is acoustically "transparent." If, in addition, the media have different densities, then the equality of the specific acoustic impedances corresponds to a condition on the velocities of sound: $c_1/c_2 = \rho_2/\rho_1$. This condition is well satisfied for some pairs of immiscible liquids, such as water and carbon tetrachloride, water and acetic anhydride, and others (see Table 4).

Amongst solids, the acoustic impedances of some solid polymers, in particular rubber, polystyrene, teflon, polyvinyl acetate, and a number of others, into which ultrasound propagates from water almost completely without significant reflection, are close to that of water. For example, at the oundary between water ($z = 15 \cdot 10^4$ $g/(cm^2 \cdot s)$) and rubber ($z = 14 \cdot 10^4$ $g/(cm^2 \cdot s)$) the amplitude reflection coefficient is only 3% and the energy reflection coefficient is only about 0.1%. Approximately 4% of the energy is reflected from polystyrene ($z = 23 \cdot 10^4$ $g/(cm^2 \cdot s)$) and about 3% is reflected from teflon. Since these materials strongly absorb ultrasonic waves, they can be regarded as almost perfect absorbers of ultrasound and can be used, for example, to deaden the walls of a vessel containing a liquid in cases when the reflected waves must be eliminated in order to perform measurements or for other purposes.

The specific acoustic impedances of metals and other solids are at least an order of magnitude higher than those of liquids (with the exception of liquid metals). Amongst metals, aluminum has the lowest acoustic stiffness ($z = 170 \cdot 10^4$ $g/(cm^2 \cdot s)$); about 30% of the energy incident from water (or vice versa) penetrates, i.e., the reflection coefficient at the water–aluminum boundary is 0.7 with respect to intensity and 0.83 with respect to amplitude. At the boundary between water and iron ($z = 46 \cdot 10^4$ $g/(cm^2 \cdot s)$) the amplitude and energy reflection coefficients are 0.94 and 0.87, i.e., only 13% of the acoustic energy penetrates through the boundary.

The specific acoustic impedances of gases are three to four orders of magnitude lower than those of liquids and solids (see Table 4). For this reason, acoustic waves are almost completely reflected at gas– liquid and gas– solid interfaces. Indeed, at the boundary between air under normal conditions ($z = 45 \cdot 10^4$ $g/(cm^2 \cdot s)$) and water ($z = 15 \cdot 10^4$ $g/(cm^2 \cdot s)$) the amplitude reflection coefficient, according to Eq. (VII.10), $\approx$ 0.999, while the energy coefficient is 0.998, i.e., only approximately 0.2% of the energy penetrates from the liquid into the air (and vice versa). The ultrasonic transmission coefficient at the boundary between a gas and a solid body is even lower, so that in practice these boundaries may be assumed to be almost perfectly reflecting.

We note that, in accordance with Eqs. (VII.8)–(VII.15) and neglecting any possible frequency dependence introduced by the dispersion of the velocity of sound in relaxing media, the reflection and transmission coefficients are practically independent of the frequency. However, dispersion is usually so small that it cannot appreciably affect the difference between the acoustic impedances, which determines the magnitude of the reflection coefficient at the boundary with a given medium. The results obtained are therefore also valid for nonmonochromatic waves with a complicated spectrum, in particular for ultrasonic pulses. Based on what was said above, the relative spectral composition, i.e., the form of the envelope of a pulse, should not change on reflection and transmission; only the absolute values of the amplitudes of the harmonics and the pulse height change in accordance with the magnitudes of the reflection and transmission coefficients. The reflection coefficient at a boundary between different media at normal incidence should evidently also be independent of the ultrasonic absorption in these media.

§ 2. Standing plane waves

We shall now study the interference of the incident and reflected waves in medium *1* with specific acoustic impedance $z_1 = \rho_1 c_1$ neglecting absorption, whose effect we shall determine later. To this end, we add the velocity potentials of the incident wave φ_1 and reflected wave φ_1' and we find the velocity potential of the resulting field:

$$\varphi(x,t) = \varphi_{1max} \exp[i(\omega t - k_1 x)] + \varphi'_{1max} \exp[i(\omega t + k_1 x)] \quad \text{(VII.16)}$$

We calculate the variable pressure and particle velocity in this field from the general definition by differentiating (VII.16) with respect to time and the coordinate (dropping the unit index, since we are concerned with only one medium) :

$$p(x,t) = \rho \frac{\partial \varphi(x,t)}{\partial t} =$$
$$= i\omega\rho\{\varphi_{max} \exp[i(\omega t - kx)] + \varphi'_{max} \exp[i(\omega t + kx)]\} \quad \text{(VII.17)}$$

$$v(x,t) = -\frac{\partial \varphi(x,t)}{\partial x} =$$

$$= i\frac{\omega}{c}\{\varphi_{max} \exp[i(\omega t - kx)] + \varphi'_{max} \exp[i(\omega t + kx)]\} \quad \text{(VII.18)}$$

We first note that the specific acoustic impedance of the medium in the presence of the reflected wave together with the incident forward wave is complex. Indeed, dividing expression (VII.17) by (VII.18), we obtain the ratio of the pressure to the velocity:

$$\frac{p}{v} = z = \rho c \frac{\varphi_{max} \exp(-ikx) + \varphi'_{max} \exp(ikx)}{\varphi_{max} \exp(-ikx) - \varphi'_{max} \exp(ikx)},$$

which for $\varphi'_{max} \neq 0$ is complex and becomes real (equal to the specific acoustic resistance of the medium ρc) only in the complete absence of the reflected wave. The complexity of the impedance, as noted in Chap. III, indicates that there is a phase shift between the pressure and velocity, i.e., part of the energy of the wave returns to the source in the form of the energy of the reflected wave.

Let us study separately the pressure and particle velocity waves, writing Eqs. (VII.17) and (VII.18) in real form, taking, for example, their imaginary parts:

$$p = p_{max} \cos(\omega t - kx) + p'_{max} \cos(\omega t + kx) \qquad \text{(VII.19)}$$

$$v = v_{max} \cos(\omega t - kx) - v'_{max} \cos(\omega t + kx) \qquad \text{(VII.20)}$$

where $p_{max} = \rho\omega\varphi_{max}$ is the amplitude of the pressure in the incident wave, etc. Adding and subtracting the term $p'_{max} \cos(\omega t - kx)$ in Eq. (VII.19) and combining like terms, we obtain:

$$p = 2p'_{max} \cos kx \cos \omega t + (p_{max} - p'_{max}) \cos(\omega t - kx). \qquad \text{(VII.21)}$$

The second term in this expression corresponds to a traveling forward wave with amplitude $p_{max} - p'_{max}$ which depends on the amplitude of the reflected wave, and the first term corresponds to a standing wave with amplitude $2p'_{max}$ equal to twice the amplitude of the reflected wave. If there is no reflected wave ($p'_{max} = 0$), then expression (VII.21) transforms into the equation for a traveling plane wave propagating in the $+x$ direction $p = p_{max} \cos(\omega t - kx)$. In the case of total reflection from a plane boundary, when $\rho_p = 1$ and $p_{max} = p'_{max}$ Eq. (VII.21) describes a pure standing pressure wave:

$$p = 2p_{max} \cos kx \cos \omega t. \qquad \text{(VII.22)}$$

The standing wave represents a sum of two traveling waves with equal amplitude, propagating in opposite directions. The amplitude of the standing wave equals twice the amplitude of the incident wave p_{max}; the **average energy density** in it is correspondingly four times higher than the energy density in the incident wave (since the energy is proportional to the square of the amplitude); the **intensity** in the field of the standing wave equals zero, because the energy flux in the incident wave is compensated by the backward flux in the reflected wave.

According to Eq. (VII.22), the pressure oscillates with maximum amplitude in planes whose coordinates satisfy the condition $kx = n\pi$, where $n = 0, 1, 2, \ldots$, i.e., for $x = n\Lambda/2$ (Λ is the wavelength of the traveling wave). These coordinates correspond to **antinodes of the pressure**. The distance between neighboring antinodes $\Delta x = \Lambda/2 = \Lambda_0$ is customarily called the *wavelength of the standing wave*. The coordinates $x = (n + 1/2)\Lambda/2$ correspond to the nodal planes in which the pressure vanishes. As is evident from the equation for the standing wave (VII.22), the pressure has the same phase at all points between nodal planes. Oscillations in neighboring standing waves occur in antiphase. In the case of incomplete reflection, i.e., for $p'_{max} < p_{max}$, a traveling wave is superposed on the standing wave, as a result of which the pressure nodes transform into minima, whose depth decreases as the amplitude of the reflected wave decreases and which completely disappear for $\rho_p = 0$ and $p'_{max} = 0$, when the wave becomes a purely traveling wave.

Performing analogous operations for Eq. (VII.20), we obtain an expression for the resulting particle velocity:

$$v = 2v'_{max} \sin kx \cos \omega t + (v_{max} - v'_{max}) \cos(\omega t - kx), \qquad \text{(VII.23)}$$

which, in this manner, can likewise be represented as a sum of a purely traveling and a purely standing waves, whose amplitudes are determined by the amplitude of the reflected wave. In this case, as is evident from a comparison of Eq. (VII.23) and (VII.21), wave parameters such as the velocity and pressure have the same in the traveling wave, while in the standing wave the phases are shifted by $\pi/2$. Therefore, in the traveling wave energy is transported toward the boundary, while in the standing wave the directed energy flux equals zero.

Thus, for incomplete reflection from a boundary, a traveling wave, whose intensity in the absence of dissipative losses in the medium is evidently equal to the intensity of the wave transmitted through the boundary into the neighboring medium, which appears to "absorb" this wave, is superposed on the standing wave formed by the interference of the reflected and incident waves. It is in this sense that one speaks about "losses due to reflection," whose magnitude with respect to energy is determined by the reflection coefficient ρ_I and with respect to amplitude by the amplitude reflection coefficient ρ_p or ρ_v.

A comparison of Eqs. (VII.21) and (VII.23) shows that nodes or antinodes of the pressure and velocity in a standing wave are displaced along the x-axis by an amount $|\Delta x| = \Lambda_0/2$. What happens in this case at the reflecting boundary? As already noted in the preceding section, the sign of the pressure or velocity in the reflected wave is reversed at the boundary, which corresponds to a jump in phase by 180° . If the wave is incident **from an acoustically stiffer medium** on the boundary with a more flexible medium, i.e., $z_1 > z_2$, then, as follows form Eq. (VII.8), a jump occurs in the phase of the pressure wave at the boundary with the more flexible medium and the wave is thereby reflected with a "loss of a half-wave". Therefore, a pressure node of the standing wave or a minimum of the total pressure, which by virtue of the equality of action and reaction is entirely transmitted to the second medium in the form of a transmitted wave, will occur at the boundary between the media in this case. Under the same conditions the sign of the velocity will not change on reflection; reflection occurs "without loss of a half-wave". Therefore, the phase of the reflected wave does not change at the boundary, where an antinode of the standing velocity wave or a maximum of the total velocity wave is thus formed. By virtue of the condition of continuity, this velocity is transmitted to particles in the adjoining medium, where a traveling (transmitted) wave forms, in which, according to Eq. (VII.9), the pressure and velocity are in-phase. In the limiting case $z_1 \gg z_2$ or $z_2 \simeq 0$, which occurs, for example, when a wave propagating in a solid or even in a liquid is reflected from the boundary with a gas, the pressure on the boundary practically equals zero and the boundary undergoes free oscillations with an amplitude equal to twice the amplitude of the incident wave. In other words, in this case, there will be a **pressure node** (compression) and an **antinode of the particle velocity** (as well as an antinode of displacement) at the boundary.

If, on the other hand, the wave is incident **from a more flexible medium** on the boundary with a stiffer medium ($z_1 < z_2$) , for example, from a gas on the boundary with a solid or liquid, then the picture is reversed: the phase of the particle velocity changes at the boundary between the media and the pressure does not undergo a jump in phase and is reflected without loss of a half-wave. Thus, for $z_1 \gg z_2$, a **pressure antinode** and a **node of the particle velocity** (displacement) form on the reflecting boundary. Physically, this corresponds to the fact that the boundary of the stiff medium must remain stationary and, therefore, the particles in the adjoining medium cannot be displaced; in addition, the displacement of neighboring particles causes the medium to be compressed at the boundary, where a pressure maximum is thereby developed, which is entirely transmitted to the second medium in the form of the transmitted wave. The pressure and velocity in this wave, according to Eq. (VII.9), are in-phase at the boundary with the pressure and velocity of the incident wave, i.e., with one another, as should occur in a traveling wave.

Figure 38 shows a schematic diagram of the formation of the standing and transmitted waves with normal incidence and almost total reflection ($\rho_l \simeq 1$, $d_l \simeq 0$) for two limiting cases $z_1 \ll z_2$ (a) and $z_1 \gg z_2$ (b), when

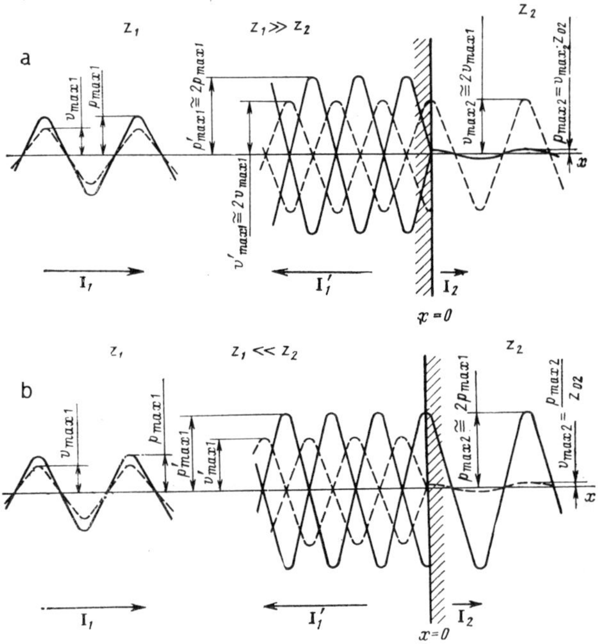

Fig. 38.

the fraction of the traveling wave in the first medium is negligibly small and almost pure nodes and antinodes form at the reflecting boundary:

for $z_1 >> z_2$ an antinode of the particle velocity wave (broken curve) forms, and for $z_1 << z_2$ an antinode of the pressure wave (solid curves) forms. Correspondingly, in the first case, the particle velocity in the wave transmitted into the second (flexible) medium is twice that in the incident wave, while in the second case, when the wave is incident on a stiff boundary, the transmitted pressure equals twice the incident pressure. The intensity of the wave transmitted into the second medium I_2 (i.e., the intensity of the traveling wave in the first medium) in both cases equals the algebraic difference between the intensity of the incident wave I_1 and that of the reflected wave I_1', i.e., their geometric sum.

§ 3. Interference of oppositely traveling waves with normal reflection in an absorbing medium

We shall now clarify the effect of ultrasonic absorption on the structure of the interference field formed by the superposition of the incident and reflected waves. Let an ideally reflecting plane boundary be located at a distance $x = +x_0$ from the origin of coordinates, where we shall assume that the parameters of the incident wave, for example, the pressure amplitude p_{max0}, are given. For $x > 0$, the amplitude of the incident wave in the absorbing medium decays exponentially

$$p_{max} = p_{max0} \exp(-\alpha_0 x) \qquad \text{(VII.24)}$$

with an amplitude absorption coefficient α_0. The amplitude of the reflected wave for $\rho_p = 1$ can be written in the form

$$p'_{max} = p_{max}(x_0) \exp[\alpha_0(x - x_0)] = p_{max0} \exp[\alpha_0(x - 2x_0)]. \qquad \text{(VII.25)}$$

The total pressure in the interference field will equal the sum of the variable pressures in the incident and reflected waves, which can be written as before in the form

$$p(x,t) = 2p'_{max}(x) \cos kx \cos \omega t + [p_{max}(x) - p'_{max}(x)] \cos(\omega t - kx), \qquad \text{(VII.26)}$$

i.e., as the sum of a traveling and standing waves. However, the amplitudes of these waves and their ratio will, in this case, be functions of the coordinate x. Then, using expressions (VII.24) and (VII.25) for the amplitudes of the standing and traveling waves we have, correspondingly,

$$2p'_{max} = 2p_{max0}e^{\alpha_0(x-2x_0)} = 2p_{max}e^{\alpha_0 x}e^{-2\alpha_0 x},$$

$$p_{max} - p'_{max} = p_{max0}[e^{-\alpha_0 x} - e^{\alpha_0(x-2x_0)}]. \quad \text{(VII.27)}$$

If the absorption is not too high, i.e., $\alpha_0 x \ll 1$, then the second expression in (VII.27) can be represented with sufficient accuracy as $p_{max} - p'_{max} = 2p_{max0}\alpha_0(x_0 - x) = p_{max0}\alpha'_0(x_0 - x)$, where $\alpha_0' = 2\alpha_0$ is the energy absorption coefficient. At $x = x_0$ (directly on the reflecting

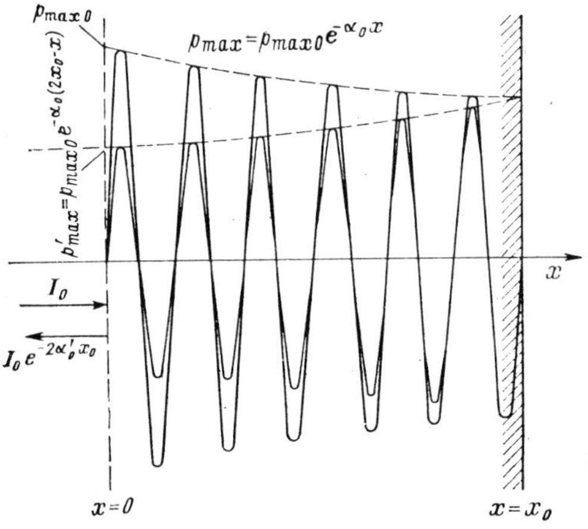

Fig. 39.

boundary) the traveling wave vanishes and only the standing wave with amplitude $2p'_{max}(x = x_0) = 2p_{max0}\exp(-\alpha_0 x_0)$, equal, as before, to twice the amplitude of the forward wave, attenuated by absorption in the medium and incident on the boundary, remains. In the case when the incident and reflected waves add, when a plane wave propagating in an acoustically stiff absorbing medium is incident normally on the boundary with a vacuum (Fig. 39), a pure pressure node forms on the reflecting boundary (the pressure is reflected with a change in sign, i.e., with a loss of a half-wave: the first term in Eq. (VII.26) in this case is negative; the phase of the reflected pressure wave is shifted by π relative to the incident wave).

As the distance from the reflecting boundary increases, the standing wave fraction decreases, while the traveling wave fraction increases. Pressure nodes in this case are smeared and transform into increasingly shallower minima. The amplitude of the traveling wave has a maximum value at $x = 0$: $(p_{max} - p'_{max})_{x=0} = p_{max0}\alpha'_0 x_0$. The energy of the traveling component with total reflection of the wave from the boundary, evidently, corresponds to the energy that is absorbed by the medium in a layer of thickness x_0, i.e., it transforms into heat irreversibly. This energy per unit area of the wave front, absorbed by the medium per unit time, can be defined as the difference of the intensity of the incident wave in the plane $x = 0$ (I_0) and the intensity of the reverse wave reflected from the boundary and traversing a total distance of $2x_0$:

$$W_{abs} = I_0 - I_0 e^{-\alpha_0 x_0} = I_0(1 - e^{-2\alpha_0 x_0}) \approx 2x_0 I_0 \alpha'_0.$$

Dividing this energy by the path traversed by the wave $2x_0$, we find the energy density absorbed by the medium, averaged over the thickness of the layer, i.e., the average energy absorbed per unit volume per unit time: $\Delta\overline{w}_{abs} = I_0\alpha'_0$, which, of course, is the same as Eq. (III.50) obtained in the calculation of the energy absorption coefficient.

We are studying here the case of normal reflection from a single plane boundary with no restrictions on the field on the side of the incident wave. In practice, however, this field is bounded on the other side by the surface of the source of plane waves or by the second boundary of the layer through which the wave penetrates from the source. In this case, multiple reflection of the plane wave from the two boundaries of the layer will lead to the formation of a standing wave, whose amplitude, energy, and other characteristics will depend on the thickness of the layer and the conditions on both of its boundaries. We shall study this situation when we analyze the transmission of a plane wave through a plane-parallel layer. We shall now study the more general case of oblique incidence of a plane wave on a plane boundary between two media.

§4. Reflection and refraction of a plane wave at oblique incidence on a plane boundary between two media

Let a plane wave be incident on a plane boundary between media *1* and 2 at an angle θ_1 to the normal (Fig. 40). For arbitrary orientation of the wave vector $\mathbf{k}$ relative to rectangular coordinate axes the equation of the plane wave, satisfying the three-dimensional wave equation (III.32), must be written in the form

$$\varphi = \varphi_{max} \exp\{i[\omega t - \mathbf{k}\cdot\mathbf{r}]\}, \qquad \text{(VII.28)}$$

where $\mathbf{r} = \mathbf{r}(x, y, z)$ is the radius vector of the point at which the value of the velocity potential φ is determined. The three-dimensional equation (VII.28) can be obtained from the one-dimensional equation (III.6) for waves propagating along the x-axis by an appropriate rotation of the coordinate axes. In doing so, we must substitute in this equation the quantity $\mathbf{n}\cdot\mathbf{r} = n_x x + n_y y + n_z z$ for the x coordinate. Keeping in mind the definition of the wave vector (see Chap. III), we obtain an equation for an arbitrarily oriented plane wave in the form (VII.28), where $\mathbf{k}\cdot\mathbf{r} = kn_x x + kn_y y + kn_z z = k_x x + k_y y + k_z z$.

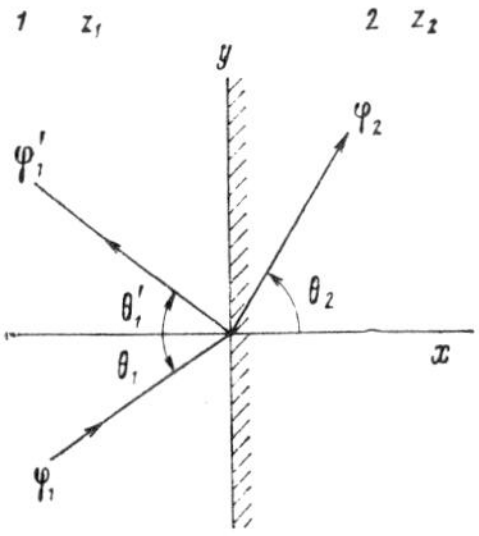

Fig. 40.

Let the boundary coincide with the yz plane so that x-axis is normal to it. If the wave vector of the incident wave lies in the xy plane and makes an angle θ_1 with the x-axis, as illustrated in Fig. 40, then its components along the coordinate axes assume the values $k_x = k\cos\theta_1$, $k_y = k\sin\theta_1$, $k_z = 0$. Therefore, the equation for the velocity potential of the incident wave φ_1 will have the form

$$\varphi_1 = \varphi_{1max} \exp\{i[\omega t - k_1(x\cos\theta_1 + y\sin\theta_1)]\}, \qquad \text{(VII.29)}$$

where, as before, $k_1 = \omega/c_1$ is the wave number for medium *1*. On the basis of the general considerations presented at the beginning of this chapter, a reflected wave and a wave passing into medium 2 must be formed at the boundary. We shall assign to the wave vector of the reflected wave an angle θ_1' relative to the x-axis and an angle θ_2 relative to the wave vector of the transmitted wave. Retaining the previous notation for the potentials and wave numbers, we obtain expressions analogous to (VII.29) for the reflected and transmitted waves, respectively:

$$\varphi_1' = \varphi_{1max}' \exp\{i[\omega t - k_1(-x\cos\theta_1' + y\sin\theta_1')]\}, \quad \text{(VII.30)}$$

$$\varphi_2 = \varphi_{2max} \exp\{i[\omega t - k_2(x\cos\theta_2 + y\sin\theta_2)]\}, \quad \text{(VII.31)}$$

where $k_2 = \omega/c_2$.

At the boundary ($x = 0$), as before, the pressures (in general, normal stresses) and the normal components of the particle velocities must be continuous. This gives the following boundary conditions:

$$\left.\begin{aligned} p_1 + p_1' &= p_2 \\ v_{1x} + v_{1x}' &= v_{2x} \end{aligned}\right|_{x=0},$$

which assume the following form for the velocity potentials:

$$\left.\begin{aligned} \rho_1\frac{\partial\varphi_1}{\partial t} + \rho_1\frac{\partial\varphi_1'}{\partial t} &= \rho_2\frac{\partial\varphi_2}{\partial t} \\ \frac{\partial\varphi_1}{\partial x} + \frac{\partial\varphi_1'}{\partial x} &= \frac{\partial\varphi_2}{\partial x} \end{aligned}\right|_{x=0}. \quad \text{(VII.32)}$$

Differentiating the velocity potentials (VII.29)—(VII.31) with respect to time and substituting the results into the first equation in (VII.32) gives at $x = 0$

$$\rho_1\varphi_{1max}\exp[i(\omega t - k_1 y\sin\theta_1)] + \rho_1\varphi_{1max}'\exp[i(\omega t - k_1 y\sin\theta_1')] =$$
$$= \rho_2\varphi_{2max}\exp[i(\omega t - k_2 y\sin\theta_2)]. \quad \text{(VII.33)}$$

The boundary conditions (VII.32) must be satisfied at any time and at all points on the boundary, i.e., for any value of the y coordinate. Therefore, all coefficients in front of y in Eq. (VII.33) in the terms determining the phases of the waves at the boundary of the media must be equal to one another, i.e.,

$$k_1 \sin\theta_1 = k_1 \sin\theta_1' = k_2 \sin\theta_2. \qquad \text{(VII.34)}$$

Of course, the same result is also obtained from the second equation in (VII.32). Indeed, differentiating the velocity potentials (VII.29)—(VII.31) with respect to x and substituting the corresponding derivatives at $x = 0$ into Eq. (VII.32), we find the second boundary condition for them

$$k_1 \cos\theta_1\, \varphi_{1max} \exp[i(\omega t - k_1 y \sin\theta_1)] -$$
$$- k_1 \cos\theta_1 \varphi_{1max}' \exp[i(\omega t - k_1 y \sin\theta_1')] =$$
$$= k_2 \cos\theta_2\, \varphi_{2max} \exp[i(\omega t - y k_2 \sin\theta_2)], \qquad \text{(VII.35)}$$

whence the equality of the coefficients in front of y likewise leads to relations (VII.34). The *laws of reflection and refraction* of acoustic waves follow from these relations:

$$\sin\theta_1 = \sin\theta_1', \quad \theta_1' = \theta_1, \qquad \text{(VII.36)}$$

i.e., the angle of incidence θ_1 equals the angle of reflection θ_1' (for this reason, in what follows, we shall not distinguish this angle with a prime):

$$\sin\theta_2/\sin\theta_1 = k_1/k_2 = c_2/c_1 \equiv n_a, \qquad \text{(VII.37)}$$

i.e., the ratio of the sine of the angle of refraction to the sine of the angle of incidence equals the ratio of the velocities of sound in the second medium to that in the first medium (a higher velocity corresponds to a larger angle). By analogy with optics, this relation can be called the *acoustic refractive index* of the two media n_a. However, the law of refraction of acoustic waves differs from the law of refraction in optics, where the ratio of the sines of the angles is inversely proportional to the ratio of the velocities of light (the case corrresponding to medium 2 having the higher velocity of light is illustrated in Fig. 40).

In this connection, it should be noted that as in optics, based on the law of refraction, it is possible to focus ultrasonic rays* with the help of a lens. But, in this case, if the lens is made of a solid material in which the velocity of sound is greater than in the surrounding medium, then the focusing lens will be a lens with concave surfaces (Fig. 41) and not convex surfaces as in the case of optics.

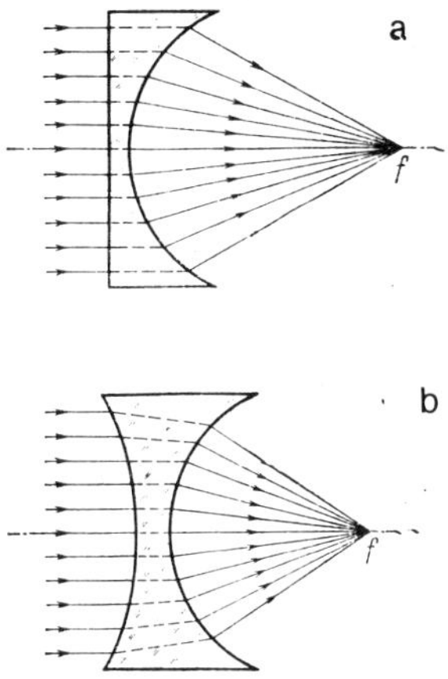

Fig. 41.

Knowing the velocity of sound in the material of a focusing lens (c_1) and in the surrounding medium (c_2), from (VII.36) it is not difficult to find a relation between the curvature of the surfaces of the lens and the focal length f of the lens for paraxial rays. In the standard approximation used in optics, we obtain

$$f = \frac{R_1 R_2}{(1 - c_1/c_2)[d(c_2/c_1 - 1) - R_1 + R_2]}, \qquad \text{(VII.38)}$$

where R_1 and R_2 are the radii of curvature of the lens surfaces and d is the thickness of the lens along the principal axis; the minus sign corresponds to the imaginary focal point of the convex lens, whose radii of curvature must be taken as negative. For a plane-concave lens ($R_1 = \infty$), with radius of curvature of the concave surface $R_2 = R$, Eq. (VII.38) gives $f = -R/(c_1/c_2 - 1) = -Rn_a/(n_a - 1)$.

To calculate the intensification of ultrasound at the focal point of a focusing lens, it is necessary to include, in addition to the acoustic impedances, such factors as the dependence of the coefficient of transmission of the wave through the lens on the angle of incidence and on the absorption of ultrasound in the lens material and the influence of

*We recall that the term 'ray' in an isotropic medium is understood to mean the normal to the wave front. A parallel bundle of rays corresponds to a plane wave, which is realized in practice in the ultrasonic frequency band.

nonlinear effects on the focusing of ultrasound. The reader is referred to the book by I. N. Kanevskii for detailed calculations of ultrasonic focusing apparatus.[60] Figure 42 shows a shadow photograph of an ultrasonic beam focused by an acoustic lens. (The shadow method of visualization of ultrasonic waves reduces to transillumination of sections of the medium with an altered optical refractive index.[12]

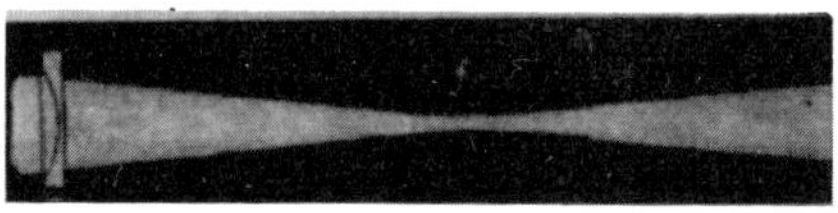

Fig. 42.

Since the latter quantity varies in phase with the density, i.e., with the pressure, a shadow photograph exposed over a period of time much longer than the period of the ultrasonic oscillations records the overall transillumination of the region of the medium "occupied" by the ultrasonic beam, thereby permitting an examination of its structure and geometry.)

We shall now calculate the reflection and transmission coefficients for an obliquely incident plane wave. We shall perform the calculation immediately for the relative intensities of the waves, without separately calculating the reflection and transmission coefficients for the pressure and particle velocity waves. The calculation of the energy coefficients is simplified by the fact that based on the boundary conditions it is sufficient to calculate only the reflection coefficient $\rho_I = I_1'/I_1$, while the transmission coefficient can be obtained by subtracting the quantity ρ_I from 1 based on the equation of energy balance (VII.13). Next, the intensity reflection coefficient can be defined as the ratio of the squares of the amplitudes of the potentials in the reflected and incident waves, because these waves propagate in the same medium, i.e., in a medium with the same specific acoustic impedance z_1 (which is not correct with respect to the transmitted and incident waves).

Introducing the notation $\varphi_1(0) \equiv \varphi_{1\max} \exp[i(\omega t - k_1 y \sin\theta_1)]$ and so on, we write the boundary conditions (VII.33) and (VII.35) for the potentials in the form

$$\rho_1\varphi_1(0) + \rho_1\varphi_1'(0) = \rho_2\varphi_2(0)\,,$$

$$k_1 \cos\theta_1 \cdot \varphi_1(0) - k_1 \cos\theta_1 \cdot \varphi_1'(0) = k_2 \cos\theta_2 \cdot \varphi_2(0)\,.$$

Multiplying the first equation by $(k_2\cos\theta_2)/\rho_2$ and subtracting the second equation from it, we obtain

$$\varphi_1'(0)\left(\frac{k_2}{\rho_2}\cos\theta_2 + \frac{k_1}{\rho_1}\cos\theta_1\right) = \varphi_1(0)\left(\frac{k_1}{\rho_1}\cos\theta_1 - \frac{k_2}{\rho_2}\cos\theta_2\right).$$

From here, taking into account the fact that $k_1 = \omega/c_1$ and $k_2 = \omega/c_2$, we obtain for the ratio of the squares of the amplitudes of the potentials, equal to the reflection coefficient,

$$\rho_l = \left[\frac{\varphi_{1\max}'}{\varphi_{1\max}}\right]^2 = \left[\frac{z_1\cos\theta_2 - z_2\cos\theta_1}{z_1\cos\theta_2 + z_2\cos\theta_1}\right]^2. \quad \text{(VII.39)}$$

Subtracting this quantity from unity, according to the equation of energy balance, we have:

$$d_l = 1 - \rho_l = \frac{4z_1z_2\cos\theta_1\cdot\cos\theta_2}{(z_1\cos\theta_2 + z_2\cos\theta_1)^2}. \quad \text{(VII.40)}$$

Thus the reflection and transmission coefficients depend on the angle of incidence of the beam on the boundary between the media. For $\theta_1 = 0$ Eqs. (VII.39) and (VII.30) transform into Eqs. (VII.14) and (VII.15), obtained for normal incidence. In the general case, however, they differ by factors of cosines of the angles of incidence and reflection. For this reason, the condition of transparency of the boundary ($\rho_l = 0$, $d_l = 1$) for an arbitrary angle of incidence will be as follows:

$$z_1\cos\theta_2 = z_2\cos\theta_1. \quad \text{(VII.41)}$$

However, the angles θ_1 and θ_2 are not independent: they are related to one another by the law of refraction (VII.37). Taking this into account, after simple transformations we obtain an expression for the angle θ_1, at which the ultrasonic wave will penetrate into the second medium without reflection:

$$\tan^2\theta_1 = \frac{z_2^2/z_1^2 - 1}{1 - c_2^2/c_1^2} = \frac{\rho_2^2/\rho_1^2 - c_1^2/c_2^2}{c_1^2/c_2^2 - 1}. \quad \text{(VII.42)}$$

It follows from this expression that the condition of transparency (VII.41) holds only for media whose densities and velocities of sound satisfy one of the inequalities $(\rho_2/\rho_1) \geqslant (c_1/c_2) \geqslant 1$ or $(\rho_2/\rho_1) \leqslant (c_2/c_1) \leqslant 1$. Under these conditions, the right side of relation (VII.42) will be positive and, therefore, for some angle of incidence in the range from 0 to $\pi/2$ the boundary will be totally transparent (including the angle of normal incidence $\theta_1 = 0$ and the angle $\theta_1 = 90°$, corresponding to propagation of a plane wave along the boundary). Examples of a pair of media for which these inequalities hold are water and diethyl phthalate, ethyl alcohol and chloroform, and others (see Table 4). In particular, for water (at 25°C), $c_1 = 149.7 \cdot 10^3$ *cm/s*, $\rho_1 = 0.997$ *g/cm*^{3}; for diethyl phthalate, $c_2 = 147.10^3$ cm/s, $\rho_1 = 1.121$ *g/cm*^{3}. Substituting these values into Eq. (VII.42), we obtain the value of the angle θ_1 for which the plane wave penetrates through the boundary of these liquids from the water side without reflection: $\theta_1 \simeq 35°$.

We shall now study the conditions for total reflection of a plane wave from the boundary between two media. Aside from the general cases $z_2 \to 0$ and $z_2 >> z_1$, corresponding to reflection from the boundary with a vacuum or from an infinite solid wall, the transmission coefficient d_l vanishes (and the reflection coefficient $\rho_l = 1$) when the cosine of one of the angles θ_1 and θ_2 vanishes. Since the condition $\cos\theta_2 = 0$ indicates propagation of the incident wave along the boundary, only the case $\cos\theta_2 = 0$, i.e., $\theta_2 = \pi/2$, is of interest. By virtue of relation (VII.37), this angle of refraction corresponds to some "critical" angle of incidence θ_{cr}, satisfying the condition

$$\sin\theta_{cr} = c_1/c_2. \tag{VII.43}$$

For this angle of incidence, the refracted ray vanishes and all of the energy in the transmitted wave for angles $\theta_1 < \theta_{cr}$, is transferred into the reflected wave. This phenomenon, known as *total internal reflection*, according to Eq. (VII.43), can occur only if $c_1 < c_2$, i.e., when the velocity of sound in the medium *2* is higher than in medium *1*, for example, for an ultrasonic wave incident from a liquid on the boundary with a solid. The magnitude of the critical angle can, in this case, be very small, for example, for a water–aluminum boundary

($c_1 \simeq 1.5 \cdot 10^5$ cm/s, $c_2 \simeq 6 \cdot 10^5$ cm/s), $\theta_{cr} \simeq 14°$. For a gas–solid boundary, on the other hand, the condition $c_1 \ll c_2$ may be assumed to hold, for which, according to (VII.43), the critical angle is close to $\pi/2$. This means that only waves incident on the boundary almost at a right angle penetrate from a gas into the solid, while the remaining waves undergo total internal reflection. It is interesting to note that in the opposite case ($c_1 \gg c_2$), according to expression (VII.37), the angle of refraction θ_2 is close to $\pi/2$ for any angles of incidence, so that waves incident, for example, from the solid onto the boundary with a gas propagate in the gas almost perpendicular to the boundary, independent of the angle of incidence.

§ 5. Interference of plane waves at oblique incidence. Quasistanding waves

We shall now study the structure of the field in the zone where the incident and reflected plane waves overlap. In so doing, we shall at first neglect ultrasonic absorption (making the assumption that it is quite small, so that the waves are not appreciably damped, at least near the reflecting boundary), nonlinear effects (for waves with low amplitudes), and "losses to reflection" (making the assumption that the reflection coefficient equals 1). After the detailed analysis performed in § 2 separately for the pressure and particle velocity, we can henceforth confine our attention to the velocity potential field only, keeping in mind that the pressure and particle velocity are found from the potential φ by differentiating it with respect to time and the coordinates, which leads to the corresponding phase shift in the reflected wave that also exists at a reflecting boundary.

Thus, using the principle of superposition and adding the potentials of the incident wave (VII.29) and reflected wave (VII.30), we obtain the total potential in the zone of overlapping of these waves, whose amplitudes are assumed to be equal

$$\varphi(x, y, t) = \varphi_1(x, y, t) + \varphi_1'(x, y, t) =$$
$$= \varphi_{max} \exp[i(\omega t - ky \sin\theta)]\,[\exp(-ikx\cos\theta) + \exp(ikx\cos\theta)] =$$

$$= 2\varphi_{max} \cos(kx \cos\theta) \exp[i(\omega t - ky \sin\theta)] =$$
$$= \Phi_{max} \exp[i(\omega t - yk_y)], \qquad \text{(VII.44)}$$

where θ is the angle of incidence (equal to the angle of reflection); $k = \omega/c$ is the wave number in the medium under study; and $k_y = k \sin\theta$ is the projection of the wave vector on the y-axis. Thus superposition of the potentials of the incident and reflected waves leads to an equation for a traveling wave propagating along the y-axis with velocity $c_y = \omega/k_y = \omega/(k \sin\theta) = c/\sin\theta$. This is essentially the velocity with which the phase of the incident wave or its "track" propagates along the boundary. For $\theta = \pi/2$ the velocity of the "track" of the wave on the boundary of the medium equals the velocity of sound in the medium, while for normal incidence, $c_y \to \infty$, all points of equal phase, forming the wave front, reach the reflecting boundary simultaneously. In the general case of an arbitrary angle of incidence, the velocity of the "track" varies in the range $c \leqslant c_y \leqslant \infty$.

As is evident from Eq. (VII.44), the traveling wave described by this equation is modulated in amplitude along the x-axis with a spatial period that also depends on the angle of incidence θ. In planes whose coordinates are given by the condition

$$x = \frac{2n+1}{2} \cdot \frac{\pi}{k\cos\theta} \quad (n = 0, 1, 2, 3, \ldots), \qquad \text{(VII.45)}$$

the amplitude of the resulting wave (VII.44) vanishes, and in planes for which $x = n\pi/(k\cos\theta)$ it assumes the maximum value, equal to twice the amplitude of the incident wave φ_{max}. These planes, parallel to the reflecting surface, form a system of nodes and antinodes of a standing wave, in which, however, the phases of the oscillations move along the y-axis with velocity c_y. Such a wave can be called a *quasistanding* wave; its amplitude is modulated in the direction of the normal to the reflecting surface. The period of this modulation, corresponding to the "wavelength of the quasistanding wave" Λ_0', is determined by the distance between the neighboring "nodes" of the planes (VII.45) and equals $\Delta x = \Lambda_0' = \Delta/(2\cos\theta) = \Lambda_0/\cos\theta$, where Λ is the wavelength of the traveling wave in the given medium at the given ultrasonic frequency.

Thus the wavelength of the quasistanding wave Λ_0' depends on the angle of incidence θ, as does the velocity of the "track" of the incident wave. For normal incidence ($\theta = 0$) $\Lambda_0' = \Lambda/2 = \Lambda_0$, and the wavelength equals the wavelength of the normal standing wave. When the incident beam is inclined away from the normal, the distance between the nodal planes increases and an energy flux corresponding to a traveling wave appears along the boundary in the direction of the projection of the wave vector of the incident wave. The vector of this flux density, i.e., the intensity of the resulting field, is oriented along the reflecting boundary for all angles θ — as the vector sum of the incident and reflected wave intensities of equal absolute magnitude (Fig. 43).

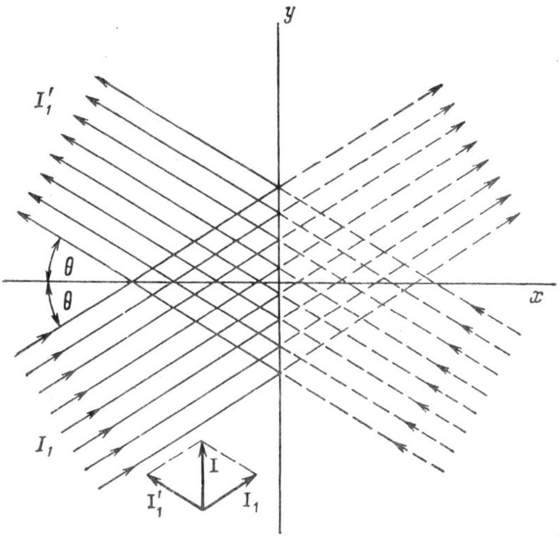

Fig. 43.

We note that, as is evident from Fig. 43, the result obtained can be interpreted as the interference of two beams of plane waves: from the actual source and from an imaginary source positioned symmetrically to it relative to the reflecting plane. Evidently, the same interference pattern with the formation of quasistanding waves is formed by the superposition of beams from two real, symmetrically positioned, coherent sources.* In both cases, the energy flux density vector in the zone of overlapping of the beams is oriented along the bisector of the angle between them and the absolute magnitude of the intensity is

*Coherence of the ultrasonic sources is achieved by exciting them with the same generator.

defined as the geometric sum of the intensities of both beams. We are, of course, talking about conditions under which the principle of superposition holds.

§ 6. Scattering of ultrasonic waves in an inhomogeneous medium

We shall briefly study the scattering of ultrasonic waves as a result of their diffuse reflection from particles with sharp boundaries and physical properties differing from those of the surrounding medium. Media containing such particles are called *heterogeneous*. Examples of heterogeneous media are suspensions (fluids containing suspended solid particles), aerosols (gases containing suspended solid particles), emulsions (liquid drops in an immiscible liquid), liquids containing gas bubbles (in particular, cavitation bubbles), as well as such media as glasses, devitrified glasses, minerals, noncrystalline metals, etc. When a primary ultrasonic wave propagates in such a medium, the wave will be reflected from the particles present in the medium, thereby exciting internal oscillations, which is what leads to the emission of secondary waves, i.e., scattered waves, by the particles. These singly scattered waves, generally speaking, can in their turn be multiply reflected by other particles. However, as soon as the single-scattered field is much smaller than the primary field, repeatedly scattered waves can be neglected, if the number of scattering centers is not too large. Neglecting multiple scattering is equivalent to the assumption that there are no "acoustic" interactions between particles, i.e., the oscillations of one particle have no effect on those of another particle. The total field scattered by the collection of particles can then be found by superposing the fields single-scattered by each particle and the problem of scattering of ultrasound in a heterogeneous medium reduces to the problem of determining the scattering by a single particle and summing the results over all particles in the scattering volume. The particles may be assumed to be spherical, especially since for particles which are much smaller than the wavelength and at sufficiently large distances from them, the deviation of the actual shape from a sphere does not play a significant role.

The acoustic field scattered by the particle, naturally, depends on the form of the primary wave. We shall study, as before, plane waves, having in mind the scattering of directed ultrasonic beams. Furthermore, scattering of ultrasound by a particle depends on its compressibility and density. It is understandable that if the the density and compressibility of the particle are the same as those of the surrounding medium, then this is equivalent to an acoustically homogeneous medium, in which there will be no scattering. If only the density of the particle differs from that of the surrounding medium and the compressibilities are the same, then in the primary acoustic field the particle will lag behind or lead the oscillatory motion of the medium, i.e., it will undergo translational-vibrational motion relative to the medium and the field scattered by the particle will be equivalent to the field radiated by an "acoustic dipole." If, on the other hand, only the compressibilities differ, then such a particle will undergo translational oscillations in phase with the acoustic oscillations of the medium, but under the action of a variable acoustic pressure the particle will pulsate relative to the medium and the field scattered by it will be equivalent to the field emitted by a pulsating sphere. In the general case, both the densities and compressibilities of the particle and medium can be different, and the fields scattered by the particles will have a complicated character. Thus the calculation of this field is closely related to the problem of the radiation of sound by a sphere undergoing different oscillations.

Scattering of ultrasound by a given particle greatly depends on the ratio of its dimensions to the wavelength of the ultrasonic wave Λ. A measure of this ratio is the so-called *scattering parameter* kR, where R is the radius of the particle and $k = 2\pi/\Lambda$ is the wave number. When $kR \gg 1$, i.e., the wavelength is much smaller than the scattering body, diffraction phenomena can be neglected and the scattering obeys athe laws of geometrical acoustics. For this reason, the region $kR \gg 1$ is called the region of *geometric scattering*. Geometric scattering is determined by the usual laws of reflection, studied previously for plane surfaces. In the case of a nonplanar but smooth surface, the surface can be subdivided into individual locally planar sections and reflection from these sections can be found according to the rule that the angle of incidence equals the angle of reflection. An acoustic shadow, whose cross-sectional area equals the cross-sectional area of the body , forms behind the scattering body. The scattered field in front of the body will be determined by all of the reflected rays and the scattered power flux will equal the incident power flux. The power flux behind the scattering body equals zero (here the scattered field seems to cancel the primary field).

Thus the scattering problem pertains to particles whose dimensions are comparable to or much less than the wavelength. The case $kR \simeq 1$ is most difficult to calculate. In optics, the theory of scattering of light for $kR \geqslant 1$ was developed by Mie and, for this reason, the region $kR \simeq 1$ for arbitrary wave processes is called the *region of Mie scattering*.

The problem of the scattering of light and sound by spherical particles with small radius was first solved by Rayleigh and forms the foundation of the classical theory of scattering of waves in inhomogeneous media. For this reason, scattering under the conditions $kR \ll 1$ is called *Rayleigh scattering*. Rayleigh's method consists of expanding the incident wave and the wave scattered by the particle in a series of spherical functions, taking into account the boundary conditions on the surface of the particle, and summing the resulting fields.

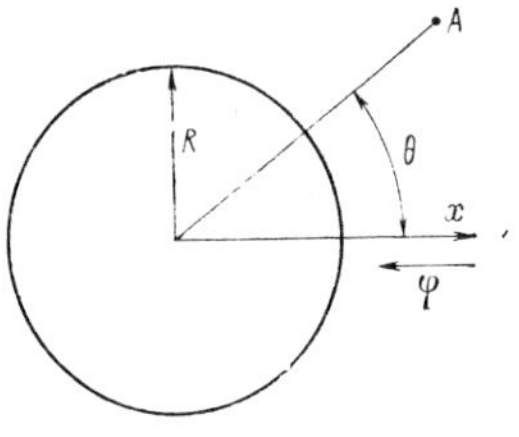

Fig. 44.

Let a spherical particle with radius R be located at the center of a spherical system of coordinates, and let the incident plane wave propagate along the negative x-axis (Fig. 44), so that its argument ikx is positive. At the point of observation A $x = r \cos\theta$. Therefore, the potential of the incident plane wave with frequency ω will have the form:

$$\varphi(r, \theta, t) = \varphi_{max} \exp[i(\omega t + kx)] = \varphi_{max} \exp(i\omega t) \exp(ikr\cos\theta).$$

Dropping the unimportant complex factor $\exp(i\omega t)$ this potential can be represented by expanding $\exp(ikr\cos\theta)$ in a series of spherical functions[61]: $\varphi = \sum_{m=0}^{\infty}(-1)^m(2m+1)P_m(\cos\theta)J_m(kr)$, where $P_m(\cos\theta)$ are Legendre polynomials and J_m is a Bessel function of order m.

The scattered wave can likewise be written as a superposition of spherical waves emanating from the origin of coordinates by expanding the potential of the scattered wave $\psi(r, \theta, t)$ in a series of spherical functions[61]

$$\psi = \sum_{m=0}^{\infty} a_m P_m(\cos\theta)\, f_m(ikr)\frac{e^{ikr}}{r}\, e^{i\omega t}\,, \qquad \text{(VII.46)}$$

where

$$f_m \equiv 1 + \frac{m(m+1)}{2ikr} + \frac{(m-1)m(m+1)(m+2)}{2\cdot 4(ikr)^2} + \dots$$

$$\dots + \frac{1\cdot 2\cdot 3\cdot \dots \cdot 2m}{2\cdot 4\cdot 6\cdot \dots \cdot 2m\,(ikr)^m}\,. \qquad \text{(VII.47)}$$

Since the scattered field results from the emission of secondary waves by particles undergoing forced oscillations induced by the incident wave, the frequency of the scattered waves remains the same. The unknown coefficients a_m in the expansion (VII.46) are found from the boundary conditions, which depend on the physical properties of the scattering particle. In this case, since the scattered field is usually observed at distances r greatly exceeding the wavelength, i.e., for $kr \gg 1$, the function $f_m(ikr)$, as is evident from expression (VII.47), can be set equal to 1.

The simplest case for calculations is the **case of an absolutely rigid, immobile spherical particle**. This model is appropriate for suspensions of solid particles in gases and liquids with small displacements in an ultrasonic wave (the source of sound in this case is the medium flowing around the particle; the flow assumed to be irrotational). The boundary condition for this model will be the vanishing of the radial components of the particle velocities in the incident and scattered waves, i.e.,

$$\left.\frac{\partial\varphi}{\partial r} + \frac{\partial\psi}{\partial r}\right|_{r=R} = 0,$$

where R is the radius of the scattering sphere. Use of this condition gives for each coefficient

$$a_m = (2m+1)\frac{kR^2 e^{ikR}}{F_m(ikR)}\, P_m\!\left(\frac{d}{d(ikr)}\right)\frac{d}{d(kR)}\,\frac{\sin kR}{kR}\,, \qquad \text{(VII.48)}$$

where

$$F_m(ikR) \equiv 1\cdot3\cdot5\cdot\ldots\cdot(2m-1)(m+1)(ikR)^{-m}\times$$
$$\times\left[1+ikR+\frac{m^2(ikR)^2}{(m+1)(2m-1)}+\ldots\right].$$

Expressing the complex function $F_m(ikR)$ in the form $F_m \equiv \alpha + i\beta = (\alpha^2 + \beta^2)/(\alpha - i\beta) = \sqrt{(\alpha^2+\beta^2)}/\exp(i\gamma)$, where $\gamma \equiv \arctan(-\beta/\alpha)$, and substituting expression (VII.48) into (VII.46), we obtain the following equation for the potential of the scattered wave for $kr \gg 1$:

$$\psi = \sum_{m=0}^{\infty}(2m+1)\frac{kR^2}{r}e^{i(\omega t-kr+kR+\gamma)}(\alpha^2+\beta^2)^{-1/2}\times$$
$$\times P_m\left(\frac{d}{d(ikR)}\right)\frac{d}{d(kR)}\frac{\sin kR}{kR}P_m(\cos\theta).$$

From here it is easy to obtain an expression for the first few spherical harmonics ($m = 0, 1, 2, \ldots$), keeping in mind that $P_0(\cos\theta) = 0$, $P_1(\cos\theta) = \cos\theta$, $P_2(\cos\theta) = 3(\cos^2\theta - 1/3)/2, \ldots$.[61] For the first three harmonics, this gives

$$m=0,\quad \alpha^2+\beta^2 = 1+(kR)^2,\quad \tan\gamma_0 = -kR,$$
$$\psi_0 = \frac{kR^2}{r}(1+k^2R^2)^{-1/2}\frac{d}{d(kR)}\frac{\sin kR}{kR}\cos(\omega t - kr + kR + \gamma_0); \tag{VII.49a}$$

$$m=1,\quad \alpha^2+\beta^2 = (kR)^2+\frac{4}{(kR)^2},\quad \tan\gamma_1 = -\frac{(kR)^3-2}{2kR},$$
$$\psi_1 = \frac{3kR^2}{r}\left(k^2R^2+\frac{4}{k^2R^2}\right)^{-1/2}\frac{d^2}{d(kR)^2}\frac{\sin kR}{kR}\cos\theta\times$$
$$\times\sin(\omega t - kr + kR + \gamma_1); \tag{VII.49b}$$

$$m=2,\quad \alpha^2+\beta^2 = (kR)^2-2+\frac{9}{(kR)^2}+\frac{81}{(kR)^4},$$
$$\tan\gamma_2 = -\frac{kR(k^2R^2-9)}{4k^2R^2-9},$$
$$\psi_2 = -\frac{45kR^2}{4r}\left(k^2R^2-2+\frac{9}{k^2R^2}+\frac{81}{k^4R^4}\right)^{-1/2}\times$$
$$\times\left[\frac{d^3}{d(kR)^3}+\frac{1}{3}\frac{d}{d(kR)}\right]\frac{\sin kR}{kR}\left(\cos^2\theta-\frac{1}{3}\right)\times$$
$$\times\cos(\omega t - kr + kR + \gamma_2). \tag{VII.49c}$$

Up to this stage, disregarding the assumption $kr \gg 1$, the problem of scattering of plane waves for the given model is solved exactly in general form. The possibility and the nature of the further development of the solution depends on the ratio of the wavelength to the radius of the particle, i.e., on the magnitude of the parameter kR. The final solution for the case $kR \ll 1$ was obtained by Rayleigh.

For $kR \ll 1$ the expression (VII.49) can be expanded in a series in increasing powers of kR, which gives

$$\psi_0 = -\frac{k^2R^3}{3r}\left(1 - \frac{3}{5}k^2R^2 + \frac{3}{7}k^4R^4 - \frac{19}{54}k^6R^6 + \ldots\right)\times$$
$$\times \cos(\omega t - kr + kR + \gamma_0);$$

$$\psi_0 = -\frac{k^2R^3}{2r}\left(1 - \frac{3}{10}k^2R^2 + \frac{3}{28}k^4R^4 - \frac{1}{27}k^6R^6 + \ldots\right)\times$$
$$\times \cos(\omega t - kr + kR + \gamma_1);$$

$$\psi_2 = -\frac{k^4R^5}{9r}\left(1 - \frac{25}{126}k^2R^2 + \frac{13}{567}k^4R^4 + \ldots\right)\times$$
$$\times\left(\cos^2\theta - \frac{1}{3}\right)\cdot\cos(\omega t - kr + kR + \gamma_2);$$

. . . .

From these expressions it is evident that the zeroth and first spherical harmonics are of the same order of magnitude relative to the small parameter kR, while the quantity ψ_2 is two orders of magnitude larger. Therefore, for $kR \ll 1$ the series for the potential of the scattered waves can be restricted to the sum of the first two spherical harmonics. They must be added taking the phase factors into account. However, according to Eqs. (VII.49a) and (VII.49b), for $kR \ll 1$ $\gamma_0 \propto 0$, while $\gamma_1 \propto \pi/2$, so that the phase factors in ψ_0 and ψ_1 are approximately equal. As a result, the following simple approximate expression is obtained for the potential of the scattered waves $\psi \propto \psi_0 + \psi_1$:

$$\psi(\theta, r, t) \propto -[k^2R^3/(3r)]\,[1 + (3/2)\cos\theta]\cos(\omega t - kr).$$

For the intensity of the scattered waves, which is proportional to the square of the amplitude, we have:

$$I_{sc} = I_0 \frac{k^4R^2}{9r^2}\left(1 + \frac{3}{2}\cos\theta\right)^2 \qquad \text{(VII.50a)}$$

or

$$I_{sc} = I_0 \frac{\omega^4R^6}{9c^4r^2}\left(1 + \frac{3}{2}\cos\theta\right)^2, \qquad \text{(VII.50b)}$$

where c is the velocity of sound in the medium.

Thus the intensity of the scattered waves is proportional to the fourth power of the frequency of the incident wave, i.e., inversely proportional to the fourth power of the wavelength Λ (*Rayleigh's law*); it is also proportional to the sixth power of the size of the scattered particle, i.e., to the square of its volume. We recall, however, that we are talking about wavelengths and particle sizes satisfying the condition $kR \ll 1$. This is the region of applicability of Eq. (VII.50) and is called the *region of Rayleigh scattering*.

According to Eq. (VII.50), the quantity $(I_{sc}/I_0)r^2$ can serve as a distance-independent measure of the angular distribution of the scattered energy (of course, for distances r satisfying the starting condition $kr \gg 1$). The corresponding curve (angular diagram) is called the *directivity function*. The directivity function for Rayleigh scattering is shown in Fig. 45. A characteristic feature of this function is the predominance of backscattering, i.e., scattering in a direction opposite to the incident wave. In accordance with Eq. (VII.50), the scattering intensity vanishes in directions for which $\cos\theta = -2/3$ ($\theta \approx 132°$ and $\theta \approx 228°$). The ratio of the scattering intensity at $\theta = 0°$ and $\theta = 180°$, according to Rayleigh's equations (VII.50), is $(5/2:1/2)^2 = 25$, i.e., backscattering is 25 times more intense than forward scattering. It is easy to calculate that about 90% of all of the scattered energy goes into the backward direction for angles ranging from 0 to 90°.

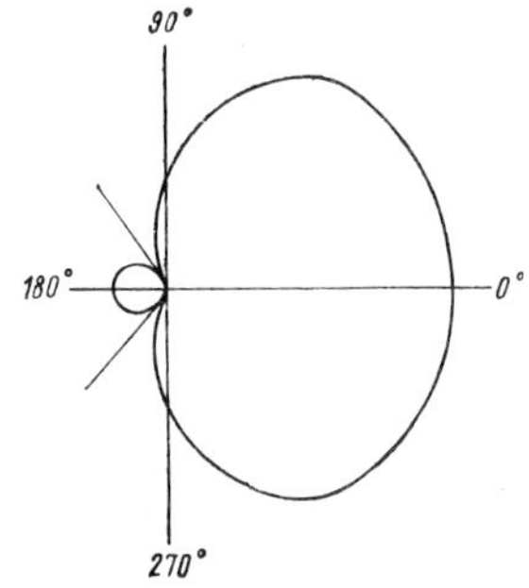

Fig. 45.

A characteristic called the *effective scattering cross section* is often used as an integral measure of scattering. The scattering cross section

σ_{eff} is defined as the ratio of the total scattered power D_{sc} to the intensity of the incident ultrasonic wave I_0:

$$\sigma_{eff} = D_{sc}/I_0, \qquad \text{(VII.51)}$$

which has the dimensions of area. In the case of geometric scattering by a sphere, the effective scattering cross section evidently equals twice the area of the diametral cross section of the sphere:

$$\sigma_{eff} = 2\pi R^2. \qquad \text{(VII.52)}$$

For Rayleigh scattering, we find the total scattered power by integrating expression (VII.50a) over the solid angle

$$D_{sc} = \int_0^{\pi} I_{sc}\, 2\pi r^2 \sin\theta\, d\theta = \frac{7}{9}\,\pi R^2 (k^4 R^4)\, I_0.$$

Dividing this expression by the intensity of the incident wave I_0, we obtain:

$$\sigma_{eff} = (7/9)\,\pi R^2 (kR)^4. \qquad \text{(VII.53)}$$

Therefore, for $kR \ll 1$ the effective scattering cross section is only a small fraction of the cross-sectional area of the sphere πR^2. Thus the quantity σ_{eff} characterizes the scattering efficiency of the given obstacle.

The problem of the scattering of plane waves by a compressible spherical particle was also first solved by Rayleigh in the approximation $kR \ll 1$. In this case the scattered field will be determined by the radiation emitted by a particle undergoing forced oscillatory and pulsational oscillations. The scheme of the solution remains the same, i.e., the scattered wave is represented as a series (VII.46) in spherical functions, but the boundary conditions from which the unknown coefficients in this series are determined will have the following form:

$$\left.\frac{\partial\varphi}{\partial r} + \frac{\partial\psi}{\partial r}\right|_{r=R} = \frac{\partial\varphi_1}{\partial r}, \qquad \left.\rho(\varphi + \psi)\right|_{r=R} = \rho_1\varphi_1.$$

Here, as before, φ and ψ are the potentials of the incident and scattered waves, respectively; φ_1 is the potential of the wave propagating in the particle; ρ_1 is the density of the particle; and, ρ is the density of the surrounding medium. The first boundary condition expresses the continuity of the normal components of the displacement velocities at the bounding surface and the second expresses the continuity of the pressures.

A simple expression is also obtained for the intensity of the scattered waves in the case $kr \gg 1$ (large distances from the point of observation):

$$I_{sc} = I_0 \frac{\omega^4 R^6}{9c^4 r^2} \left(\frac{K_1 - K_2}{K_1} + 3 \frac{\rho_1 - \rho}{2\rho_1 + \rho} \cos\theta \right)^2, \qquad \text{(VII.54)}$$

where K_1 and K_2 are the elastic moduli of the particle and of the medium surrounding it. In the limit $K_1 \to \infty$ and $\rho_1 \gg \rho$ this expression transforms into the equation for scattering by an incompressible immobile sphere (VII.50b). The model of a compressible spherical particle is more applicable to emulsions, suspensions of liquid drops in the atmosphere, and gas bubbles in a liquid. It follows from expression (VII.54), however, that for $kR \ll 1$ all the basic characteristics of Rayleigh scattering obtained for a rigid particle are valid even in the case of a compressible sphere: dependence on the fourth power of the frequency and the sixth power of the particle radius. The difference lies only in the details of the directivity function; in particular, the angle θ at which the scattering intensity vanishes depends on the ratio between the density and the compressibility of the particle and that of the surrounding medium.

The case $kR > 1$, corresponding to Mie scattering, involves very laborious calculations (a large number of spherical harmonics must be summed) which yield cumbersome results. The simplest expression is obtained for the intensity of waves scattered by an **incompressible sphere** for $kR \gg 1$ (and $kr \gg 1$)[2]:

$$I_{sc} = I_0 \frac{R^2}{4r^2} \left[1 + \cot^2 \frac{\theta}{2} J_1^2 (kR \sin\theta) \right].$$

where J_1 is a first-order Bessel function. Although this expression applies to the case $kR \gg 1$, it contains the basic characteristics of Mie scattering. In contrast to Rayleigh scattering, the frequency dependence is expressed here in a complicated manner in terms of the Bessel function in the second term; the dependence on the particle size is also different.

different. Mie phase functions have a different form for different values of the scattering parameter kR. Figure 46 presents three directivity functions for scattering by a rigid sphere, calculated for different values of kR. For small kR, the directivity function is close to the pattern for Rayleigh scattering. As kR increases, the directivity function begins to stretch out in the direction of the incident wave and a number of peaks appear in it. For $kR > 20$–30, the form of the directivity function stabilizes on the whole, with the exception of the "diffraction nose" where the scattered field, adding to the incident wave, forms the shaded region.

Analysis of scattering by a **compressible sphere** for $kR > 1$ shows that even in this case the basic characteristics of Mie scattering studied

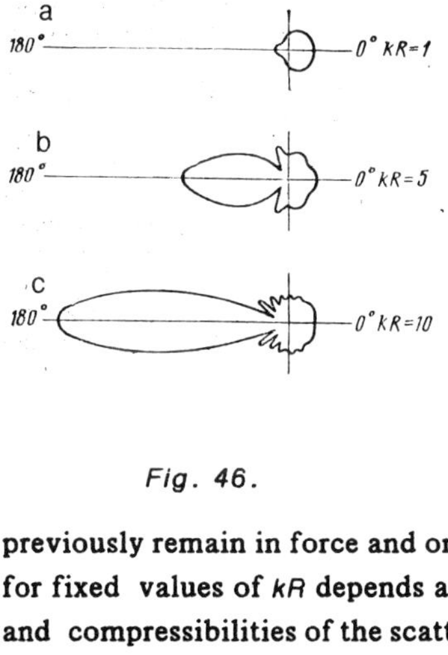

Fig. 46.

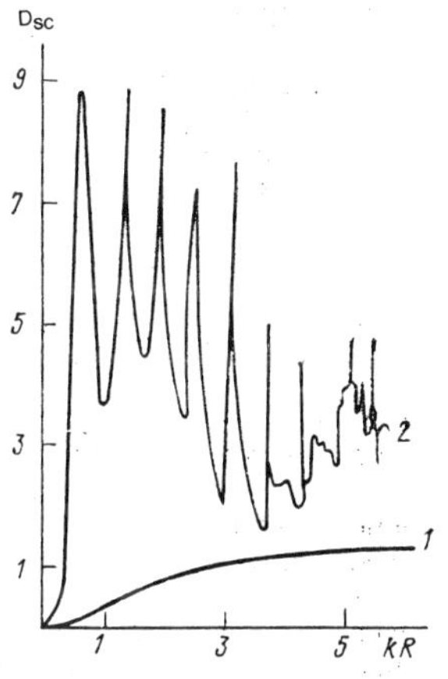

Fig. 47.

previously remain in force and only the form of the directivity functions for fixed values of kR depends additionally on the ratio of the densities and compressibilities of the scattering particle and of the medium.

Amongst the factors that distinguish compressible and incompressible particles, the most important is the possibility of resonant excitation of different characteristic oscillations of the elastic scatterers. In this case resonance peaks corresponding to excitation of some characteristic oscillatory modes of the scattering particles can appear in the frequency dependence of the scattering. As an example, Fig. 47 [62] shows computed curves illustrating the dependence of the reduced scattering power D_{sc} on the parameter kR for a rigid sphere (1) and for a compressible spherical particle, in which the velocity of sound c_1 and the density ρ_1 are half the velocity of sound c and density ρ of the surrounding medium (2). Of course, such scattering peaks for particles with fixed physical properties appear in a well-defined ("critical") region of frequencies (values of kR), corresponding to its characteristic resonant oscillations. From the position of the scattering peaks it is possible to determine the resonance frequencies and the nature of these oscillations.

The intensification of scattering at resonance is explained by the fact that, as already mentioned, the scattered field is formed by radiation of ultrasound by particles undergoing forced oscillations in the field of the primary wave. The amplitude of the forced oscillations, however, sharply increases at resonance; the scattering intensity also increases by a factor equal to the Q-factor of the oscillatory system (see Chap. VIII). For pulsational oscillations of an air bubble in water, for example, this increases the effective scattering cross section by approximately 12 orders of magnitude, whence follows the strong scattering of ultrasound accompanying the appearance of cavitation in the liquid, when, as we have seen, bubbles with resonant sizes are always present or are formed in the liquid. Resonant scattering is used successfully in hydroacoustic sonar location of fish schools; the resonant bubbles in this case are the bubbles produced by the fish. The sharp increase in scattering at resonance (including also the backscattering recorded by the echo-locator) makes it possible to determine reliably the size of both the fish and the school.

Attenuation of ultrasonic waves as a result of scattering. Since the scattered energy is removed from the energy of the primary ultrasonic wave, additonal attenuation (aside from absorption and other factors) of ultrasonic waves as they propagate in a medium occurs owing to scattering by clusters of particles and other inhomogeneities in the medium. A measure of this attenuation, contributed by a single particle, is the effective scattering cross section ("transverse cross section") σ_{eff}, which, according to its definition (VII.51), is precisely the fraction of ultrasonic power which is removed by scattering from the specific power (i.e., intensity) of the incident ultrasonic wave. For clusters of acoustically noninteracting particles the overall scattering will equal the sum of the scattering from single particles. If we are talking about microinhomogeneities in media with close-lying obstacles, which are much smaller than the ultrasonic wavelength, then this collection of inhomogeneities can be represented by a regular (uniform) distribution on which concentration fluctuations are superposed. A uniform distribution of inhomogeneities is equivalent to a three-dimensional diffraction grating, and it will not lead to diffuse scattering. An analogous situation occurs in optics in the case of the propagation of light in a regular crystal: the light waves scattered by each molecule cancel one another in all directions, except the direction of propagation of the primary wave. This means that incoherent scattering by concentration fluctuations will occur, and if the concentration fluctuations in different volumes are independent, then the powers scattered by them likewise simply sum.

In any case, neglecting secondary scattering, an inhomogeneous medium can be characterized by the specific scattering cross section, defined as the product of the effective transverse cross section of each scatterer and the number of independent scatterers per unit volume n_0. Then, according to the definition of σ_{eff}, the primary traveling ultrasonic plane wave with intensity I will lose in the form of scattered waves an intensity of $n_0\sigma_{eff} I$ per unit path length, i.e., $dI/dx = -n_0\sigma_{eff} I$, whence we find by integration: $I = I_0 \exp(-n_0\sigma_{eff} x) = I_0 \exp(\alpha'_{sc} x)$.

Thus scattering, like absorption, leads to exponential attenuation of plane ultrasonic waves with an energy scattering coefficient $\alpha'_{sc} = n_0\sigma_{eff}$, determined by the effective scattering cross section σ_{eff}. Based on its general definition (VII.51), we can write $\alpha'_{sc} = (n_0/I_0)\oint I_{sc}\, dS$. In the case of Rayleigh scattering, according to (VII.53), we obtain

$$\alpha'_{sc} = (7/9)n_0\pi k^4 R^6 = 7\pi^3 n_0 V^2 \nu^4/c^4,$$

i.e., the attenuation coefficient due to Rayleigh scattering is proportional to the fourth power of the frequency and the square of the volume of the scattering particle V_0. In the case of geometric scattering, according to (VII.52), $\alpha'_{sc} = 2\pi R^2 n_0$ and $\alpha_{sc} = \pi R^2 n_0$, i.e., the scattering coefficient is independent of the frequency. In all cases it is proportional to the density of the scattering centers n_0. This is a result of neglecting the interaction of particles. A more accurate theory must take into account secondary scattering, as well as such factors as ultrasonic absorption in the particle material and the friction force on the surface of a particle.

VIII. Transmission of Plane Waves Through Layers. Electroacoustical Analogies. Radiation of Plane Waves.

§1. Transmission of ultrasonic plane waves through a plane-parallel layer

We shall now study the transmission of ultrasonic plane waves through a layer with plane-parallel boundaries. We denote the acoustic impedance of the layer by $z = \rho c$ and the acoustic impedance of the medium on both sides of the layer by $z_1 = \rho_1 c_1$. We orient the x-axis perpendicular to the boundaries of the layer, to which we assign the coordinates $x = 0$ and $x = d$ (d is the thickness of the layer), and we consider at the outset the general case of oblique incidence of ultrasonic waves at an arbitrary angle θ to the x-axis (Fig. 48). Reflected and refracted waves will appear at each boundary; in addition, because of the symmetry, the wave passing through the layer will emerge from the layer at the angle of incidence θ_1. By direct analogy with Eqs. (VII.29)–(VII.31) we have the following expressions for the potentials of these waves: for the incident wave,

$$\varphi_1 = \varphi_{1\max} \exp\{i[\omega t - k_1(x \cos\theta_1 + y \cos\theta_1)]\};$$

for the wave reflected at the boundary $x = 0$,

$$\varphi_1' = \varphi_{1\max}' \exp\{i[\omega t + k_1(x \cos\theta_1 - y \cos\theta_1)]\};$$

for the wave transmitted through the first boundary,

$$\varphi_2 = \varphi_{2\max} \exp\{i[\omega t - k(x \cos\theta + y \cos\theta)]\};$$

for the wave reflected from the second boundary (by analogy with the the wave reflected from the first boundary),

$$\varphi_2' = \varphi_{2max}' \exp\{i[\omega t + k(x\cos\theta - y\cos\theta)]\};$$

and, for the transmitted wave,

$$\varphi_3 = \varphi_{3max} \exp\{i[\omega t - k_1(x\cos\theta_1 + y\cos\theta_1)]\}.$$

Thus we have four unknown quantities φ_1', φ_2, φ_2', and φ_3 which are related by four equations following from the boundary conditions

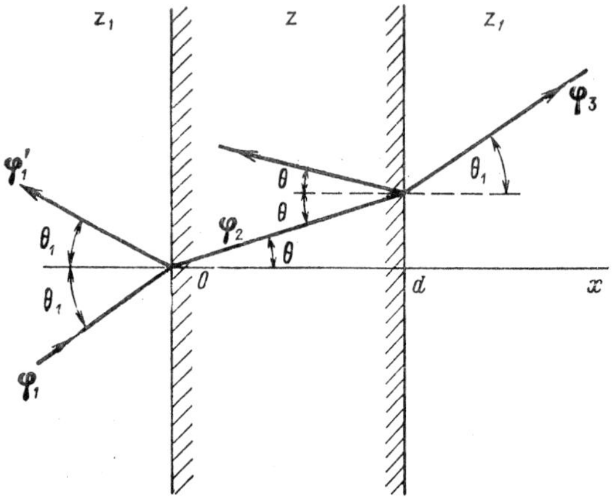

Fig. 48.

requiring continuity of the pressure and of the normal components of the velocities on both sides of the boundary. For the potentials this gives at $x = 0$:

$$\left.\begin{aligned} \rho_1(\varphi_1 + \varphi_1') &= \rho(\varphi_2 + \varphi_2') \\ k_{1x}(\varphi_1 - \varphi_1') &= k_x(\varphi_2 - \varphi_2') \end{aligned}\right|_{x=0}, \qquad \text{(VIII.1)}$$

where $k_{1x} = k_1 \cos\theta$ and $k_x = k\cos\theta$. The condition (VIII.1) holds only at $x = 0$, so that $\varphi_1 = \varphi_1|_{x=0} \equiv \varphi_1(0) = \varphi_{1max}\exp[i(\omega t - k_1 y \sin\theta_1)]$, etc. At $x = d$ the potentials will differ from the previous values by the factors $\exp(\pm ik_x d)$ and $\exp(-ik_1 xd)$:

$$\rho[(\varphi_2(0)e^{-ik_xd} + \varphi_2'(0)e^{ik_xd}] = \rho_1\varphi_3(0)e^{-ik_{1x}d};$$

$$\text{(VIII.2)}$$

$$k_x[\varphi_2(0)e^{-ik_xd} - \varphi_2'(0)e^{-ik_xd}] = k_{1x}\,\varphi_3(0)e^{-ik_{1x}d}.$$

Eliminating now the intermediate waves φ_2 and φ_2' from Eq. (VIII.1) and (VIII.2), we obtain for the pressure reflection coefficient

$$\rho_p = \frac{\varphi_1'}{\varphi_1} = \frac{k_x\rho_1/(k_{1x}\rho) - k_{1x}\rho/(k_x\rho_1)}{\sqrt{[k_x\rho_1/(k_{1x}\rho) + k_{1x}\rho/(k_x\rho_1)]^2 + 4\cot^2(d\cdot k_x)}}, \quad \text{(VIII.3)}$$

where $k_x\rho_1/(k_{1x}\rho) = k\cos\theta\cdot\rho_1/(k_1\cos\theta_1\cdot\rho) = z_1\cos\theta/(z\cos\theta_1)$. Aside from the trivial case $d = 0$ or $\rho_1 = 0$ and $c_1 = 0$ (no layer), the reflection coefficient vanishes at $z_1\cos\theta = z\cos\theta_1$, which corresponds to the condition (VII.41) for the transparency of an interface between two media at oblique incidence. In addition, as is evident from expression (VIII.3), the reflection coefficient of the layer vanishes if $\cot(d\cdot k_x) = \infty$, i.e., $d\cdot k_x = n\pi$, where $n = 0, 1, 2, \ldots$. Since $d\cdot k_x = d\cdot k\cos\theta = (2\pi d/\Lambda)\cos\theta$, the condition of transparency of the layer $(\rho_p = 0)$ assumes the following form:

$$d = (n\Lambda/2)/\cos\theta, \quad n = 0, 1, 2, 3, \ldots, \quad \text{(VIII.4)}$$

i.e., a layer with arbitrary acoustic impedance becomes acoustically transparent when the projection of the thickness of the layer on the ray refracted in it equals an integral number of half-waves, corresponding to the material of the layer, i.e., it is determined by the velocity of sound in the layer $c = \Lambda v$. The condition (VIII.4) can be used when the velocity of sound in the layer is greater than in the surrounding medium $(\Lambda > \Lambda_1)$, for example, in the case of a plane-parallel plate consisting of a solid material in a liquid or gas. By rotating this plate relative to the front of the incident ultrasonic wave it is always possible to achieve acoustic transparency of the plate. This technique, for example, was used in Ref. 63 to filter the harmonics of a finite-amplitude wave. It should be kept in mind, however, that when a wave is incident obliquely on a plate made of a solid material, aside from longitudinal waves, shear waves also appear in the plate. Being transformed at the second boundary, the shear waves will also generate longitudinal waves in the liquid (see below), so that the process of transmission of ultrasound through a solid plate is quite complicated.

In the case of normal incidence $\theta_1 = \theta = 0$, $k_{1x} = k_1$, and $k_x = k$, and the equation for the amplitude reflection coefficient (VIII.3) assumes the form

$$\rho_p = \frac{k\rho_1/(k_1\rho) - k_1\rho/(k\rho_1)}{\sqrt{[k\rho_1/(k_1\rho) + k_1\rho/(k\rho_1)]^2 + 4\cot^2(d\cdot k)}}, \quad \text{(VIII.5)}$$

This result is also applicable to a solid plate, because wave transformation does not occur at normal incidence. The intensity reflection coefficient equals, correspondingly, $\rho_I = \rho_p^2$, while the transmission coefficient $d_i = 1 - \rho_I = 1 - \rho_p^2$. According to Eq. (VIII.5), $\rho_p = \rho_I = 0$, and $d_I = 1$ if $d\cdot k = n\pi$, i.e.,

$$d = n\Lambda/2, \quad n = 1, 2, 3, \ldots, \quad \text{(VIII.6)}$$

when an integral number of half-waves (integral number of standing waves) fits into the thickness of the layer (plate). It is not difficult to understand the physical significance of this result. Due to multiple reflections, a standing wave is established in the plate. If $z < z_1$, this corresponds to pressure antinodes at its boundaries, and if $z > z_1$, it corresponds to velocity antinodes. In any case, the plate oscillates in resonance and itself becomes a source of plane waves with the same amplitude ($\varphi_3 = \varphi_1$, if, of course, damping in the material of the layer is neglected), propagating further along the positive x-axis.

Thus, when the condition (VIII.6) holds $\rho_{I\min} = 0$ and $d_{I\max} = 1$ for any characteristic impedance of the layer. The maximum value of the reflection coefficient, however, occurs when $\cot^2(d\cdot k) = 0$, i.e.,

$$d = (2n + 1)\Lambda/4, \quad \text{(VIII.7)}$$

when an odd number of quarter-waves fits into the thickness of the layer ($d = \Lambda/4, 3\Lambda/4, 5\Lambda/4$, etc.). According to Eq. (VIII.5)

$$\rho_{I\max} = \left[\frac{(k\rho_1)^2 - (k_1\rho)^2}{(k\rho_1)^2 + (k_1\rho)^2}\right]^2 = \left[\frac{z_1^2 - z_L^2}{z_1^2 + z_L^2}\right]^2 = \left[\frac{\imath^2 - 1}{\imath^2 + 1}\right]^2, \quad \text{(VIII.8)}$$

where $\imath \equiv z/z_1$. In this case the transmission coefficient has a minimum value

$$d_{I\min} = 1 - \rho_{I\max} = 4\,\imath^2/(\imath^2 + 1)^2. \quad \text{(VIII.9)}$$

At $z_1 = z$, $\rho_{l\max} = 0$ and $d_{l\min} = 1$, i.e., the layer is transparent for any value of the ratio d/Λ. If, on the other hand, $\iota \to 0$ or $\iota \to \infty$, then $\rho_{l\max} \to 1$ and $d_{l\min} \to 0$. In other words, if the characteristic impedances of the layer and of the medium differ considerably, then under the condition (VIII.7) the layer will completely reflect the ultrasonic wave.

Intermediate values of ρ_l and d_l for fixed x, z, d, and Λ can be calculated using Eq. (VIII.5). The results of such a calculation are presented in Fig. 49 for three specific values of the parameter ι: $\iota = 0.094$ (aluminum plate in water), $\iota = 0.454$ (plexiglass plate in water), and $\iota = 1$ (layer with characteristic impedance z equal to the

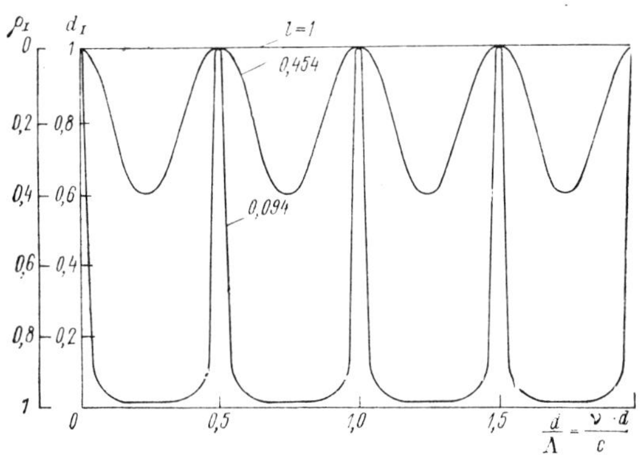

Fig. 49.

characteristic impedance of the medium z_1). The ratio d/Λ or vd/c, where v is the ultrasonic frequency and c is the velocity of sound in the material of the layer, are measured along the abscissa axis. When the frequency v varies and the thickness d is fixed, the transmission coefficient of the plane-parallel plate will exhibit peaks at multiple resonance frequencies satisfying the condition $v_n = nc/(2d)$, which is equivalent to the condition (VIII.6). The peaks will be all the sharper the larger the deviation of the ratio of the characteristic impedances ι from unity; for $\iota = 1$, the peaks disappear entirely.

The dependence of the transparency of the layer on its thickness at a fixed ultrasonic frequency is illustrated in Fig. 50, which shows the transmission of ultrasonic beams through wedge-like layers with different angles. Maximum transmission is observed in sections where the wedge thickness satisfies the condition (VIII.6). It is interesting that narrow beams passing through the wedge have very good directionality, which is apparently attributable to the nature of the distribution of the amplitudes of the oscillations on the surface of the wedge.

As follows from Eq. (VIII.5), the layer (plate) likewise becomes transparent for $d \ll \Lambda$. It should be kept in mind, however, that this is correct when the boundary conditions (VIII.1) and (VIII.2), which presume continuity of the particle velocity at both boundaries of the plate, are satisfied. In this case the acoustic transparency of the thin plate is determined by the fact that the faces of the plate oscillate almost with the same phase ($d \ll \Lambda$), emitting an ultrasonic wave along the positive x-axis, like a plate with resonant thickness. If, however, the plate is rigidly fixed to a mounting, then the boundary conditions used are not satisfied for the plate. For example, a perfectly rigid, immobile, fixed plate will give 100% reflection regardless of how thin it is. However, even an unclamped thin plate can give considerable reflection if the velocity of sound in the plate material and the density of the material differ considerably from the corresponding values for the surrounding medium. Indeed, let $c/c_1 \gg 1$ (steel plate in air). Then Eq. (VIII.5) assumes the form

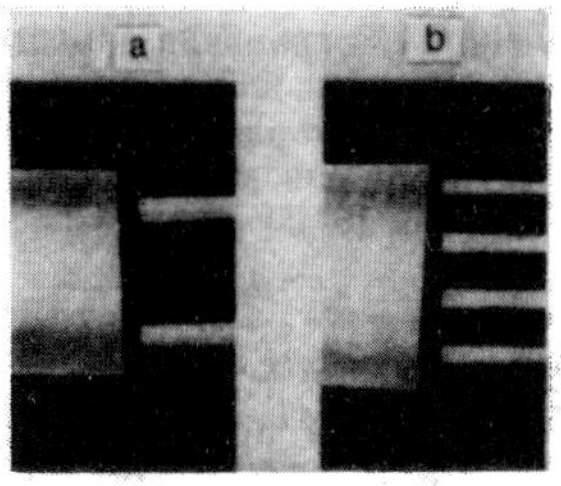

Fig. 50.

$$\rho_l \approx \frac{[\pi \rho d/(\rho_1 \Lambda)]^2}{1 + [\pi \rho d/(\rho_1 \Lambda)]^2},$$

whence it is evident that for small d/Λ the reflection coefficient is determined by the density ratio $|\rho/\rho_1|$, i.e., by the inertial properties of the layer, which can give almost total reflection in this case also. With these stipulations the curves in Fig. 49 can be interpolated down to zero values.

§2. "Antireflection" (impedance-matched) layers

The possibility of "acoustic matching" of two media with different acoustic impedances, in the sense that the coefficient of reflection from the boundaries of these media is close to zero for different ultrasonic frequencies, is very important for various applications in applied ultrasonics. We shall analyze on this level an intermediate layer of thickness d with acoustic impedance z, placed between media with acoustic impedances z_1 and z_2. In other words, we shall study the transmission of plane ultrasonic waves through two boundaries separating three media with different acoustic impedances. We shall limit the analysis to normal incidence ($\theta_1 = 0$), which is also useful for solids. The problem is solved in exactly the same manner as before, so that we shall only present the final results for the transmission coefficient, which has the following form[64]:

$$d_l = \frac{4z_2/z_1}{(z_2/z_1 + 1)^2 - (z_2^2/z^2 - 1)(z^2/z_1^2 - 1)\sin^2 kd}. \quad \text{(VIII.10)}$$

If $kd \ll 1$, i.e., $d \ll \Lambda$, as well as for $kd = n\pi$ when $\sin kd = 0$, this equation assumes the form $d_l \simeq (4z_2/z_1)/(z_2/z_1 + 1)^2 = 4z_1z_2/(z_1 + z_2)^2$, which agrees with the expression (VII.15) obtained for the case of the transmission of plane waves through the boundary between two media, so that the transmission coefficient under these conditions does not depend on the properties of the intermediate layer and this layer has no effect. If $\sin kd = 1$, i.e., $d = (2n + 1)\Lambda/4$, then it follows from Eq. (VIII.10) that $d_l = 4z_1z_2/[z^2(1 + z_1z_2/z^2)^2])$. From here it is evident that for

$$z = \sqrt{z_1z_2} \quad \text{(VIII.11)}$$

the transmission coefficient equals 1.

An analogous principle is used in optics to develop antireflection coatings. In ultrasonics, the use of an intermediate quarter-wave layer made of a material satisfying the conditions (VIII.11) greatly improves the transmission of ultrasonic energy from one medium into another. In particular, it greatly increases the efficiency of radiation of ultrasound by a solid transducer into a liquid. The best practical results are obtained with a combination of quarter-wave layers.[65,66]

Unfortunately, layers of this kind are frequency-tuned, and other methods must be found in order to prepare wide-band impedance matched layers, which would give a transmission coefficient close to unity in a more or less wide frequency band. One such method is the creation and utilization of transitional layers whose acoustic properties vary over the thickness of the layer. Theoretically, the problem of finding the corresponding conditions reduces to determining how and which acoustic characteristics must vary along the thickness of the layer in order that its reflection coefficient vanish in a given frequency band. Based on the results obtained in preceding sections, one can imagine that the reflection coefficient of a layer with a gradient of the acoustic impedance, which varies monotonically over the thickness and at the boundaries of the layer equals the characteristic impedance of the adjoining media, must equal zero. However, these results follow from the wave equation (III.4), which was obtained for media with constant acoustic properties, and for an inhomogeneous layer the previous scheme is no longer applicable. In this case it is necessary to use the equation of propagation of acoustic waves in an inhomogeneous medium and to solve it with appropriate boundary conditions. This is not a simple problem, but it is of practical importance in modern ultrasonics. It is therefore worthwhile to devote some attention to it.

To obtain the equation sought for a continuously inhomogeneous medium, we shall use the continuity equation in the general form (II.10): $d\rho/dt + \rho \operatorname{div} \mathbf{v} = 0$. For the total derivative $d\rho/dt$ we can write:

$$\frac{d\rho}{dt} = \frac{\partial\rho}{\partial p}\frac{dp}{dt} = \frac{1}{c^2}\frac{dp}{dt} = \frac{1}{c^2}\left(\frac{\partial p}{\partial t} + \mathbf{v}\operatorname{grad} p\right),$$

thereby reducing the continuity equation to the form

$$\partial p/\partial t + \mathbf{v}\operatorname{grad} p + \rho c^2 \operatorname{div} \mathbf{v} = 0. \qquad \text{(VIII.12)}$$

We shall set, as before, $\rho = \rho_0 + \Delta\rho$, and we shall assume that the equilibrium density ρ_0 is a function of the coordinates. Linearizing Eq.

(VIII.12), i.e., neglecting second-order infinitesimals, we obtain:

$$\partial p/\partial t + \rho_0 c_0^2 \operatorname{div} \mathbf{v} = 0. \qquad \text{(VIII.13)}$$

The equation of motion can be used directly in linearized form (II.5), i.e.,

$$-\operatorname{grad} p = \rho_0 \partial v/\Delta t, \qquad \text{(VIII.14)}$$

because neither the acoustic pressure p nor the particle velocity $\mathbf{v}$ depend on the inhomogeneity of the medium. Eliminating the velocity $\mathbf{v}$ from Eqs. (VIII.13) and (VIII.14) and dropping once again the zero indices, we have $\operatorname{div}(\operatorname{grad} p/\rho) - 1/(\rho c^2)\cdot\partial^2 p/\partial t^2 = 0$. For a monochromatic wave, $\partial/\partial t = -i\omega$, so that $\operatorname{div}[(1/\rho)\operatorname{grad} p] + k^2 p/\rho = 0$ or

$$\Delta p + k^2 p - (1/\rho)\operatorname{grad}\rho\cdot\operatorname{grad} p = 0, \qquad \text{(VIII.15)}$$

where $k = \omega/c$ is the wave number. Equation (VIII.15) can be put into the form of a wave equation. For this, we introduce a new function Ψ defined as $\Psi \equiv p\sqrt{\rho}$. Then, after simple transformations and confining our attention to the one-dimensional problem, based on (VIII.15) we obtain an equation which has the same form as the wave equation (III.4):

$$\partial^2\Psi(x)/\partial x^2 + k'^2\Psi(x) = 0, \qquad \text{(VIII.16)}$$

where

$$k'^2 = k'^2(x) = k^2 + \frac{1}{2\rho}\frac{\partial^2\rho}{\partial x^2} - \frac{3}{4}\left(\frac{1}{\rho}\frac{\partial\rho}{\partial x}\right)^2. \qquad \text{(VIII.17)}$$

Here all acoustic quantities — the density ρ, the wave number k, the velocity of sound c, i.e., the modulus of elasticity ρc^2 also, and the acoustic impedance of the medium ρc — are assumed to be functions of the coordinate x. If they are constant, then $k'^2(x) = k^2 = \text{const}$ and Eq. (VIII.16) transforms into the standard wave equation (III.4).

It is difficult to solve Eq. (VIII.16) in general form. However, several solutions of this equation are known for the simplest forms of the function $k'^2(x)$.[64] In particular, the problem of reflection from an inhomogeneous layer was solved by Rayleigh[1] for the case when the velocity of sound varies monotonically over the thickness of the layer from a value c_1, equal to the velocity of sound in one of the media adjoining to the layer, to the value c_2 equal to the velocity of sound in the second adjoining medium. If the layer has a thickness d and the coordinates of its boundaries are $x_1 = -d/2$ and $x_2 = +d/2$, then for the function $k'^2(x)$ this corresponds to the conditions:

$$k'^2(x) = \begin{cases} k_1 = \omega/c_1 = \text{const} & \text{for } x \leqslant -d/2, \\ k_0'^2(x) & \text{for } -d/2 < x < d/2, \\ k_2 = \omega/c_2 = \text{const} & \text{for } x \geqslant d/2, \end{cases}$$

where k_1 and k_2 are the wave numbers for the homogeneous media. We introduce the relative refractive index defined as follows*: $n_a(x) = k_0'(x)/k_1$. Then, for the first boundary $n_a(-d/2) \equiv n_1 = 1$, and for the second boundary $n_a(d/2) \equiv n_2 = k_2/k_1 = c_1/c_2$, while inside the layer $n_a(x)$ varies according to the law: $n_a(x) = M[M + (x/d + 1/2)]^{-1}$, where $M \equiv n_2(1 - n_2)^{-1}$. Therefore, the function $k_0'^2(x)$ for the layer will have the following form: $k_0'^2(x) = n_a^2(x)k_1^2 = k_1^2M^2[M + (x/d + 1/2)]^{-2}$. This problem was solved by Rayleigh for a medium with a constant modulus of elasticity, for which $\rho c^2 = \text{const}$, i.e., $c \sim (\rho c)^{-1}$. If n_a is interpreted as the ratio of the specific acoustic impedances rather than the ratio of the velocities of sound, then the results obtained by Rayleigh are generalized to the case of arbitrary media. The energy reflection coefficient in this case equals

$$\rho_1 = \frac{\sin^2(\mu_0 \ln n_2)}{4\mu_0^2 + \sin^2(\mu_0 \ln n_2)} \qquad \text{for } |k_1 d \cdot M| > {}^1/2$$

and

$$\rho_l = \frac{\sinh^2(i\mu_0 \ln n_2)}{-4\mu_0^2 + \sinh^2(i\mu_0 \ln n_2)} \qquad \text{for } |k_1 d \cdot M| < {}^1/2,$$

where $\mu_0 \equiv \sqrt{[(k_1 d \cdot M)^2 - 1/4]}$. Since the parameter $k_1 d \cdot M$ is

*In acoustics, according to condition (VII.37), the refractive index at the boundary between two media is defined as the ratio of the velocities in these media, i.e., the inverse ratio of the wave numbers.

proportional to the thickness of the layer d, the condition $|k_1 d \cdot M| > 1/2$ can be satisfied for any k_1 and k_2. If, in this case, $|k_1 d \cdot M| \gg 1/2$, then the equation for the reflection coefficient assumes the simpler form:

$$\rho_I = \frac{\sin^2(|k_1 d \cdot M| \ln n_2)}{4(k_1 d \cdot M)^2}. \tag{VIII.18}$$

Thus, even in the case of an inhomogeneous layer, we again obtain oscillations of the reflection coefficient as a function of the thickness of the layer d or frequency of sound $\omega = k_1 c_1$: for $|k_1 d \cdot M| \ln n_2 = m\pi$, where $m = 0, 1, 2, \ldots$, the reflection coefficient vanishes, and it in-

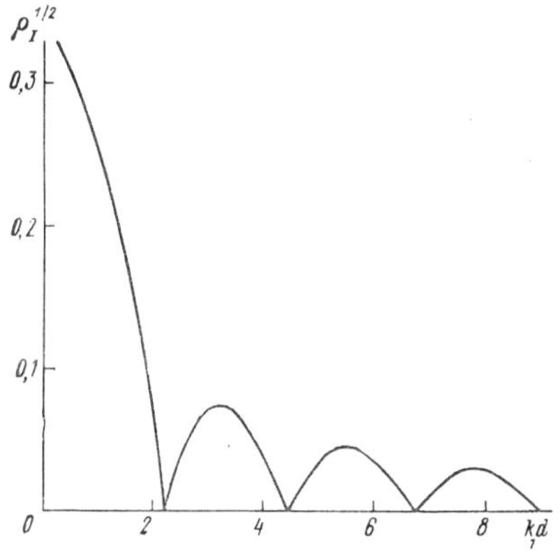

Fig. 51.

creases intermediate points. In contrast to a homogeneous layer, however, the presence of the same parameter $k_1 d \cdot M$ in the denominator in Eq. (VIII.18) leads to damping of the amplitude of these oscillations with increasing layer thickness d, as a result of which the reflection coefficient of such a layer can always be made less than some fixed magnitude in a fixed frequency band.

Let us consider a numerical example. Let a plane ultrasonic wave with frequency ω propagate out of a solid body through an inhomogeneous layer with thickness d into a liquid. Let us assume that the specific characteristic impedances of the media external to the layer differ by a factor of two, i.e., $n_2 = 2$, and the velocity of sound in the solid is $c_1 = 5 \cdot 10^3\ m/s$. Then $M = -2$; $\ln n_2 = \ln 2 \approx 0.7$; and,

$$\rho_l = \frac{\sin^2(0.7\mu_0)}{\sin^2(0.7\mu_0) + 4\mu_0}; \quad \mu_0 = \sqrt{4(k_1 d)^2 - 1/4}. \quad \text{(VIII.19)}$$

The graph of the function (VIII.19) for a fixed value of μ_0 and $n_2 = 2$ is shown in Fig. 51. It is evident from the graph that a limit on the ultrasonic frequency $\omega = k_1 c_1$ for fixed $\rho_{l\max}$ exists only at low frequencies (values of k_1). For $(k_1 d)^2 >> 1/4$, we can set $\mu_0 \approx 2k_1 d$, which gives $\rho_l \approx [\sin^2(1.5k_1 d)]/(4k_1 d)^2$. Assume that the maximum magnitude of the reflection coefficient in a given frequency band must not exceed, say, 1%, i.e., $\rho_l \leqslant 0.01$, which gives $dk_1 > 25$. At an ultrasonic frequency of 1 MHz, this corresponds to a layer with a thickness $d > 0.4\Lambda_1$, where Λ_1 is the wavelength of the incident ultrasonic wave, i.e., for a velocity of sound $c_1 = 5 \cdot 10^3\ m/s$, $d > 2\ mm$, and for a thicker layer the reflection coefficient will be even smaller. Therefore, the layer will be practically "transparent" at all frequencies above 1 MHz.

Thus, from the theoretical viewpoint, the problem is completely solvable. As far as the practical realization is concerned, however, one possibility is to develop special glasses in which the impurity concentration varies with thickness. This possibility is based on the fact that for some glasses the velocity of sound and specific acoustic impedance depend strongly on the impurity concentration.[67] As an example Fig. 52 shows the concentration dependences of the specific acoustic impedances for for longitudinal ultrasonic waves in glassy boron anhydride and silicate glasses doped with lead oxide.[68] By creating a gradient of the concentration of such impurities over the thickness of a glass plate

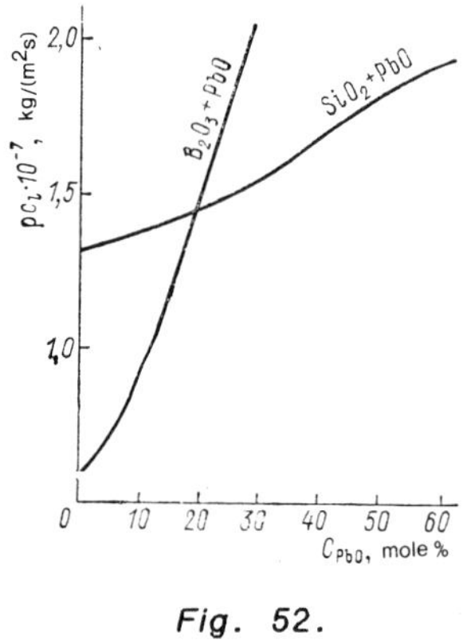

Fig. 52.

the acoustic transparency of the plate can be increased considerably over a wide frequency band. Another method for obtaining broad-band transparency is based on the same principle, employing composite materials in which the average acoustic impedance varies over the thickness.

§ 3. Characteristic acoustic oscillations of plates

We shall now study the conditions under which ultrasonic waves can propagate in homogeneous layers from the viewpoint of the possible frequencies and the structure of the field. For this, the wave equation (III.4) must be solved with the appropriate boundary conditions. For greater clarity and convenience, we shall write the wave equation for the harmonic **displacements** ξ along the x-axis, perpendicular to the boundaries of a flat layer with thickness d. In this case, for the amplitudes of the displacements we have

$$\partial^2 \xi_{max}(x) / \partial x^2 + k^2 \xi_{max}(x) = 0, \qquad \text{(VIII.20)}$$

where $k = \omega / c$ is the wave number and c is the velocity of sound in the layer. Since both forward and backward waves are present, we write the solution of Eq. (VIII.20) in the general form (see § III.1)

$$\xi_{max} = A \sin kx + B \cos kx. \qquad \text{(VIII.21)}$$

The complete solution of the problem will be the time-dependent displacement, which we shall obtain by multiplying the displacement amplitude ξ_{max} by the time-dependent factor $\sin \omega t$: $\xi(x,t) = (A \sin kx + B \cos kx) \sin \omega t$. The coefficients A and B in these solutions are found from the boundary conditions.

We shall first examine two limiting conditions: free boundaries and a rigidly clamped plate. We shall consider the intermediate cases by introducing the method of electroacoustical analogies.

Layer with immobile boundaries. This model corresponds, for example, to a layer of gas between two parallel solid walls. The boundary conditions, in this case, will be the absence of displacements at the boundaries of the layer, i.e., $\xi_{max} = 0$ at $x = 0$ and $x = d$. The first boundary condition gives $A \sin k \cdot 0 + B \cos k \cdot 0 = 0$, i.e., $B = 0$, and the second condition gives $A \sin kd = 0$, which means that $kd = n\pi$, $n = 0, 1, 2, 3, \ldots$, whence

$$k_n = n\pi/d. \tag{VIII.22}$$

From the mathematical viewpoint, the condition (VIII.22) defines the eigenvalues of the problem. Substituting the condition (VIII.22) into the solution (VIII.21), we obtain the eigenfunctions in the form

$$\xi_{n\,\max} = A_n \sin k_n x = A_n \sin(n\pi x/d), \tag{VIII.23}$$

which are the particular solutions of Eq. (VIII.20) and describe the characteristic oscillations of the layer. The general solution including the time factor will have the form $\xi(x,t) = \xi_{\max} \sin \omega t = \sum_n A_n \sin(n\pi x/d) \sin \omega_n t$. This solution represents a superposition of standing waves, i.e., harmonics with the frequencies

$$\omega_n = k_n c = n\pi c/d. \tag{VIII.24}$$

The value $n = 0$ corresponds to the absence of oscillations. The frequency of the fundamental tone (first harmonic) is

$$\omega_1 = \pi c/d \quad \text{or} \quad \nu_1 = \omega/2\pi = c/(2d), \tag{VIII.25}$$

and the wavelength $\Lambda_1 = c/\nu_1 = 2d$ (the thickness of the layer equals one-half the wavelength of the traveling wave), i.e., a displacement standing wave is formed with nodes at the boundaries and antinodes at the center.

The frequency of the second tone (first overtone or second harmonic) is $\omega_2 = 2\pi c/d$; $\nu_2 = c/d = 2\nu_1$; $\Lambda_2 = c/\nu_2 = d = \Lambda_1/2$; etc.

The entire set of characteristic oscillations of the layer (VIII.23), determined by the condition (VIII.22), which, naturally, is the same condition as the one we obtained previously for the transparency of a layer to external plane waves (VIII.6), can in this case be excited in the layer.

Layer with free boundaries. This model corresponds, for example, to oscillations of a solid plate in a gas. The boundary condition in this case will be the absence of stress on the boundaries, i.e.,

$$\left.\frac{\partial \xi_{max}}{\partial x}\right|_{x=0} = 0, \quad \left.\frac{\partial \xi_{max}}{\partial x}\right|_{x=d} = 0.$$

In this variant, the problem essentially reduces to the preceding problem. Indeed, differentiating Eq. (VIII.20) with respect to x, $(\partial^2/\partial x^2)(\partial \xi_{max}/\partial x) + k^2 (\partial \xi_{max}/\partial x) = 0$, and introducing the new variable $y \equiv \partial \xi_{max}/\partial x$, we obtain the previous boundary conditions and the previous solution $y = \partial \xi_{max}/\partial x = A_n' \sin k_n x$ or the solution for ξ_{max}

$$\xi_{n\,max} = A_n \cos k_n x, \tag{VIII.26}$$

which differs from the preceding solution only in phase, while the wave numbers k_n and characteristic frequencies ω_n and ν_n, determined by Eqs. (VIII.22) and (VIII.24), respectively, remain the same. We find the position of the displacement nodes in this case by assuming that

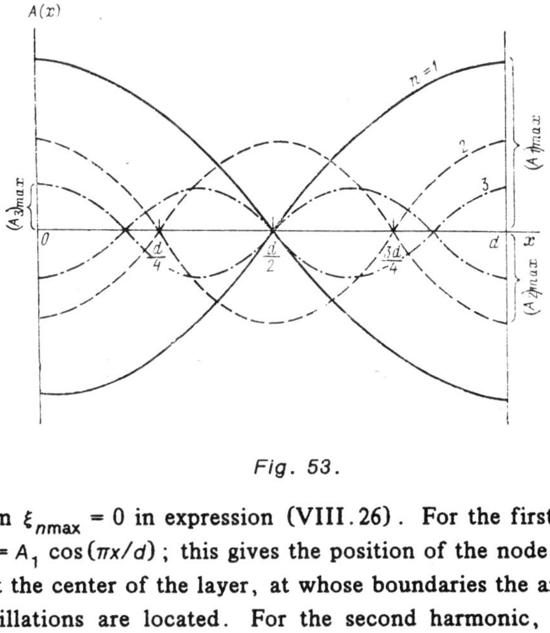

Fig. 53.

for them $\xi_{nmax} = 0$ in expression (VIII.26). For the first harmonic, $\xi_{1max} = A_1 \cos(\pi x/d)$; this gives the position of the node at $x = d/2$, i.e., at the center of the layer, at whose boundaries the antinodes of the oscillations are located. For the second harmonic, $\xi_{2max} = A_2 \cos(2\pi x/d) = 0$, for which for the coordinates of the nodes we have: $x_1 = d/4$ and $x_2 = 3d/4$. For the third harmonic we find, analogously,

$x_1 = d/6$, $x_2 = 3d/6$, $x_3 = 5d/6$, etc. Figure 53 shows the distribution of the displacement amplitude A_n for the first three harmonics of the characteristic oscillations of a free plate (for fixed boundaries, the entire pattern is displaced by $d/2$). It is evident from the figure that clamping the plate in the plane $x = d/2$ has no effect on its characteristic oscillations for the odd harmonics, because for these harmonics the displacements have a node in this plane. Therefore, a plate with thickness $d' = d/2$ and one fixed boundary can oscillate with the frequency

$$\omega_0 = \pi c/d = \pi c/(2d') \qquad \text{(VIII.27)}$$

and at all **odd** overtones. The frequency ω_0 is one-half the fundamental frequency with which a plate with thickness d' can oscillate with both boundaries remaining free. Therefore, for any thickness d, Eq. (VIII.27) determines the frequency of the "subharmonic" ω_0 with which this plate can still undergo characteristic oscillations with one of its boundaries fixed. The remaining characteristic frequencies of such a plate will be odd-number multiples of the fundamental frequency. This corresponds to the condition that an odd number of quarter-wavelength traveling waves must "fit into" the thickness of the plate.

§ 4. Method of electroacoustical analogies

The method of electroacoustical analogies is based on the fact that the characteristics of an acoustic oscillatory system can be compared to well-defined "equivalent" parameters of an oscillatory electric circuit, and the well-known equations and results from electrodynamics can be used to solve problems in ultrasonics.[69,70] This method greatly simplifies, for example, the analysis of free and forced acoustic oscillations of a layer (plate) when it radiates ultrasound into an adjacent medium with a finite characteristic impedance. Because, however, electroacoustical transducers, in which electric energy is converted directly into acoustic energy and vice versa (for example, based on the direct and reverse piezoelectric effect), are primarily used for radiation and reception of ultrasound, the method of electroacoustical analogies is generally widely and successfully used in ultrasonics to develop such transducers, and for this reason it is worthwhile to become acquainted with this method.

We have already encountered electroacoustical analogies in Chap. III when we introduced the concept of the characteristic impedance of a medium. The term "impedance" in the physical sense actually means the cause-to-effect ratio of some phenomenon. In electrodynamics the cause of the motion of charge along a conductor is the potential difference (voltage) and the effect is the current. The ratio of the voltage U to the current strength I is the impedance of the corresponding section of the circuit $R_e = U/I$. In acoustics, the cause of the oscillatory motion of particles in the medium is the variable pressure p and the effect is the particle velocity v. The ratio of these quantities in a plane wave is called the specific acoustic impedance of the medium $z = \rho c$, while the total acoustic impedance is $z = \rho c S = F_p/v$, where F_p is the pressure force acting on the area S. Thus in acoustics the analog of the electric voltage is the pressure force and the analog of the current is the particle velocity. The same ratio in mechanics – in the form of the ratio of the force of friction to the velocity of a body in a viscous medium – determines the coefficient of friction or the resistance to motion $r = F_f/|v|$. We note that both the electric impedance and the characteristic acoustic impedance can in general be complex. Here, in any case, the real (active, "ohmic") part of the impedance R_e determines the loss of power to Joule heating $D_e = I^2_{eff} R_e$, while the active acoustic impedance of the medium (radiation resistance) determines the loss of acoustic power radiated into this medium $D_a = v^2_{eff} \rho c S$. In all cases, as already noted in Chap. III, the active impedance determines the power irreversibly lost by the source.

Thus these circumstances already reveal the analogy between electric and acoustic systems and enable generalization to oscillatory systems. Moreover, they can be extended to the case of an arbitrary oscillatory system, including a mechanical system, and we can talk about electromechanical-acoustical analogies. We shall use the expression *electroacoustical* or *electromechanical analogies* for all three oscillatory systems: acoustic, mechanical, and electric. The acoustic system will consist of an oscillating plate (although, in general, this can be any system characterized by characteristic oscillations), the mechanical system will consist of a mass on a spring, and the electrical system will consist of an oscillatory circuit. The last two systems, as an ideal, can be represented as systems with *lumped parameters*, i.e., each characteristic of the system is concentrated in its element, for example, stiffness (elasticity) in a spring, mass in a material point, capacitance in a capacitor, etc. The acoustic oscillatory system, however, is a system with *distributed parameters*. In this case we cannot assign, for example, the mass to one element and the stiffness to another; these charcteristics are distributed over the volume of the system. Any oscillatory system, however, is characterized by a set of normal modes. In a system of N material points there are $3N$ normal modes; for example, in a crystal, N equals the total number of atoms (sites) in the lattice. One normal mode corresponds to one material point. We shall associate this normal mode with one of the normal oscillations of the plate at one of its characteristic frequencies, for example, the fundamental frequency.

We shall now study the analogies between these systems, starting with simple cases and progressing to more complicated cases.

§ 5. Oscillatory systems without damping

Consider a mechanical oscillatory system consisting of a material point with mass m_0 suspended on a spring with stiffness K from a stationary wall. Let us compare its oscillations with the oscillations of the free boundary of a plate at the fundamental frequency. To make the analogy especially clear, we shall examine a quarter-wave plate with thickness d', one of whose boundaries is fixed to the same stationary wall (Fig. 54). The oscillations of the free boundary of the plate will be sinusoidal in time with some amplitude which we shall denote by the letter A: $\xi(t) = A \sin \omega_0 t$, where ω_0 is the fundamental frequency of characteristic oscillations of such a plate, defined by Eq. (VIII.27), i.e., $\omega_0 = \pi c/(2d')$. Here c is the velocity of sound in the plate material.

The oscillations of the mechanical system (Fig. 54a) are determined by Newton's law:

$$m_0 d^2\xi/dt^2 = -K\xi \quad \text{or} \quad d^2\xi/dt^2 + (K/m_0)\xi = 0. \quad \text{(VIII.28)}$$

The solution of this equation is $\xi(t) = A \sin \omega_0 t$, i.e., a harmonic oscillation with frequency

$$\omega_0 = \sqrt{K/m_0}. \qquad \text{(VIII.29)}$$

Thus the oscillations of the faces of the plate and of the mechanical system will be completely identical if the mass m_0 and the stiffness K are assigned well-defined equivalent values. The masses, naturally, can be simply equated, expressing the mass of the plate m' in terms of its density ρ: $m_0 = m' = Sd'\rho$. We then find the equivalent stiffness K by equating the frequencies (VIII.27) and (VIII.29): $K = m'\pi^2c^2/(4d'^2)$.

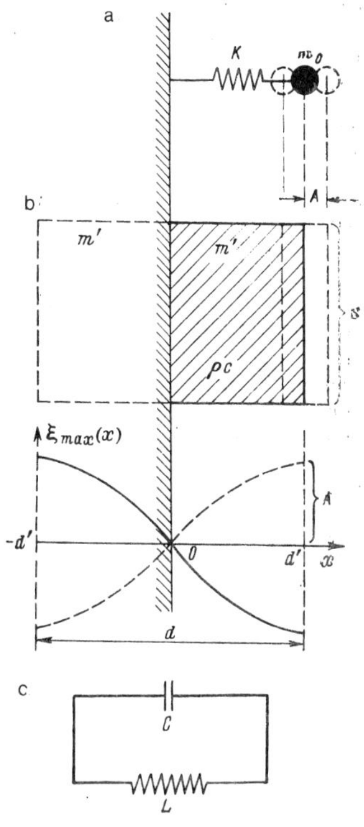

Fig. 54.

We shall be interested primarily in symmetrical oscillations of the plate. This case is completely equivalent to the preceding case with the exception that we must replace d' by $d/2$ and m' by $m/2$, where d and m are the thickness and mass of the symmetrical plate. Then the frequency remains the same, but we shall keep the zero index to denote any frequency of characteristic oscillations. Thus,

$$\omega_0 = \pi c/d, \qquad \text{(VIII.30)}$$

and for the equivalent masses and stiffnesses we shall have:

$$m_0 = m/2 = \rho dS/2, \qquad \text{(VIII.31)}$$

$$K = m\pi^2c^2/(2d^2) \qquad \text{(VIII.32)}$$

or

$$K = S\pi^2 E/(2d), \qquad \text{(VIII.33)}$$

where $E = \rho c^2$ is the effective modulus of elasticity for ultrasonic waves of this type. For definiteness, we shall consider longitudinal waves, regardless of the material of the plate.

Thus the oscillations of a mechanical system whose mass and stiffness are determined by Eqs. (VIII.31)–(VIII.33) will be completely identical to the oscillations of the acoustic system (Fig. 54b) at its fundamental frequency (VIII.30). For the harmonics, on the other hand, leaving the equivalent mass unchanged, we obtain the equivalent stiffness $K = n^2 S\pi^2 E/(2d)$, i.e., the equivalent stiffness for harmonic frequencies increases as the square of the number of the harmonic.

We shall now examine an electric oscillatory system (Fig. 54c) consisting of a circuit with a capacitor with capacitance C and a coil with inductance L connected in series. Ohm's law for such a circuit (the sum of the voltage drops equals zero: there is no emf)

$$L dI/dt + q/C = 0 \qquad \text{(VIII.34)}$$

gives a differential equation for the charge $q(t)$: $Ld^2q/dt^2 + q/C = 0$ or for the current $I = dq/dt$: $Ld^2I/dt^2 + I/C = 0$. The solutions of these equations are likewise sinusoidal functions of time $q = q_{max} \sin\omega_0 t$ and $I = I_{max} \sin\omega_0 t$, which describe harmonic oscillations with frequency

$$\omega_0 = (LC)^{-1/2}. \qquad \text{(VIII.35)}$$

Since, on the basis of the above discussion, the current is equivalent to the particle velocity $v = d\xi/dt$, the displacement ξ will be equivalent to the variable charge q. Oscillations of the electric circuit will be equivalent to oscillations of the mechanical or acoustic system, if appropriate equivalent values are assigned to the inductance and capacitance. In a conservative mechanical oscillatory system with lumped parameters, the mass carries the kinetic energy and the potential energy is stored in the spring. The inductance L and the capacitance C fulfill analogous functions in the oscillatory circuit. Therefore, comparing Eqs. (VIII.29), (VIII.30), and (VIII.35) we obtain the equivalent inductances and capacitances:

$$L \rightarrow m_0 \rightarrow \rho d S/2, \qquad \text{(VIII.36)}$$

$$C \rightarrow 1/K \equiv k_0 \rightarrow 2d/(S\pi^3 E), \qquad \text{(VIII.37)}$$

i.e., the capacitance C is equivalent to the "elasticity" k_0 of the mechanical system.

Thus the oscillations of a mechanical or an acoustic system can be described with the help of an oscillatory circuit, if the parameters of the circuit are assigned equivalent values, defined by Eq. (VIII.36) and (VIII.37).

§ 6. Characteristic oscillations of electric, mechanical, and acoustic oscillatory systems with damping

We shall now study the oscillations of real systems in the presence of losses. The equivalent mechanical, electric, and acoustic systems for this case are illustrated in Fig. 55. A piston in a viscous medium with mechanical resistance r serves as the element in which the dissipation of energy in the mechanical system is concentrated (Fig. 55a). The friction force is given by

$$F_{fr} = -rv = -r\,d\xi/dt, \qquad \text{(VIII.38)}$$

where the minus sign indicates that the force is oriented opposite to the velocity. The ohmic resistance R_e plays the same role in an electric circuit. The voltage drop across the resistance, according to Ohm's law, equals $U_R = IR_e = R_e dq/dt$. In an acoustic system the role of the acoustic resistance is played by the radiation resistance, which for symmetric oscillations of the plate in the surrounding medium with the specific characteristic impedance $z_1 = \rho_1 c_1$ is determined by the total area of contact with this medium, i.e., twice the cross-sectional area of the plate $z = \rho_1 c_1 2S$, which corresponds to symmetric bilateral radiation of ultrasound into this medium. For unilateral radiation (from one side of the vacuum) $z = \rho_1 c_1 S$; for radiation into different media with specific characteristic impedances z_1 and z_2, $z = z_1 S$

$+ z_2S$, etc. For the time being, we shall study the case of symmetric oscillations. The "internal friction" r_0, determined by the ultrasonic absorption of the plate, must be added to the radiation resistance in an acoustic system. Thus, for an acoustic system the coefficient of resistance in Eq. (VIII.38) must be taken to be

$$r = r_0 + 2\rho_1 c_1 S \qquad \text{(VIII.39)}$$

We now add the friction force (VIII.38) to Eq. (VIII.28), which assumes the form

$$m_0 \frac{d^2\xi}{dt^2} + r\frac{d\xi}{dt} + K\xi = 0 \qquad \text{(VIII.40)}$$

or with the velocity $v = d\xi/dt$

$$m_0 \frac{d^2v}{dt^2} + r\frac{dv}{dt} + Kv = 0. \qquad \text{(VIII.41)}$$

To the expression for Ohm's law (VIII.34) we add the voltage drop across the resistance R_e, and we obtain for the charge

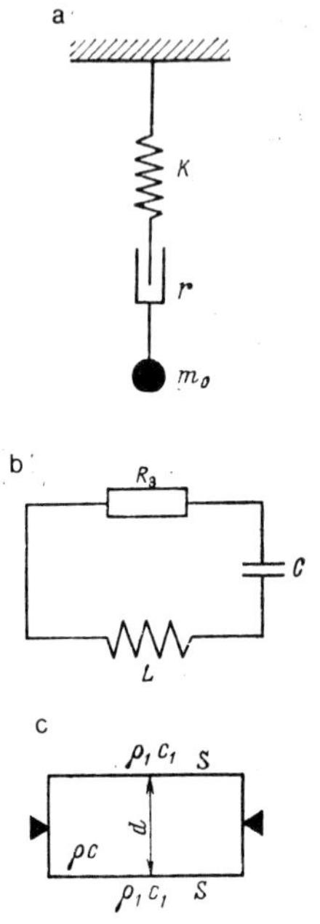

Fig. 55.

$$L\frac{d^2q}{dt^2} + R_e\frac{dq}{dt} + \frac{q}{C} = 0 \qquad \text{(VIII. 42)}$$

or for the current

$$L\frac{d^2I}{dt^2} + R_e\frac{dI}{dt} + \frac{I}{C} = 0. \qquad \text{(VIII.43)}$$

We introduce the general variable x, which can denote the displacement, velocity, charge, or current (the acceleration $a = dv/dt$, etc.). Then Eqs. (VIII.40–(VIII.43) can be represented in the unified form:

$$\frac{d^2x}{dt^2} + 2\delta_0\frac{dx}{dt} + \omega_0^2 x = 0, \qquad \text{(VIII.44)}$$

where

$$\delta_0 = r/(2m_0) \qquad \text{(VIII.45)}$$

in the case of a mechanical or an acoustic system and

$$\delta_0 = R_e/(2L) \qquad \text{(VIII.46)}$$

in the case of an electric system, while the quantity ω_0 is determined by expressions (VIII.29), (VIII.30), or (VIII.35), respectively. The well-known solution of the homogeneous differential equation (VIII.44) $x = x_{max0} \exp(-\delta_0 t) \sin \omega' t$, where $\omega' = \sqrt{(\omega_0^2 - \delta_0^2)}$, describes damped oscillations whose amplitude decreases with time exponentially

$$x_{max} = x_{max\,0} \exp(-\delta_0 t) \qquad \text{(VIII.47)}$$

with the *damping coefficient* δ_0. Strictly speaking, this process can no longer be called harmonic: it is characterized by a spectrum of frequencies, but for small δ_0 the previous terminology can be retained and we can speak about "damped harmonic" oscillations with the period $T' = 2\pi/\omega'$.

We have already introduced the time constant (see §III.5) as one of the characteristics of a damped wave process in an infinite medium. According to the definition (III.43),

$$\delta_0 = \alpha_0 c, \qquad \text{(VIII.48)}$$

where α_0 is the small-amplitude absorption coefficient and c is the velocity of sound in the medium. Using the definition (VIII.48), we can relate the "internal friction" in the plate to the previous characteristics of ultrasonic damping in it. For this, we set the radiation resistance in the external medium in Eq. (VIII.39) equal to zero and compare expressions (VIII.45) and (VIII.48). This gives

$$r_0 = \alpha_0 \rho c d S = \alpha_0 \rho c V, \qquad \text{(VIII.49)}$$

where V is the volume of the plate and ρc is the characteristic impedance of the plate material.

The radiation resistance is usually much higher than the internal losses, i.e., $2\rho_1 c_1 S \gg r_0$. Then, we can set in Eq. (VIII.39) $r = 2\rho_1 c_1 S$ (for bilateral radiation), and for the damping coefficient of the acoustic system we obtain:

$$\delta_0 = 2\rho_1 c_1/(\rho d). \qquad \text{(VIII.50)}$$

In Chap. III we also introduced the inverse characteristic: the damping time constant $\tau_0 = 1/\delta_0$, which, according to expression (VIII.47), defines the time interval $t = \tau_0$ over which the amplitude of the oscillations decreases e-fold.

The next characteristic of the damping of the oscillatory system is the logarithmic decrement v, which is defined as the logarithm of the ratio of the amplitudes of two "neighboring" oscillations separated by a time interval T': $v = \ln(x_{max1}/x_{max2}) = \delta_0 T'$. Therefore, the logarithmic damping decrement

$$v = 2\pi\delta_0/\omega', \qquad \text{(VIII.51)}$$

where $\omega' = \sqrt{(\omega_0^2 - \delta_0^2)}$. From here, incidentally, it is evident that the oscillatory nature of the process persists up to values $\delta_0 < \omega_0$ (for $\delta_0 = \omega_0$, the decrement $v \to \infty$ and $x_{max} \to 0$), i.e., if $\omega_0 t_0 > 1$ or $\tau_0 > T_0/(2\pi)$. In the opposite case, the amplitude of the oscillations decays over a time less than one period: the system, displaced from the equilibrium position, strives to return to this position or passes through it only once.

The logarithmic decrement, related to the characteristic ultrasonic absorption of the material, was already introduced above (see Chap. III): by definition, $v_0 = \alpha_0\Lambda$, where Λ is the wavelength of the traveling wave. Expression (VIII.49), which expresses the internal friction r_0 in terms of the decrement of this characteristic damping, will have the form $r_0 = \rho c S v_0/2$. In the case of bilateral radiation of ultrasound by the plate $(2\rho_1 c_1 S >> r_0)$ and the standard, for a more or less long oscillatory process, condition $\delta_0 << \omega_0$, according to the expressions (VIII.51) and (VIII.50), we obtain $v = 4\pi\rho_1 c_1/(\rho\omega_0 d)$ or, substituting (VIII.30),

$$v = 4\rho_1 c_1/(\rho c). \qquad \text{(VIII.52)}$$

Thus the damping decrement of a "loaded" plate is entirely determined by the ratio of the specific acoustic impedance of the external medium to that of the plate material. We should not forget, however, that our simplified scheme applies to the case when there is an antinode of displacements and velocities at the boundaries of the plate and this, according to the analysis performed in § VII.2, corresponds to the condition $\rho c > \rho_1 c_1$, which corresponds, for example, to oscillations of a solid plate in a liquid or gas. In the opposite case, the equivalent parameters of the system will be different, because the system itself is different.

One of the most important characteristics of oscillatory systems is their *Q-factor*. There are various definitions of this factor. By analogy with electric circuits, we shall define the Q-factor as the ratio of the reactive impedance of the oscillatory circuit to the active impedance, i.e.,

$$Q_e = \omega_0 L/R_e = 1/(\omega_0 C R_e),$$

where $\omega_0 = (LC)^{-1/2}$ is the resonance frequency of the circuit. Since, according to the condition (VIII.46), $\delta_0 = R_e/(2L)$, for any oscillatory system

$$Q \simeq \omega_0/(2\delta_0) = \pi/\upsilon, \qquad \text{(VIII.53)}$$

i.e., the Q-factor is inversely proportional to the damping decrement. This approximate equality indicates that, as usual, we assume $\delta_0 \ll \omega_0$ and $\omega_0 \simeq \omega'$. Since $\delta_0 = 1/\tau_0$, from Eq. (VIII.53) we also obtain $Q = \omega_0\tau_0/2$, i.e., the Q-factor is proportional to the damping time constant. The minimum value of the Q-factor, corresponding to the minimum value $\omega_0\tau_0 = 1$ $(\delta_0 = \omega_0)$, is $Q_{min} = 1/2$. According to expression (VIII.53), the Q-factor is approximately determined by the number of possible free oscillations of the system before their amplitudes decrease by factor of e. The lower the losses of the energy of oscillations in the system, the higher their Q-factor is. For example, electric circuits have a Q-factor of about 50–100; for a tuning fork, a Q-factor of about 3000 is characteristic; and, the Q-factor of a quartz plate oscillating in a vacuum (i.e., without an external load $\rho_1 c_1 S$) reaches values of the order of 10^6, i.e., the quartz plate can complete this many free oscillations before their amplitudes decrease approximately by a factor of three.

We shall relate the Q-factor of an acoustic system to the characteristics of its internal losses, i.e., in the absence of ultrasonic radiation into an external medium $(\rho_1 c_1 S) = 0$. According to (VIII.53) and (VIII.43),

$$Q_a = \omega_0/(2\delta_0) = \omega_0/(2\alpha_0 c) = \pi/(\alpha_0 \Lambda).$$

Since the ultrasonic absorption coefficient α_0 usually increases with frequency as ω_0^2, the characteristic acoustic Q-factor of a plate, as a rule, decreases with frequency, i.e., it is lower at the harmonics than at the fundamental frequency. We note that in the literature a quantity that is inversely proportional to the Q-factor is sometimes used as the characteristic damping of ultrasound in the plate material: $Q_a^{-1} = 2\alpha_0 c/\omega_0$, called the *coefficient of internal friction*. This term definition differs from our definition of the internal friction r_0, expressed by Eq. (VIII.49).

If, however, radiation losses exceed internal losses, i.e., $\rho_1 c_1 S \gg r_0$,* then we obtain the following expression for the acoustic Q-factor of a "loaded" plate, using Eq. (V.53) and (VIII.50) or (VIII.52) with bilateral radiation,

$$Q_a = \frac{\pi}{} \propto \frac{\pi \rho c}{4\rho_1 c_1} \propto \frac{\rho c}{\rho_1 c_1} = \iota^{-1}, \qquad \text{(VIII.54)}$$

where $\iota \equiv z_1/z$ is the notation introduced in Eq. (VIII.8) in §VIII.1.

Thus the Q-factor of a loaded acoustic system is determined simply by the ratio of the specific characteristic impedances of this system and of the external medium into which the ultrasound is radiated. For example, the Q-factor of a quartz plate $(\rho c = 1.5 \cdot 10^6\ g/(cm \cdot s))$ for oscillations in water $(\rho_1 c_1 = 1.5 \cdot 10^5\ g/(cm \cdot s))$ is $Q_a \simeq 10$, while for oscillations in air $(\rho_1 c_1 = 4.5\ g/(cm \cdot s))$, $Q_a \simeq 3 \cdot 10^5$. We must, however, make two remarks regarding the acoustic Q-factor of real systems. First, a real plate is placed on a base, in a "holder" into which radiation is also emitted, so that the Q-factor of a clamped plate can be substantially lower. For this reason, in systems in which it is necessary to maintain a high Q-factor the plate is clamped along the nodal (central) plane (as indicated in Fig. 55c). Second, in Eq. (VIII.54) an ideal acoustic contact is presumed between the plate and the external medium, which is realized, for example, with a good contact between a solid body and a liquid in contact with it.

---

*For plates consisting of high-Q materials, such as quartz, corundum, etc., radiation losses exceed internal losses already for oscillation in air.

Experience shows, however, that when a plate consisting of a solid material has a bilateral contact, even with the same kind of material, then its Q-factor still constitutes several units. The point is that this contact is realized through transitional layers that increase the Q-factor. For this reason, obtaining a low Q-factor is another technical problem in ultrasonics associated with the extension of the transmission band (see below).

§7. Forced oscillations. Resonance

Let an external force F, varying in time sinusoidally with frequency ω, act on a mechanical system: $F = F_{max} \sin \omega t$. Then the equation of motion of a material point on the spring will have the form:

$$m_0 \frac{d^2\xi}{dt^2} + r\frac{d\xi}{dt} + K\xi = F_{max} \sin \omega t$$

or

$$\frac{d^2\xi}{dt^2} + 2\delta_0 \frac{d\xi}{dt} + \omega_0^2\xi = F'_{max} \sin \omega t, \qquad \text{(VIII.55)}$$

where $\delta_0 = r/(2m_0)$, $\omega_0^2 = K/m_0$, and $F' = F/m_0$ is the force per unit mass. The same type of equation with equivalent parameters r, K, and m_0 can also be written down for an acoustic system, expressing the force F in terms of the mechanical stress (pressure) and the area on which it acts: $F = pS$. Differentiating (VIII.5)) with respect to t, we obtain an analogous equation for the displacement velocities $v = d\xi/dt$. But, since the derivative of $\sin \omega t$ is $\omega \cos \omega t$, the quantity F'_{max} in this case will equal $\omega F_{max}/m_0$. As far as the initial phase of the force is concerned, it can always be set equal to zero, without distinguishing between the functions $\sin \omega t$ and $\cos \omega t$.

The role of the force in an electric circuit is played by the electromotive force (emf) $E(t) = E_{max} \sin \omega t$. The equivalent electric circuit for the case of forced oscillations is shown in Fig. 56. Ohm's law for it gives the equation

$$L\frac{d^2q}{dt^2} + R_e\frac{dq}{dt} + \frac{q}{C} = E_{max} \sin \omega t$$

or

$$\frac{d^2q}{dt^2} + 2\delta_0 \frac{dq}{dt} + \omega_0^2 q = E'_{max} \sin \omega t \qquad \text{(VIII.56)}$$

where $\delta_0 = R_e/(2L)$, $\omega_0 = (LC)^{-1/2}$, and $E'_{max} = E_{max}/L$. An analogous equation is obtained for the current I, only with $E'_{max} = \omega E_{max}/L$. Thus Eqs. (VIII.55) and (VIII.56) can again be combined into one equation with the arbitrary variable x:

$$\frac{d^2x}{dt^2} + 2\delta_0 \frac{dx}{dt} + \omega_0^2 x = F'_{max} \sin \omega t, \qquad \text{(VIII.57)}$$

keeping in mind, however, that the quantity F'_{max} will assume different values depending on the meaning of the parameter x: if $x \equiv q$, then $F'_{max} \equiv F_{max}/L$; if $x \equiv I$, then $F'_{max} = \omega F_{max}/L$; if $x \equiv \xi$, then $F'_{max} \equiv F_{max}/m_0$; if $x \equiv v$, then $F'_{max} \equiv \omega F_{max}/m_0$; if $x = a$ (acceleration), then $F'_{max} \equiv \omega^2 F_{max}/m_0$, etc. Of course, if, for example, the displacement amplitude $\xi_{max} = A$ is obtained from the equation for ξ, then the velocity amplitude will equal $v_{max} = \omega A$, i.e., the result will be the same.

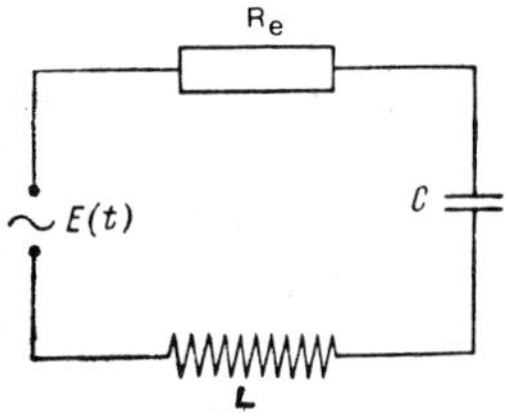

Fig. 56.

The solution of the inhomogeneous differential equation (VIII.57) is the sum of the general solution for the corresponding homogeneous equation ($F'_{max} \sin\omega t = 0$) and the particular solution of the inhomogeneous equation $x_1(t)$: $x(t) = x_1(t) + A_0 \exp(-\delta_0 t) \cdot \sin\omega' t$. The first solution describes forced oscillations and the second one describes free oscillations determined only by the initial action, and then by the parameters of the system. Free oscillations damp out sooner or later (depending on the magnitude of δ_0).

We shall now analyze only forced **stationary** oscillations, which are established in the system after a time interval $t > T_0 = \delta_0^{-1}$, when the characteristic oscillations vanish. We shall seek the solution for $x_1(t)$ in the form of a harmonic function

$$x_1(t) = x_{max} \sin(\omega t + \beta_0), \qquad \text{(VIII.58)}$$

and we shall find the unknown quantities x_{max} and β_0, i.e., the amplitude and the intial phase of the forced oscillations, by substituting Eq. (VIII.58) into Eq. (VIII.57), which yields

$$x_{max} = F'_{max} / \sqrt{4\omega^2\delta_0^2 + (\omega_0^2 - \omega^2)^2}\,, \qquad \text{(VIII.59)}$$

$$\beta_0 = \tan^{-1} 2\delta_0\omega / (\omega^2 - \omega_0^2)\,, \qquad \text{(VIII.60)}$$

Thus the solution for forced oscillations will have the following form:

$$x_1(t) = \frac{F'_{max}}{\sqrt{4\omega^2\delta_0^2 + (\omega_0^2 - \omega^2)^2}} \sin\left(\omega t + \tan^{-1} \frac{2\delta_0\omega}{\omega^2 - \omega_0^2}\right).$$

It follows from this general solution that when the frequency of the driving force (ω) is equal to the frequency of characteristic oscillations of the system (ω_0), the amplitude of forced oscillations will reach a maximum value $(x_{max})_{res} = F'_{max}/(2\omega_0\delta_0)$, corresponding to the *condition of resonance*. The resonance amplitude depends on the damping coefficient δ_0 and in the limit $\delta_0 \to 0$, $(x_{max})_{res} \to \infty$. This, of course, cannot occur, because even if the internal friction (resistance) is very small, for high displacement velocities (or currents) it will increase due to nonlinear effects. As far as the initial phase of forced oscillations is concerned, it makes sense to compare these oscillations with others, for example, displacement oscillations. At resonance ($\omega = \omega_0$), the phase difference between the driving force and the displacement (or the emf and oscillations of the charge on the capacitor) equals exactly 90° independent of the magnitude of δ_0, while as the frequency ω varies from zero to infinity, the phase difference changes by 180°. A graph of β_0 as a function of ω is shown in Fig. 57 for the idealized ($\delta_0 = 0$) and real ($\delta_0 \neq 0$) cases. In the case $\delta_0 = 0$, according to Eq. (VIII.60), at $\omega = \omega_0$ the phase jumps from zero to π. In the real case, the phase varies over the same limits in a more or less wide (depending on δ_0) frequency band, though the

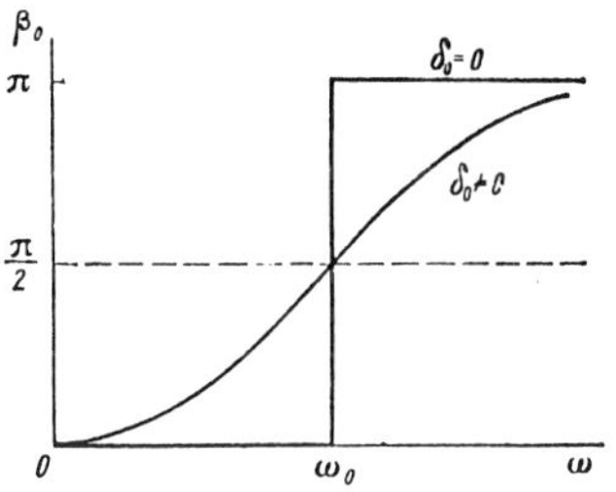

Fig. 57.

basic change occurs near the resonance frequency. The phase difference between the driving force and the displacement velocity (or the emf and the current) equals zero at resonance and the phase difference between the force and the acceleration is again $\pi/2$, etc., but in any case, it changes by π over a quite wide frequency band. All this must be kept in mind when comparing the **phase of ultrasonic oscillations with the phase of the electric voltage** exciting the ultrasonic transducer at different frequencies near the resonance frequency. In other cases, the initial phase of the forced oscillations β_0 is not important and it can be set equal to zero.

We shall now analyze the frequency dependence of the amplitude of forced oscillations for different values of the variable x.

1. Let $x \equiv q$. Then $F'_{max} = F_{max}/L$ and Eq. (VIII.59) gives

$$q_{max} = E_{max}/\{\omega\sqrt{R_e^2 + [1/(\omega C) - \omega L]^2}\}. \quad \text{(VIII.61)}$$

As $\omega \to \infty$, $q_{max} \to 0$; as $\omega \to 0$, $q_{max} \to E_{max}C$, which corresponds to a static charge q_{stat} on the capacitor. For $\omega = \omega_0$, $(q_{max})_{res} = E_{max}/(\omega_0 R_0)$, and the ratio $(q_{max})_{res}/q_{stat} = 1/(\omega_0 R_e C)$, i.e., it equals the Q-factor of the circuit.

2. Let $x = \xi$. Then $F'_{max} = F_{max}/m_0$. The result for the displacement amplitude A can be obtained by replacing in Eq. (VIII.61) all quantities by their equivalent quantities:

$$A = F_{max}/\{\omega\sqrt{[r^2 + (K/\omega - \omega m_0)^2]}\}.$$

As $\omega \to \infty$, $A \to 0$; as $\omega \to 0$, $A \to A_{stat} = F_{max}/K$ (Hooke's law). At $\omega = \omega_0$, $A_{res} = F_{max}/\omega_0 r$ and the ratio $A_{res}/A_{stat} = (K/\omega_0 r)Q_m$, i.e., for a plate loaded on two sides

$$A_{res}/A_{stat} = Q_a \simeq \rho c/(\rho_1 c_1).$$

Thus the amplitude of oscillations of a plate radiating ultrasound under resonant conditions can be easily calculated if its Q-factor and the static deformation, for example, due to the inverse piezoelectric effect, are known. The acoustic Q-factor, however, of the radiating plate is determined simply by the ratio of its specific acoustic impedance to that of the external medium. The general form of the frequency dependences of the quantitites q and A is shown in Fig. 58 for different Q-factors (for low Q-factors, the peak x_{max} is displaced somewhat to the left of the resonance frequency ω_0).

3. Let $x \equiv I$. Then $F'_{max} = F_{max}\omega/L$ and $I_{max} = E_{max}/\sqrt{\{R_e^2 + [1/(\omega c) - \omega L]^2\}}$. The expression in the denominator is the electric impedance of the circuit, which determines the current strength in it with a fixed emf. At resonance ($\omega = \omega_0$), $(I_{max})_{res} = E_{max}/R_e$, the current strength is determined only by the resistance. As $\omega \to 0$ and $\omega \to \infty$, $I_{max} \to 0$.

4. Analogously, if $x \equiv v$ ($F_{max} \equiv F_{max}\omega/m_0$), then

$$v_{max} = \omega A = F_{max}/\sqrt{r^2 + (K/\omega - \omega m_0)^2}. \quad \text{(VIII.62)}$$

The quantity $z_m = \sqrt{[r^2 + (k/\omega - \omega m_0)^2]}$, which determines the amplitude of the particle velocity for a fixed amplitude of the driving force, can be called the *mechanical impedance*. With the help of the equivalent parameters r, K, and m_0 it can be reduced to the acoustic impedance of the plate (layer) z_0, which we used above. At the resonance frequency of the plate

$$(v_{max})_{res} = \frac{F_{max}}{r} \propto \frac{F_{max}}{2\rho_1 c_1 S} = \frac{p2S}{2\rho_1 c_1 S} = \frac{p}{\rho_1 c_1};$$

the amplitude of the particle velocity with $r = 2\rho_1 c_1 S$ is determined by the specific acoustic resistance of the medium, a result which we obtained previously by another method from an analysis of the solutions of the wave equation.

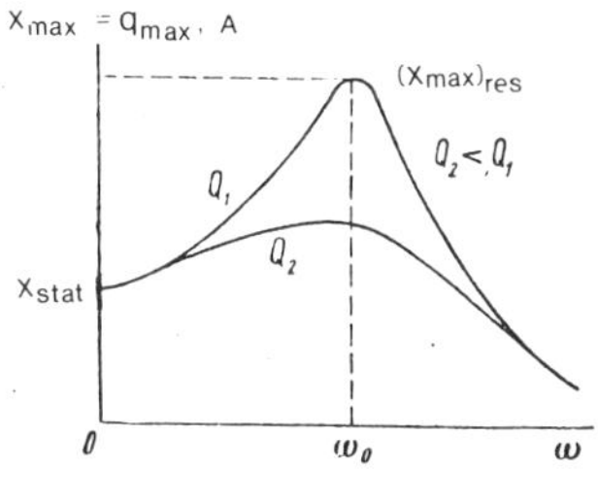

Fig. 58.

Far from the resonance frequency ($\omega \to 0$ and $\omega \to \infty$) $v_{max} \to 0$, and the graphs of the frequency dependence of the particle velocity (like the currents in an electric circuit) whose qualitative form is illustrated in Fig. 59 are obtained.

From Eq. (VIII.62) it is clear that an increase in the acoustic resistance not only decreases the resonance amplitude, but also increases the widths of the resonance curve. Using expression (VIII.62) we can, however, also find a quantitative relation between the parameters of this curve and the damping characteristics. The simplest, but very important

J

relation, which is convenient for practical applications, is obtained between the Q-factor of the system and the width of the energy curve of the frequency dependence, i.e., the resonance curve of the square of the amplitude of the particle velocity v^2_{max}. We shall construct the reduced curve for the ratio $v^2_{max}/(v^2_{max})_{res}$ (Fig. 60). Since $(v_{max})_{res} = F_{max}/r$,

$$\frac{v^2_{max}}{(v^2_{max})_{res}} = \frac{r^2}{r^2 + (K/\omega - \omega m_0)^2}. \qquad \text{(VIII.63)}$$

Using the definition of the Q-factor $Q = \omega_0 m_0/r = K/(\omega_0 r)$, Eq. (VIII.63) can be put into the form

$$\frac{v^2_{max}}{(v^2_{max})_{res}} = \frac{(\omega/\omega_0)^2}{(\omega/\omega_0)^2 + Q[(\omega/\omega_0)^2 - 1]^2}. \qquad \text{(VIII.64)}$$

Let us select a frequency ω_1 on the resonance curve at which

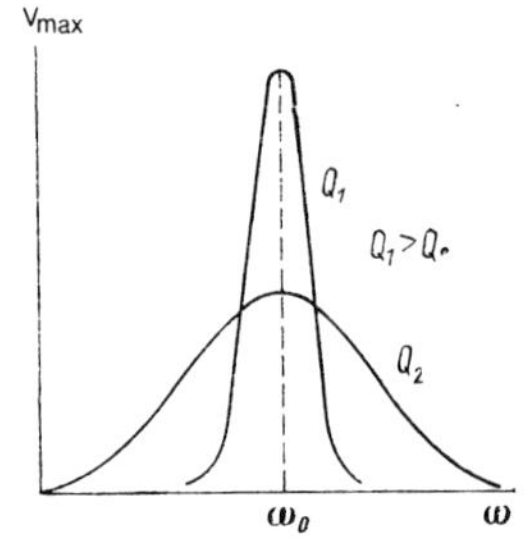

Fig. 59.

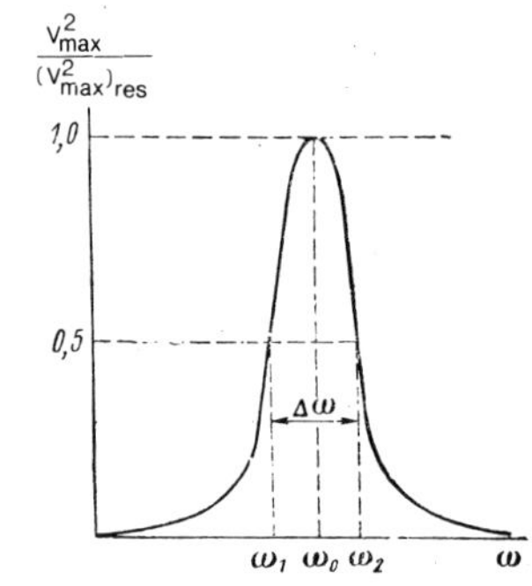

Fig. 60.

$v^2_{max}/(v^2_{max})_{res} = 1/2$. For this frequency, we have from Eq. (VIII.64)

$$Q = \left|\frac{\omega_1\omega_0}{\omega_1^2 - \omega_0^2}\right| \qquad \text{(VIII.65)}$$

(the absolute value is taken because $Q > 0$). It is evident from expression (VIII.64) that if its right side equals 1/2 (or a different constant value) at the frequency ω_1, then it also equals this value at another frequency ω_2 such that $\omega_2/\omega_0 = (\omega_1/\omega_0)^{-1}$, whence $\omega_1\omega_2 = \omega_0^2$.

Substituting this value of ω_0^2 into Eq. (VIII.65), we obtain

$$Q = \frac{\omega_0}{|\omega_1 - \omega_2|} = \frac{\omega_0}{|\Delta\omega|} = \frac{\nu_0}{|\Delta\nu|}. \qquad \text{(VIII.66)}$$

The frequency band $\Delta\omega$ (or for cyclical frequencies $\Delta\nu$), in which by definition the energy of oscillations equals one-half the energy at the resonance frequency (i.e., at the frequency ω_0), is called the *width of the resonance curve*. Thus the Q-factor of an oscillatory system equals the ratio of its characteristic frequency to the width of the energy resonance curve, whence the Q-factor (and together with it the other damping characteristics also) is easily determined experimentally from the frequency dependence of any acoustic quantity. If the ultrasonic intensity is measured (energy density, power, etc.), then the Q-factor is found directly from the curve of the frequency dependence obtained. If, on the other hand, the measured quantity is, for example, the amplitude of the pressure (particle velocity, displacement, etc.), then in order to use Eq. (VIII.66) the frequency dependence obtained for the given quantity must first be rescaled to the frequency dependence of the square of this quantity. In its turn, the Q-factor of the system determines its frequency selectivity or the transmission band, i.e., the frequency range in which the energy of forced oscillations is less than 50% of the energy at the resonance frequency. This means, for example, that a plate with a Q-factor Q_a, used as a transducer, can radiate ultrasound with intensity exceeding 50% of the maximum in the frequency band $\Delta\nu = \nu_0/Q_a$. This also means that the transmission coefficient d_l of a plane-parallel layer on which plane ultrasonic waves are incident exceeds 0.5 of the maximum in the frequency band ν_0/Q_a. Since the Q-factor of the loaded layer at its fundamental oscillation frequency is determined by the ratio of the characteristic impedance of the layer to that of the external medium $\rho c/(\rho_1 c_1)$, for the transmission band of the layer near the fundamental frequency this gives $\Delta\nu_l = \nu_0 \rho_1 c_1/(\rho c) = \nu_0 \ell$. We would also have obtained the same result, of course, for the other curves shown in Fig. 49 by analyzing Eq. (VIII.5) for different frequencies. This method, however, is more laborious.

§ 8. Radiation of plane waves. The field of a real plane ultrasonic radiator

Up to now we have been studying ideal plane waves excited by harmonic oscillations of an infinite flat surface. Real radiators of plane ultrasonic waves have finite dimensions, and this leads to an interference structure of the field in the near zone of such radiators and to diffraction of an ultrasonic beam.

Let a circular piston radiator with radius R, surrounded by an infinite screen, radiate in the direction of the positive x-axis, passing through the center of the radiator (Fig. 61). The velocity potential φ_A at an arbitrary point of observation $A(x, y, z)$, located at a distance r from the element of the surface dS of an arbitrary source with area S, can be calculated with the help of the well-known Rayleigh equation[1]:

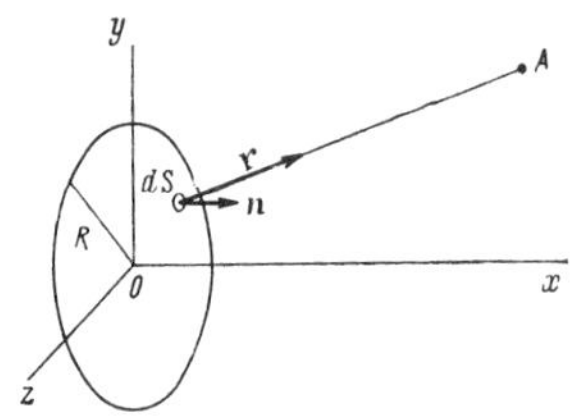

Fig. 61

$$\varphi_A = -\exp(i\omega t)/(2\pi)\int_S(\partial\varphi/\partial n)_S[\exp(-ikr)/r]\,dS$$

where **n** is the unit normal to the surface of the emitter; $\partial\varphi/\partial n$ is the distribution of the amplitudes of the particle velocities on the surface S. Assuming this distribution to be uniform, which corresponds to the boundary conditions

$$-(\partial\varphi/\partial n)_{x=0} = v_{\max 0} = \text{const for } y, z \leqslant R, \qquad \text{(VIII.67)}$$
$$\partial\varphi/\partial n = 0 \qquad \text{for } y, z > R,$$

we obtain,

$$\varphi_A = \frac{v_{\max 0}}{2\pi}\int_S \frac{\exp(-ikr)}{r}\,dS\,e^{i\omega t} \qquad \text{(VIII.68)}$$

The quantity $d\varphi = v_{\max 0}/(2\pi)\,[\exp(-ikr)/r]\,dS$ is the potential of a point source emitting into a solid angle 2π. Thus Eq. (VIII.68) indicates that the potentials $d\varphi$ at the point A from separate point sources distributed over the surface area S are summed taking into account the phase delay (the factor exp $(-ikr)$), i.e., it expresses the Huygens–Fresnel principle. According to this principle, as $S \to 0$, at any distance x from the source an ideal plane wave is formed with a uniform distribution of amplitudes. In the case of a finite area S, to which the integral (VIII.68) refers, the distribution of amplitudes and phases of oscillations in the yz plane at different distances x will be nonuniform, though it is clear from general considerations that the larger the source as compared to the wavelength of the wave which it emits, the closer the wave front will be to an ideal plane wave front.

Equation (VIII.68) with boundary conditions (VIII.67) refers to ideal "piston" oscillations of a flat source, surrounded by an infinite stationary flat screen. Real sources of ultrasound can radiate without a screen. The distribution of the amplitudes of a real source, however, is not as a rule strictly uniform for various reasons, including the nonuniformity of the electric voltage applied to the piezoelectric transducer, the nonuniformity of the mechanical properties of the transducer material and its support, resonances of parasitic transverse or bending oscillations, etc. In addition, Eq. (VIII.68) does not take into account damping of the amplitude of oscillations at a distance r from the point of observation, i.e., it applies to an ideal medium. However, even under the idealized conditions to which this equation refers,

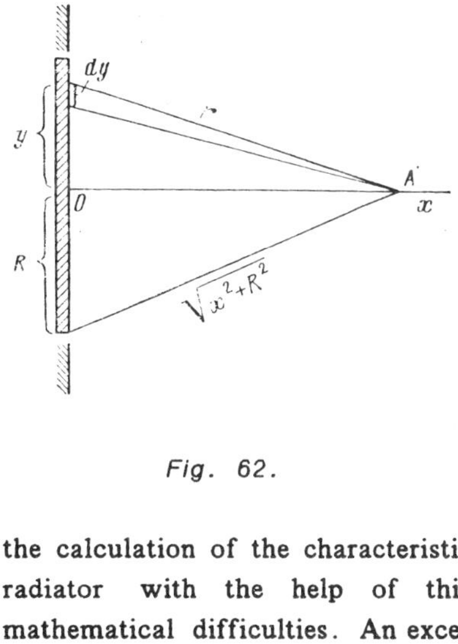

Fig. 62.

the calculation of the characteristics of the near field of the piston radiator with the help of this equation entails considerable mathematical difficulties. An exception is the problem of finding the field on the axis of a circular radiator, which shows the basic characteristics of the structure of the field of a real plane radiator. Introducing the instantaneous coordinate y and choosing a ring of radius y and width dy for the surface element of the radiator dS (Fig. 62), based on Eq. (VIII.68) we immediately obtain for the pressure on the x-axis, dropping the time-dependent factor,

$$p(x) = \rho\frac{\partial\varphi_A}{\partial t} = i\omega\rho\frac{v_{\max 0}}{2\pi}\int_0^R \frac{\exp(-ik\sqrt{x^2+y^2})}{\sqrt{x^2+y^2}}2\pi y\,dy =$$

$$= v_{\max 0}\,\rho c\,[\exp(-ikx) - \exp(-ik\sqrt{x^2+R^2}), \qquad \text{(VIII.69)}$$

where ρ and c are, respectively, the density of and the velocity of sound in the medium.

At the center of the radiator on its surface ($x = 0$) $p_{\max 0} = \text{Re}\,p(0) = 2\rho c v_{\max 0}\,|\sin(kR/2)|$, i.e., the amplitude of the pressure at the center of the radiator can vary, depending on the value of $kR = 2\pi R/\Lambda$, from zero to $2\rho c v_{\max 0}$, corresponding to twice the magnitude of the pressure in the plane wave with velocity amplitude $v_{\max 0}$. If $kR = 2\pi n$, i.e., $R = n\Lambda$ ($n = 0, 1, 2, 3, \ldots$), then $p_{\max 0} = 0$; if $kR = (2n+1)\pi$, i.e., $R = (2n+1)\Lambda/2$, then $p_{\max 0} = 2\rho c v_{\max 0}$.

For $x \neq 0$, we introduce the notation

$$\alpha \equiv \sqrt{\alpha^2 + R^2} - x = x(\sqrt{1 + R^2/x^2} - 1) \approx R^2/(2x) \quad (\text{for } x \gg R). \qquad \text{(VIII.70)}$$

Then, according to expression (VIII.69), for the amplitude of the pressure for $x \gg R$ we shall have:

$$p_{\max}(\alpha) = \text{Re}\,p(x) = \text{Re}\,v_{\max 0}\rho c\,|(1 - \cos k\alpha + i\sin k\alpha)\exp(ikx)|$$

$$= 2\rho c v_{\max 0}\,|\sin(k\alpha/2)|. \qquad \text{(VIII.71)}$$

The amplitude of the pressure $p_{\max}(x)$ along the x-axis, as follows from Eq. (VIII.71), will also reach a series of maximum values equal to $2\rho c v_{\max 0}$ if

$$k\alpha = (2n+1)\pi \quad \text{or} \quad \alpha = (2n+1)\Lambda/2, \qquad \text{(VIII.72)}$$

and minimum values equal to zero if

$$k\alpha = 2n\pi \quad \text{or} \quad \alpha = n\Lambda. \qquad \text{(VIII.73)}$$

In all cases the minimum magnitudes of the pressure, evidently, correspond to the fact that an even number of ring-shaped Fresnel zones fits on the surface of the radiator and they cancel at the point of observation, while the maximum magnitudes of the pressure correspond to an odd number of Fresnel zones.

From relations (VIII.72), (VIII.73), and (VIII.70), we find the position of the maxima and minima relative to the center of the radiator:

$$x_m = R[R/(m\Lambda) - (m/4)(\Lambda/R)], \qquad \text{(VIII.74)}$$

where $m = 2n + 1$ for maxima, $m = 2n$ for minima, and $n = 0, 1, 2, 3, \ldots$. The most distant maximum is obtained with $m = 1$, i.e., $n = 0$, when $x_1 = R^2/\Lambda - \Lambda/4$.

At ultrasonic frequencies the condition $R \gg \Lambda$ is almost always satisfied. In addition, the position of the last interference maximum in the field of a circular piston radiator is determined by the simple relation

$$x_1 = R^2/\Lambda. \qquad \text{(VIII.75)}$$

There are still other maxima and minima located in front of this last maximum. It is evident from Eq. (VIII.74) that positive values of x_m for the minima are obtained if $R^2/(2n\Lambda) - 2n\Lambda/4 > 0$, i.e., $n < R/\Lambda$, from which it follows that the number of minima in the interference zone equals the nearest integer less than R/Λ. The distance between them (and the maxima) gradually increases as the distance from the radiator increases, as is evident in Fig. 63, which shows the distribution of the amplitudes of the pressure on the axis of a circular piston radiator for $x_1/\Lambda = 30$ as a function of the relative distance x/Λ.

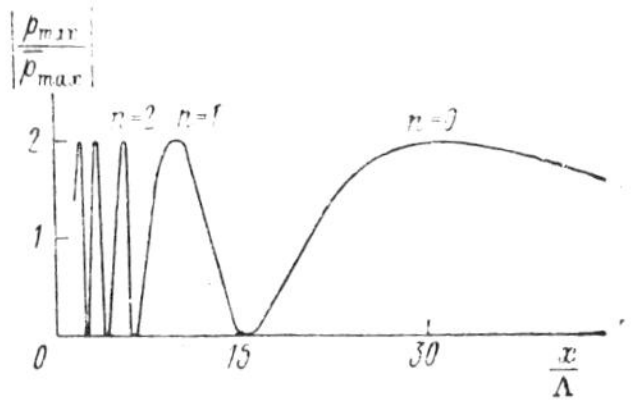

Fig. 63.

The relation (VIII.75) determines the length of the interference *near zone* of a plane radiator, which is also called the *Fresnel zone*. In this zone the ultrasonic beam is nearly cylindrical, i.e., the wave front remains nearly planar. The region of the beam for $x > x_1$ is called the *far zone* or the *Fraunhofer zone*. The calculation of the acoustic pressure in this zone likewise does not present any special difficulties, because for it we can set in the denominator of Eq. (VIII.68) r = const for any point of observation at an angle υ. Simple calculations then lead to the following expression for the pressure amplitude as a function of the angle υ [3]:

$$p_{max}(\upsilon) = \pi\rho\frac{\pi R^2 v_{max0}}{2\pi r}\left[2\frac{J_1(kR\sin\upsilon)}{kR\sin\upsilon}\right], \qquad \text{(VIII.76)}$$

where J_1 is a first-order Bessel function. The expression in square brackets in Eq. (VIII.76) has a maximum equal to 1 at $\upsilon = 0$, i.e. in the axial direction, and first vanishes when the argument $kR\sin\upsilon_0 = 3.83$, i.e., when

$$\sin\upsilon_0 = 3.83/(kR) = 0.61\Lambda/R. \qquad \text{(VIII.77)}$$

The first secondary maximum of the quantity $p_{max}(\upsilon)$, whose amplitude is only $\approx$ 13% of the main maximum, while the intensity is approximately 60 times lower than the intensity on the axis, occurs when $kR\sin\upsilon = 5.33$. The next secondary maxima will appear farther along with an even smaller amplitude. On the whole the angular pattern of the amplitude distribution will be identical to the well-known pattern of the Fraunhofer diffraction pattern for light diffracted by a circular aperture. All of the energy of the wave in the Fraunhofer zone is concentrated in a cone whose angle is determined by Eq. (VIII.77). The front of the wave in this zone is nearly spherical and its amplitude decreases along the x-axis in accordance with the law governing the propagation of spherical waves, which is studied in the next chapter. On the whole, the pattern of the ultrasonic field radiated by a circular piston radiator can be schematically illustrated as shown in Fig. 64. The length of the near zone x_1 is determined by Eq. (VIII.75), while the angle of divergence of the beam in the far zone is determined by Eq. (VIII.77). Analogous equations are also obtained for a rectangular radiator. Thus, for a square radiator with sides $2R$ the angle of divergence is determined by the relation $\sin\upsilon_0 = 0.5\Lambda/R$. In any case the length of the Fresnel zone will increase and the angle of divergence will decrease as the ratio of the

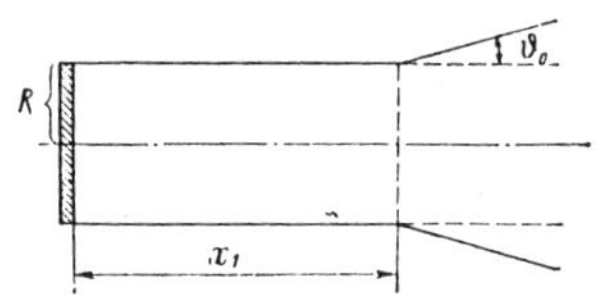

Fig. 64.

transverse dimensions of the radiator to the wavelength increases. At ultrasonic frequencies this ratio is usually not less than several tens or hundreds of units. In this case, for example at a frequency of 10 MHz, when the ultrasound is radiated by a piezoelectric plate with a diameter of 2 cm in water ($R/\Lambda \simeq 100$), the length of the near zone, according to Eq. (VIII.75), is $x_1 \simeq 200$ cm, which greatly exceeds the scales of interest in physical ultrasonics.

In the near zone, as we have seen, the structure of the ultrasonic field is characterized by a strong interference inhomogeneity, which, of course, occurs not only along the axis of the ultrasonic beam, but also in any cross section of the beam. Calculation of the integral (VIII.68) for arbitrary points in sections of the near field is a very difficult problem, which is solved with the help of complicated series that can be calculated numerically for specific values of kR. Such a calculation shows that in any section of the beam in the near field the pressure amplitude and the intensity also pass through a series of maxima and minima, though the wave fronts, i.e., the surfaces of constant phase, are nearly planar. In addition, the average pressure on different sections of the beam remains quite stable. Real ultrasonic receivers, however, usually record precisely the average pressure, defined mathematically as $\bar{p} = (1/S')\int_{S'} p_A dS'$, where S' is the effective area of the receiver and p_A is the local pressure at the point of observation A, lying in the S' plane. The calculation shows that for $S' = S$ the average pressure in the near field differs from the pressure in an ideal uniform plane wave by not more than 10–15%. As S' decreases the interference effects will naturally increase, but if receivers with sufficiently large surface area are used, such effects will practically not appear. This is confirmed experimentally, and justifies the use of ideal plane waves for the analysis of real ultrasonic beams in ultrasonics. beams, performed in ultra-acoustics, in terms of ideal plane waves. In measurements of the absorption, and even of the velocity of sound, however, the interference and diffraction of an ultrasonic beam can lead to considerable errors, and they must be taken into account by the introduction of appropriate corrections.[71]

IX. Spherical Waves

§ 1. Wave equation for spherical waves

In ultrasonics, spherical waves are often encountered in addition to plane waves. We have already encountered such waves in our analysis of the scattering of ultrasound by spherical particles, cavitation processes, and radiation pressure. Spherical waves are formed in the far zone of real plane ultrasonic radiators, as well as in the near zone of spherical radiators. For this reason, in this chapter we shall study separately the characteristics and singularities of the propagation of spherically symmetric waves, i.e., waves whose acoustic parameters depend on the distance from a center.

We shall obtain the wave equation for spherical waves from the general wave equation (II.32), writing the Laplacian of the velocity potential $\Delta\varphi$ in spherical coordinates. Since φ is, in this case, a function of the polar coordinate r only, in expression (II.36) for the Laplacian of φ in spherical coordinates only the first term will be different from zero, and the linearized equation (II.32) for this case will have the form

$$\frac{1}{r^2}\frac{\partial}{\partial r}\left(r^2\frac{\partial\varphi}{\partial r}\right) = \frac{1}{c_0^2}\frac{\partial^2\varphi}{\partial t^2} \qquad \text{(IX.1)}$$

(the zero subscript corresponds to the linear approximation). Rewriting the left side as

$$\frac{1}{r^2}\frac{\partial}{\partial r}\left(r^2\frac{\partial\varphi}{\partial r}\right) = \frac{1}{r}\left(r\frac{\partial^2\varphi}{\partial r^2} + 2\frac{\partial\varphi}{\partial r}\right) = \frac{1}{r}\left(\frac{\partial^2 r\varphi}{\partial r^2}\right),$$

Eq. (IX.1) can be written in the form

$$\frac{\partial^2 (r\varphi)}{\partial r^2} = \frac{1}{c_0^2}\frac{\partial^2 (r\varphi)}{\partial t^2}. \tag{IX.2}$$

In this form Eq. (IX.2) is the same as the wave equation (II.37) for one-dimensional plane waves, with the exception that the coordinate x is replaced here by the coordinate r and the velocity potential φ is replaced by the product $r\varphi$. The solution of Eq. (IX.2) will therefore have a form analogous to (II.41), i.e.,

$$r\varphi = f_1(c_0t - r) + f_2(c_0t + r)$$

or

$$\varphi(r, t) = (1/r)f_1(c_0t - r) + (1/r)f_2(c_0t + r).$$

The first term of this solution describes an outgoing wave, propagating with velocity c_0 in all directions from a center at $r = 0$, and the second term describes an incoming wave propagating toward this center. We shall study one of these waves – the outgoing wave – for which the velocity potential is

$$\varphi = (1/r)f(c_0t - r). \tag{IX.3}$$

§ 2. Monochromatic spherical waves

The function f in (IX.3) represents an arbitrary disturbance. For a sinusoidal disturbance with frequency ω, the expression for the velocity potential in the outgoing spherical wave will have the form

$$\varphi = \varphi_{\max 0}\sin\omega(t - r/c_0) = \frac{\varphi_{\max 0}}{r}\sin(\omega t - kr) \tag{IX.4}$$

or, in the complex notation,

$$\varphi = \varphi_{\max 0}\exp[i(\omega t - kr)], \tag{IX.5}$$

where $k = \omega/c_0 = 2\pi/\Lambda$ is the wave number and $\varphi_{\max 0}$ is the starting amplitude of the velocity potential, which is determined by the boundary conditions. We shall find the pressure and particle velocity in the spherical wave from the velocity potential, based on the relations (II.7) and (II.9) which are valid in three-dimensional case, i.e. $p = \rho_0 \partial\varphi/\partial t$ and $v = -\mathrm{grad}\varphi = -\partial\varphi/\partial r$. Differentiating Eq. (IX.4) with respect to time and the coordinate, we obtain

$$p = \frac{\rho_0 \varphi_{\max 0} \omega}{r} \cos(\omega t - kr) = \frac{p_{\max 0}}{r} \cos(\omega t - kr), \qquad \text{(IX.6)}$$

$$v = -\frac{\partial\varphi}{\partial r} = \frac{\varphi_{\max 0}}{r^2} \sin(\omega t - kr) + \frac{\varphi_{\max 0} k}{r} \cos(\omega t - kr). \qquad \text{(IX.7)}$$

Comparing relations (IX.6) and (IX.7), we find

$$v = \frac{\varphi_{\max 0}}{r^2} \sin(\omega t - kr) + \frac{p}{\rho_0 c_0} =$$
$$= \frac{\varphi_{\max 0}}{r^2} \sin(\omega t - kr) + \frac{p_{\max 0}}{\rho_0 c_0} \cos(\omega t - kr). \qquad \text{(IX.8)}$$

In a plane wave the relation between the pressure and the particle velocity has the form (see Chap. II): $v = p_{\max}/(\rho_0 c_0) \cos(\omega t - kx) = p/(\rho_0 c_0)$, which is almost the same as the second term in the expression (IX.7), the only difference being that the amplitude of the pressure in the spherical wave $p_{\max}$ decreases with distance as $1/r$. Here there is a new term which vanishes for large r (since it decreases with distance as $1/r^2$), when the form of the spherical wave front approaches a planar form. The presence of this term indicates that there is a phase difference between the particle velocity and the pressure in the spherical wave. If the coefficients in both terms of Eq. (IX.7) had been identical, then these terms would have differed in phase by $\pi/2$ as the sine and cosine with identical arguments. Because the coefficients are different, however, the phase difference will lie between $\pi/2$ and 0, and will vary with the distance since the coefficients depend differently on r.

Using the fact that $\varphi_{\max} = p_{\max}/(\rho_0 \omega)$, expression (IX.8) can be written as follows:

$$v = \frac{p_{\max}}{\rho_0 c_0 \cos\beta} \cos(\omega t - kr - \beta) = v_{\max} \cos(\omega t - kr - \beta), \qquad \text{(IX.9)}$$

where

$$\cos\beta = kr(1 + k^2r^2)^{-1/2}, \ \sin\beta = (1 + k^2r^2)^{-1/2}, \ \tan\beta = (kr)^{-1},$$

and

$$v_{max} = p_{max}/(\rho_0 c_0 \cos\beta) . \qquad \text{(IX.10)}$$

If $kr \ll 1$, i.e., $r \ll \Lambda$, then $\tan\beta \to \infty$, $\beta \to \pi/2$, $\cos\beta \to kr$, and expression (IX.9) assumes the form

$$v = \frac{p_{max}}{\rho_0 c_0 kr}\cos(\omega t - kr - \frac{\pi}{2}) = \frac{p_{max\,0}}{\rho_0 c_0 kr^2}\cos(\omega t - kr - \frac{\pi}{2}) .$$

It follows from here that for small r the particle-velocity wave lags behind the pressure wave in phase by $\pi/2$, the velocity amplitude decreases with distance as $1/r^2$, and the pressure amplitude decreases as $1/r$.

If, on the other hand, $kr \gg 1$, i.e., $r \ll \Lambda$, then $\tan\beta \to 0$, $\cos\beta \to 1$, and we have $v = p_{max}/(\rho_0 c_0)\cos(\omega t - kx)$, $p_{max} = v_{max}\rho_0 c_0$, i.e., we obtain the relations characteristic for a plane wave, the only difference being that the amplitude values of all acoustic quantities decrease inversely as $1/r$: $p_{max} = p_{max0}/r$, $v_{max} = v_{max0}/r$, $A = A_0/r$, etc.

§ 3. The intensity of a spherical wave

The phase difference between the pressure and velocity in a spherical wave leads to singularities in the expressions for the intensity of the wave. The intensity can be calculated as the average work performed by acoustic pressure forces per unit area per unit time, i.e., $I = A = pv$. We have

$$v = p_{max\,0}/(r\rho_0 c_0 \cos\beta)\cos(\omega t - kr - \beta) ;$$

$$p = (p_{max\,0}/r)\cos(\omega t - kr) .$$

Thus

$$A = pv = \frac{p_{\max 0}^2}{r^2 \rho_0 c_0 \cos\beta} \cos(\omega t - kr)\cos(\omega t - kr - \beta) .$$

Introducing the notation $\gamma \equiv \omega t - kr$ and $c \equiv p_{\max 0}^2/(\rho_0 c_0 r^2)$, this expression can be rewritten as follows:

$$A = C\frac{\cos\gamma}{\cos\beta}(\cos\gamma\cos\beta + \sin\gamma\sin\beta) = C(\cos^2\gamma + \sin\gamma\cos\gamma\tan\beta) .$$

Using the relations $2\cos^2\gamma = 1 + \cos 2\gamma$ and $2\sin\gamma\cos\gamma = \sin 2\gamma$, we obtain

$$A = \frac{C}{2} + \frac{C}{2}\cos 2(\omega t - kr) + \frac{C}{2}\sin 2(\omega t - kr)\tan\beta . \qquad \text{(IX.11)}$$

After averaging over a period the second and third terms in this equation vanish, so that

$$I = A = p_{\max 0}^2/(2r^2\rho_0 c_0) \qquad \text{(IX.12)}$$

or, since $p_{\max 0}/r = p_{\max}$,

$$I = p_{\max}^2/(2\rho_0 c_0) . \qquad \text{(IX.13)}$$

Comparing this result with Eq. (III.21), we see that the intensity of the spherical wave is expressed in terms of the amplitude of the pressure in the wave in the same way as in a plane wave, except that the amplitude of the pressure in the spherical wave decreases with distance as $1/r$ and, therefore, the intensity of the spherical wave decreases inversely as the square of the distance. This is as it should be, since the total power remains constant while the area of the front of the spherical wave, over which this power is distributed, increases as $4\pi r^2$.

The amplitude of the pressure in a spherical wave is related to the amplitude of the particle velocity by the relation (IX.10), i.e., $p_{\max} = v_{\max}\rho_0 c_0 \cdot kr(1 + k^2r^2)^{-1/2}$. Substituting this into expression (IX.13), we obtain

$$I = \frac{\rho_0 c_0}{2}\frac{k^2r^2}{1 + k^2r^2}v_{\max}^2, \qquad \text{(IX.14)}$$

where $v_{\max} = (v_{\max 0}/r)$. Thus the expression for the intensity of a spherical wave in terms of the particle velocity differs from the same expression for a plane wave, since in the spherical wave the pressure and velocity differ in phase. Taking this into account, expression (IX.14) can be rewritten in the form

$$I = (p_{max} v_{max} \cos\beta)/2, \qquad \text{(IX.15)}$$

which is the same as the equation for the ac power in a circuit with a reactive resistance. In the spherical wave, however, this phase shift is a function of distance, i.e., $\cos\beta = f(r)$, and it vanishes for $kr >> 1$. Here, $\cos\beta = 1$ and Eqs. (IX.15) and (IX.14) transform into the expression for the intensity of a plane wave with decreasing amplitude.

We shall estimate the distances from the center of the spherical wave at which the phase shift between the pressure and the particle velocity in the wave may be large. We assume that $r = \Lambda$. Then $kr = 2\pi$, $\beta = \tan^{-1}(2\pi)^{-1} \simeq 9°$, and $\cos\beta \simeq 0.988$. Thus the "angle" β decreases very rapidly with distance, and the spherical and plane waves differ considerably only in the near field of the radiator $(r < \Lambda)$, whose dimensions, in this case, do not exceed the wavelength of the radiated wave. In the megahertz frequency band the wavelengths of ultrasonic waves are millimeters or fractions of a millimeter and the dimensions of real radiators are much larger. In this connection, the singularities of the near field of a real radiator of high-frequency spherical ultrasonic waves are not important. At low ultrasonic frequencies, however, the condition $kr < 1$ can be met; it is also met when ultrasound is reradiated by small suspended particles and cavitation bubbles. For this reason, in the next section we shall briefly study the ultrasonic radiation from pulsating spheres.

§ 4. Radiation of spherical waves from a pulsating sphere

Consider a sphere with radius R, whose surface undergoes small, coherent, radial (pulsational) oscillations with constant amplitude. The acoustic field of this pulsating sphere will evidently consist of symmetric uniform spherical waves without interference nodes. Such radiators are called *zero-order radiators*.

Let the velocity of radial displacements be given on the surface of the sphere, i.e., at $r = R$. We write this velocity in the complex form

$$v(R) = v_{\max 0} \exp(i\omega t). \tag{IX.16}$$

We shall find the velocity potential, writing it in the general complex form (IX.5) with some initial phase β_0, $\varphi = (\varphi_{\max 0}/r)\exp[i(\omega t - kx) - kx + \beta_0)]$, taking into account the fact that $v(r) = -(\partial\varphi/\partial r)_{r=R}$. Differentiating this potential and setting $r=R$, we obtain $\beta_0 = kR$ and

$$\varphi(R) = (1/R)[R^2 v_{\max 0}/(1 + ikR)]\exp(i\omega t). \tag{IX.17}$$

We now calculate the sound pressure arising on the surface of the sphere due to the reaction of the medium to the motion of the sphere. This gives the initial pressure amplitude $p_{\max 0}$, which then decreases with distance. We find the pressure from the general definition

$$p(R) = \rho_0 \frac{\partial\varphi}{\partial r}\bigg|_{r=R},$$

where ρ_0 is the density of the surrounding medium. Differentiating the expression (IX.17) with respect to time, we find

$$p(R) = \rho_0 \frac{i\omega R v_{\max 0}}{1 + ikR}\exp(i\omega t) = \rho_0 c_0 \frac{ikR}{1 + ikR} v_{\max 0}\exp(i\omega t). \tag{IX.18}$$

The pressure force acting on the sphere, i.e., the reaction force of the medium equals the product of the pressure $p(R)$ and the area of the sphere $S_0 = 4\pi R^2$. The sphere obviously exerts the same force on the medium, producing there an oscillatory process with the displacement velocity $v(R)$. Independent of the relationship between the pressure and velocity, the ratio of the pressure force to the velocity determines the total acoustic impedance of the medium (in contrast to the specific impedance, which is referred to unit area) Z. Dividing expression (IX.18) by (IX.16) and multiplying by the area of the sphere, we obtain

$$Z = \frac{S_0 p}{v} = \rho_0 c_0 S_0 \frac{ikR}{1 + ikR}. \tag{IX.19}$$

Of course, the same expression will be obtained for the impedance in the field of spherical waves at any distance from the center of the source $r > R$: $Z = \rho_0 c_0 S ikr/(1 + ikr)$, where $S = 4\pi r^2$ is the area of a spherical wave front with radius r. For greater clarity, however, we shall analyze the result obtained relative to the surface of the ultrasonic source, whose radius R is the lowest value that r can assume. Separating the expression (IX.19) into real and imaginary parts, we obtain

$$\frac{ikR}{1+ikR} = \frac{k^2R^2}{1+k^2R^2} + i\frac{kR}{1+k^2R^2} \equiv X + iY,$$

Equation (IX.19) can then be written in the form $Z = \rho_0 c_0 S_0 (X + iY)$. The quantity Re $Z = \rho_0 c_0 S_0 X$ represents the active part of the characteristic impedance and Im $Z = \rho_0 c_0 S_0 Y$ is the reactive part. Both depend on kR, but they behave differently as a function of kR, i.e., as a function of the ratio of the size of the sphere to the wavelength of the sound. For $kR \gg 1$, i.e., for $R \gg \Lambda$, $X = 1$, $Y = 0$, and $Z = Z_0 = \rho_0 c_0 S_0$, and only the active part of the impedance, which represents the total resistance to radiation, remains. For $kR \ll 1$, $X = 0$ and $Y = 0$. For $kR < 1$, the quantity Y increases more rapidly with kR than does X. For $kR = 1$, $X = Y$ and thereafter, for $kR > 1$, the first term, which increases to 1 while Y decreases to 0, dominates. The general dependence on the parameter kR of the real and imaginary parts of the impedance of a pulsating sphere is illustrated graphically in Fig. 65.

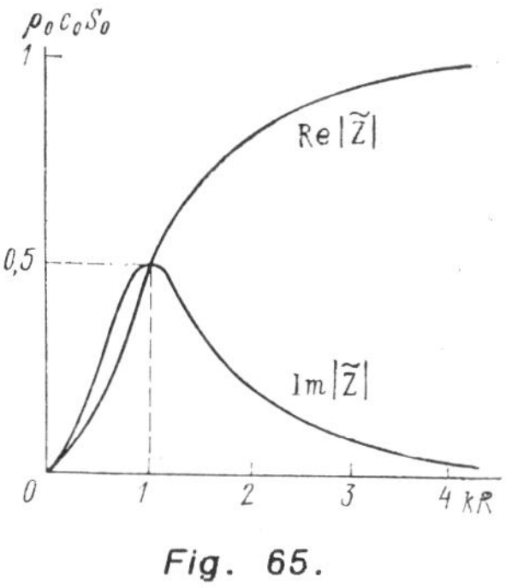

Fig. 65.

To clarify the significance of the result obtained, we shall calculate the power radiated from a pulsating sphere. For this, we multiply the previously obtained expression for the intensity of a spherical wave (IX.14) by the area of the sphere $4\pi R^2$:

$$D = 4\pi R^2 \left[\rho_0 c_0 \frac{k^2 r^2}{1+k^2 r^2} \frac{v_{max}^2}{2}\right]_{r=R},$$

But $v_{max}|_{r=R} = v_{max0}$, so that

$$D = \rho_0 c_0 S_0 X v_{\max 0}^2 / 2. \tag{IX.20}$$

Since $S(r) \sim r^2$ and $v_{\max}(r) \sim r^{-2}$, this power remains constant for any r (neglecting, of course, absorption losses in the medium). The magnitude of this power, according to expression (IX.20), is proportional to the active impedance of the medium $\rho_0 c_0 S_0 X$, which depends on kR. Therefore, the radiation efficiency of a pulsating sphere depends on the ratio of the radius of the sphere to the wavelength of the radiated wave, i.e., on the ultrasonic frequency. For small kR, the radiation efficiency is low, independently of the amplitude of oscillations of the source. In this case, the reactive part of the impedance, which, as always, determines the relative fraction of the energy of the source returned by the medium during a half-period of the oscillations, plays a large role. This is easily seen by integrating expression (IX.11) over part of the period instead of over the entire period. Then, the first and third terms in this expression will differ from zero and will give the additional term $(1/2) M v_{\max 0}^2$ in the power of radiation over one-fourth of the period, where M is some constant with the dimension of mass, corresponding to the mass of the medium displaced by the pulsating sphere, called the *associated mass*. Over the next one-fourth period the additional power will have the same magnitude but opposite sign. This means that the kinetic energy stored in the associated mass over one-fourth the period is then returned to the radiator.

Thus the power related to the reactive part of the impedance is analogous to the power consumed by the inductance in an ac circuit, and the reactive part Im Z is itself anlogous to the inductive impedance of the coil. The active part Re $Z = \rho_0 c_0 S_0 R$, on the other hand, determines the power irreversibly lost by the source to radiation into the medium, and it is equivalent to the active impedance of an electric circuit. The equivalent circuit of the acoustic impedance of a pulsating sphere can therefore be represented by a coil and a resistor connected in parallel.

As kR increases, the reactive part of the impedance decreases rapidly while the active part and, concomitantly, the radiation efficiency of the pulsating sphere increase. As is evident from Fig. 65, already at $kR = 1$ the active part of Z equals the reactive part, and for $kR = 3$–4 the reactive part almost completely vanishes. The value $kR = 1$, however, is attained at a frequency of 1 MHz, for example, when the ultrasonic radiation is emitted into water ($\Lambda = c_0/\nu = 1.5\ mm$), at $R = 1/k = \Lambda/2\pi \simeq 0.25\ mm$. For this reason, as already noted, the associated mass and the reactive impedance of a pulsating sphere usually do not play an important role at ultrasonic frequencies; we therefore do not study this question in detail, and instead refer the interested reader to the literature.[72]

X. Propagation of Ultrasound in an Isotropic Solid

§ 1. Wave equation for an infinite solid

Unlike liquids and gases, which in practice exhibit only bulk elasticity, solids also exhibit shear elasticity ("elasticity of shape"). The presence of shear elasticity, which, as always, we shall first assume to be ideal, leads to the fact that in a solid, together with the longitudinal elastic waves studied in the preceding chapters, shear deformation waves can also propagate in the form of so-called *transverse* (*shear*) *waves*. The laws of propagation of both types of waves in an **infinite** isotropic solid are exactly the same as the laws studied in the preceding chapters, which concerned ideal media with ideal elasticity, so that most of the previous results are equally valid for transverse waves. The distinctive features of the propagation of elastic waves in isotropic solids are, however, manifested primarily at the boundaries as different types of surface waves, mixed deformations, transformation of waves with reflection from boundaries, etc. For this reason, in this chapter, after deriving and analyzing the wave equation for an isotropic solid, we shall study only the basic problems concerning the propagation of ultrasonic waves, as well as some of the distinctive features of the propagation of finite-amplitude ultrasonic waves, in infinite solids.

The laws of propagation of elastic waves in solids follow from the general equations of motion derived in Chap. I. In linearized form, valid for waves with infinitesimal amplitude, these equations have the form of expression (I.11), i.e.,

$$\partial \sigma_{ik}/\partial x_k = \rho \partial^2 u_i/\partial t^2, \qquad (X.1)$$

where σ_{ik} are the components of the stress tensor (I.6); u_i are the components of the displacements along the coordinate axes $x_k = x,\ y,\ z$

(i, k = 1, 2, 3). The zero subscript for the density, corresponding to the linear approximation, is dropped. Here, and in what follows, as in the preceding chapters ρ is the equilibrium density. To reduce Eq. (X.1) to a single variable the stresses σ_{ik} can be expressed in terms of the corresponding deformations ϵ_{ik} by means of Hooke's law for an isotropic solid (I.15):

$$\sigma_{ik} = \lambda\Theta\delta_{ik} + 2\mu\epsilon_{ik}, \qquad \text{(X.2)}$$

where Θ is the volume expansion, equal to the sum of longitudinal extensions; δ_{ik} is the Kronecker delta; and, λ and μ are the Lamé constants. The latter constants are two independent elastic moduli, completely characterizing the elastic properties of an isotropic solid.

The volume expansion Θ is, by definition, given by

$$\Theta = \epsilon_{11} + \epsilon_{22} + \epsilon_{33} = \frac{\partial u_1}{\partial x_1} + \frac{\partial u_2}{\partial x_2} + \frac{\partial u_3}{\partial x_3} = \operatorname{div}\mathbf{u}. \qquad \text{(X.3)}$$

We recall that the relation (X.3) is the mathematical expression of the continuity of the medium and represents the linearized equation of continuity. Using this relation, as well as the definition of the components of the small deformation $\epsilon_{ik} = (\partial u_i/\partial x_k + \partial u_k/\partial x_i)$, and differentiating Eq. (X.2) with respect to x_k, after combining like terms we obtain $\partial\sigma_{ik}/\partial x_k = (\lambda + \mu)(\partial\Theta/\partial x_i) + \mu\Delta u_i$, where Δ is the Laplace operator (the sum of the second derivatives with respect to the coordinates). Substituting this result into the equation of motion (X.1), we obtain three equations for the three components of the displacement u_i: $(\lambda + \mu)(\partial\Theta/x_i) + \mu\Delta u_i = \rho(\partial^2 u_i/\partial t^2)$, which can be combined into a single vector equation for the displacement vector $\mathbf{u}$:

$$(\lambda + \mu)\operatorname{grad}\operatorname{div}\mathbf{u} + \mu\Delta\mathbf{u} = \rho\partial^2\mathbf{u}/\partial t^2. \qquad \text{(X.4)}$$

With an arbitrary orientation relative to the coordinate axes the vector $\mathbf{u}$ can be expressed as a sum of two vectors:

$$\mathbf{u} = \mathbf{u}_l + \mathbf{u}_T, \qquad \text{(X.5)}$$

one of which (u_l) corresponds to the longitudinal deformation and the other (u_T) corresponds to a purely shear deformation. The longitudinal deformation is characterized by the absence of tangential components and, therefore, for it

$$\text{curl } \mathbf{u}_l = 0. \qquad \text{(X.6)}$$

For the shear deformation, however, curl $\mathbf{u}_T \neq 0$ but

$$\text{div } \mathbf{u}_T = 0. \qquad \text{(X.7)}$$

Substituting the expressions (X.5)–(X.7) and applying in succession the operations curl and div to Eq. (X.4), we obtain:

$$\Delta \mathbf{u}_l (\lambda + 2\mu)/\rho = \partial^2 \mathbf{u}_l / \partial t^2 \qquad \text{(X.8)}$$

for the longitudinal displacements and

$$\Delta \mathbf{u}_T \mu/\rho = \partial^2 \mathbf{u}_T / \partial t^2 \qquad \text{(X.9)}$$

for the shear (tangential) displacements.

Thus Eq. (X.4) separates into two identical equations (X.8) and (X.9), which have the familiar form of wave equations. The first of these equations describes the propagation of purely longitudinal waves with the velocity

$$c_l = [(\lambda + 2\mu)/\rho]^{1/2}. \qquad \text{(X.10)}$$

and the second one describes the propagation of purely transverse waves with the velocity

$$c_T = (\mu/\rho)^{1/2}. \qquad \text{(X.11)}$$

The difference between the propagation velocities of these waves, as we can see, is related only to the presence of elastic characteristics which determine the stiffness of the medium with respect to the given type of dynamic deformations. We can therefore introduce the generalized concept of *effective stiffness* E, related to the velocity of propagation of the corresponding wave by the relation

$$E = \rho c^2, \qquad \text{(X.12)}$$

and

$$c = (E/\rho)^{1/2}. \qquad \text{(X.13)}$$

For a transverse wave propagating with velocity c_T, E equals the shear modulus

$$E = \rho c_T^2 = G = (c_{11} - c_{12})/2, \qquad \text{(X.14)}$$

Table 10

Representation of the Effective Stiffnesses for Longitudinal and Shear Waves in Terms of Different Elastic Constants

Effective stifness	c_{11}, c_{12}	λ, μ	μ, K	μ, E	ν_0, E	ν_0, K
ρc_l^2 (longitudinal waves)	c_{11}	$\lambda + 2\mu$	$K + \frac{4}{3}\mu$	$\frac{\mu(4\mu - E)}{3\mu - E}$	$\frac{E(1-\nu_0)}{(1+\nu_0)(1-2\nu_0)}$	$\frac{3K(1-\nu_0)}{1+\nu_0}$
ρc_T^2 (shear waves)	$\frac{c_{11} - c_{12}}{2}$	μ	μ	μ	$\frac{E}{2(1+\nu_0)}$	$\frac{3K(1-2\nu_0)}{2(1+\nu_0)}$

and for a longitudinal wave

$$E = \rho c_l^2 = \lambda + 2\mu = c_{11}, \qquad (X.15)$$

where c_{11} and c_{12} are the elastic moduli (see Table 1). Together with these moduli and the Lamé constants, we introduced in § I.6 other characteristics of elasticity: Young's modulus E, Poisson's ratio ν_0, and the modulus of hydrostatic compression K. Using the relations presented there as well as Eqs. (X.14) and (X.15), the effective stiffnesses $E = \rho c^2$ for longitudinal and shear waves can be expressed in terms of different pairs of independent moduli. The corresponding expressions are summarized in Table 10, whence it follows, in particular, that

$$c_T/c_l = \sqrt{(1 - 2\nu_0)/(2 - 2\nu_0)}\,. \qquad (X.16)$$

Thus Poisson's ration ν_0 is a measure of the ratio of the velocities of transverse and longitudinal waves in a given medium. . The maximum value $\nu_0 = 0.5$ corresponds to a liquid for which $c_T = 0$, and the effective stiffness is the bulk modulus of elasticity K, which determines the velocity of the longitudinal wave. The value $\nu_0 = 0$ corresponds to the maximum ratio of the velocities $(c_T/c_l)_{max} = 2^{-1/2}$. Therefore, in

Table 11

Acoustic Characteristics of Some Isotropic Solids

Material	T, °C	$\rho \cdot 10^{-3}$, kg/m^3	ν_0	$c_l \cdot 10^{-3}$, m/s	$c_T \cdot 10^{-3}$, m/s	$\rho c_l \cdot 10^{-5}$, $kg/(m^2 \cdot s)$	$\rho c_T \cdot 10^{-5}$, $kg/(m^2 \cdot s)$
Aluminum	20	2.7	0.34	6.26	3.08	169	83.2
Bismuth	20	9.8	0.33	2.18	1.10	214	108
Tungsten	20	19.1	0.35	5.46	2.62	1042	500
Iron	20	7.8	0.28	5.85	3.23	456	252
Gold	20	19.3	0.42	3.24	1.20	626	232
Cadmium	20	8.6	0.30	2.78	1.50	240	129
Constantan	20	8.8	0.33	5.24	2.64	460	232
Brass	20	9.1	0.35	4.43	2.12	361	172
Manganese	20	8.4	0.33	4.66	2.35	393	197
Copper	20	8.9	0.35	4.70	2.26	418	201
Nickel	20	8.8	0.31	5.63	2.96	495	260
Tin	20	7.3	0.33	3.32	1.67	242	122
Platinum	20	21.4	0.39	3.96	1.67	846	357
Lead	20	11.4	0.44	2.16	0.70	246	80
Silver	20	10.5	0.38	3.60	1.59	380	167
Zinc	20	7.1	0.25	4.17	2.41	296	171
Glass silicate (fused quartz, SiO_2)	17	2.21	0.17	6.02	3.78	133	83
borate (B_2O_3)	17	1.8	—	3.47	1.25	62	23
germanate (GeO_2)	17	3.63	—	3.61	2.21	130	80
chalco-genide							
As_2S_3	17	3.27	—	2.58	1.49	85	50
As_2Se_3	17	4.62	—	2.23	1.29	103	100
glassy selenium	17	4.28	—	1.84	0.96	79	41
beryllium fluoride	17	4.70	—	4.70	3.90	221	183
crown glass	20	2.5	0.22	5.66	3.42	141	86
flint	20	3.6	0.22	4.26	2.56	154	92
heavy flint	20	4.6	0.24	3.76	2.22	173	102
Plexiglass	20	1.18	0.35	2.67	1.12	32	13
Gypsum	20	2.26	0.34	4.79	2.37	110	58
Ice	0	1.0	0.33	3.98	1.99	32	20
Polystyrene	20	1.06	0.32	2.35	1.12	23	12
Porcelain	20	2.41	0.23	5.34	3.12	129	75
Ebonite	20	1.2	—	2.40	—	29	—

any medium the propagation velocity of longitudinal waves exceeds the propagation velocity of shear waves by not less than a factor of $\sqrt{2} \simeq 1.4$. The value of ν_0 for solids usually lies in the range 0.3–0.25 and the velocities c_l and c_t differ by 50–70%. The values of c_l and c_T for some infinite isotropic solids at room temperature are presented in Table 11. The same table shows Poisson's ratio ν_0, the density, and the acoustic impedances ρc_l and ρc_T for these materials.

Returning to our analysis of the wave equation, we note that using the relations (X.10)–(X.13), Eqs. (X.8) and (X.9) can be represented in a unified form:

$$\Delta \mathbf{u} = c^{-2}\partial^2\mathbf{u}/\partial t^2. \tag{X.17}$$

Therefore, the laws of propagation of shear waves in an infinite isotropic body are identical to the general laws of propagation of longitudinal waves studied in the preceding chapters. In addition, the wave equation in the form (X.17) describes the propagation of either a purely longitudinal wave with velocity c_l or a purely shear wave with velocity c_T. Equation (X.4), however, refers to an arbitrary orientation of the displacement vector $\mathbf{u}$, in which it is in general possible to separate both longitudinal and shear components and, in addition, these components $u_i = u_x$, u_y, and u_z are mutually perpendicular. The solution of Eq. (X.4) in a rectangular coordinate system $x_i = x$, y, and z is thus a plane wave with arbitrary orientation of the displacement vector $\mathbf{u}$ relative to **these** coordinates:

$$\mathbf{u} = \mathbf{u}_{max}\exp\{i[\omega t \mp \mathbf{k}\cdot\mathbf{r}]\}, \tag{X.18}$$

where $\mathbf{u}_{max}$ is its vector amplitude, which is independent of the coordinates and time; ω is the circular frequency fixed by the source; $\mathbf{r}(x_i)$ is the radius vector; $\mathbf{k}$ is the wave vector; the minus sign refers to the forward wave; and, the plus sign refers to the backward wave. By definition (§ III.1) $\mathbf{k} = k\mathbf{n} = \mathbf{n}\omega/c$, where $\mathbf{n}$ is a unit vector which determines the direction of propagation, normal to the wave front and $\mathbf{k}\cdot\mathbf{r} = k\mathbf{n}\cdot\mathbf{r} = k(xn_x + yn_y + zn_z)$, where $n_i = n_x$, n_y, n_z are the projections of the unit normal on the coordinate axes, i.e., the direction cosines. In accordance with what was said above, a wave with arbitrary displacement (X.18) can be represented as a sum of two waves: a longitudinal wave with the displacement lying along the wave normal $\mathbf{n}$ ($\mathbf{u}_l \parallel \mathbf{n}_l$, i.e., $\mathbf{u}_l \times \mathbf{n}_l = 0$), propagating with velocity c_l and a transverse wave with the displacement lying in the plane perpendicular to $\mathbf{n}$ ($\mathbf{u}_T \perp \mathbf{n}_T$, i.e., $\mathbf{u}_T \cdot \mathbf{n}_T = 0$, i.e., in the plane of the wave front. The velocity of propagation of this wave is c_T, and in an isotropic solid it does not depend on the direction of the displacement of the particles in the plane of the front. In other words, we can say that in an isotropic solid two waves can propagate in any direction: one longitudinal wave with velocity c_l and one transverse wave with velocity c_T, independent of the polarization.

In one-dimensional problems, for an unbounded isotropic solid the x-axis of a Cartesian coordinate system can always be oriented along the wave vector $\mathbf{k}$. Then, $n_y = n_z = 0$, $|n_x| = 1$, and the longitudinal displacement will be a displacement along the x-axis, i.e., $u_x = \xi$, and the transverse displacement vector $\mathbf{u}_T$ will lie in the yz plane and will have the components $u_y = \eta$ and $u_z = \zeta$. In this case, the wave equation (X.4) separates into two one-dimensional equations: one for the longitudinal wave

$$\frac{\partial^2 \xi}{\partial x^2} = \frac{1}{c_l^2} \frac{\partial^2 \xi}{\partial t^2}$$

and one for the components of the displacement vector of the shear wave $u_T(\eta, \zeta)$

$$\frac{\partial^2 \eta}{\partial x^2} = \frac{1}{c_T^2} \frac{\partial^2 \eta}{\partial t^2}; \qquad \frac{\partial^2 \zeta}{\partial x^2} = \frac{1}{c_T^2} \frac{\partial^2 \zeta}{\partial t^2}.$$

The solution of these equations is a one-dimensional plane wave of the form $u_i = u_{i\max} e^{(\omega t - kx)}$, where $k = k_l = \omega/c_l$ for the longitudinal wave and $k = k_T = \omega/c_T$ for the shear wave.

Analogous solutions can be obtained in spherical and cylindrical coordinate systems, but this does not give anything new.[73]

§ 2. Reflection, refraction, and transformation of ultrasonic waves at the boundaries of solids.

Unlike the reflection and refraction of ultrasound in liquids and gases, where, as we saw in Chap. VII, only longitudinal waves can exist, at the boundaries of solids the character of the wave changes. A purely longitudinal or purely shear wave incident on the boundary between two solids, in general, creates both longitudinal and tangential displacements at the boundary. As a result, both longitudinal and shear waves, which propagate at different velocities and are therefore reflected and refracted at different angles, arise in both media. Thus, at the boundary between solids, waves of one type are transformed into waves of another type, as a result of which, in general, two reflected and two refracted waves propagating in different directions arise at the boundary. There are only two exceptions: the case of normal incidence of a plane wave of either type on a plane boundary and the case of arbitrary incidence of a transverse wave polarized perpendicular to the plane of incidence, i.e., with the displacement parallel to the boundary.

Let the boundary between two solid media *1* and *2* lie in the yz plane perpendicular to the x-axis at $x = 0$. In a plane wave propagating along the x-axis, the character of the displacement remains unchanged at the boundary and only the velocity of propagation c, i.e., the wave number $k_x = \omega/c$, changes. As a result, the reflection coefficient of an arbitrary **normally incident** wave is determined by the previous relation (VII.14): $\rho_i = [(\rho_2 c_2 - \rho_1 c_1)/(\rho_2 c_2 + \rho_1 c_1)]^2$, which, as already noted in Chap. VII, remains valid for both longitudinal and shear waves. In the latter case, the quantity $\rho c_T = z_T$ represents the specific characteristic impedance of the medium relative to the shear wave. If this medium is stiff, then for it the shear modulus $G = \mu = 0$, $c_T = 0$, i.e., $z_T = 0$ and $\rho_z = 1$: the shear wave is completely reflected from the boundary with a liquid (gas). In this case, the reflection coefficient of the longitudinal wave at the boundary with the same liquid can be small.

Let a shear wave with displacement along the z-axis be incident on the same boundary ($x = 0$) in the xy plane *at* **an angle** θ_1 to the x-axis. Such a wave also does not create other components of displacement at the boundary, and it is therefore reflected and refracted as two pure shear waves with the same displacement. The relation (VII.39) remains in force for the energy reflection coefficient of such a wave:

$$\rho_l = \left[\frac{z_{T1}\cos\theta_2 - z_{T2}\cos\theta_1}{z_{T1}\cos\theta_2 + z_{T2}\cos\theta_1}\right]^2, \qquad (X.19)$$

where θ_1 is the angle of incidence (equal to the angle of reflection θ_1'); θ_2 is the angle of refraction (see Fig. 40); z_{T1} and z_{T2} are the characteristic impedances with respect to the shear waves in media *1* and *2*, respectively. If the second medium is a liquid, then for it $z_{T2} = 0$ and the shear wave is completely reflected from the boundary for any angle of incidence. If the second medium is a solid, then a refracted shear wave, propagating at the angle of refraction θ_2 determined by relation (VII.37) $\sin\theta_2/\sin\theta_1 = c_{T2}/c_{T1} \equiv \eta_T$, arises in it.

We shall now study the **more general cases of oblique incidence** of a longitudinal wave or arbitrarily polarized shear wave. For simplicity, we shall consider a two-dimensional problem, making the assumption that the wave is incident on the boundary at $x = 0$ in the xy plane at an angle θ_1 to the x-axis (Fig. 66), i.e., the wave vector has nonzero components $k_x = kn_x = k\cos\theta_1$ and $k_y = kn_y = k\sin\theta_1$, while $k_z = 0$. The displacement $\mathbf{u}_l$ in the longitudinal wave in this case has only two components $u_x = \xi$ and $u_y = \zeta$, and the displacement $\mathbf{u}_T$ in an arbitrarily polarized shear wave, in general, has all three components: $u_x = \xi$, $u_y = \eta$. and $u_z = \zeta$. Since, however, the component ζ refers to the previously studied case of a shear wave polarized perpendicular to the plane of incidence, without loss of generality we can study a transverse wave with displacement $\mathbf{u}_T$ in the plane of incidence (the xy plane) with displacement components ξ and η. The value of the component ζ in the reflected and refracted waves can therefore be found from Eq. (X.19) and added to the values of the displacement in the corresponding shear waves.

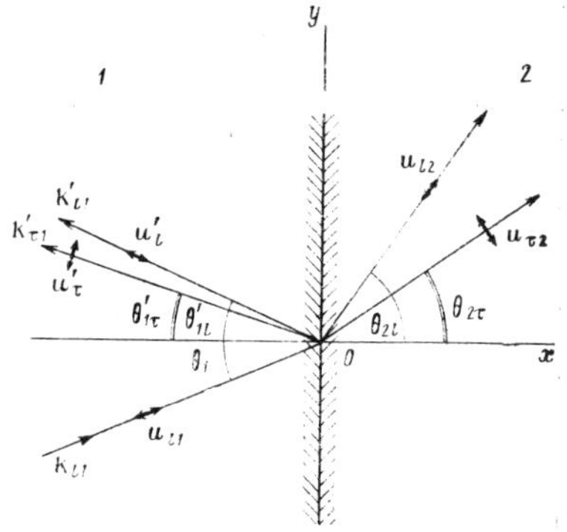

Fig. 66.

The components ξ and η create normal and tangential stresses, respectively, at the boundary between solid media. As a result, longitudinal and transverse waves, which propagate in both media with different velocities, i.e., at different angles of reflection and refraction, appear on both sides of the boundary. The general condition for finding these angles is that the projection of the wave vector **k** on the boundary yz, in this case, along the x axis, i.e., the component k_y for waves propagating on both sides of the boundary, must be constant – the "tracks" of these waves must be equal at the boundary. This condition, studied in detail and already used in Chap. VII to determine the angles of reflection and refraction and representing the condition that the tracks of the waves must be equal on both sides of the boundary, is an obvious requirement in order for there not to be any discontinuities in the medium. On another level, this condition follows from the fact that the boundary is oriented, in this case, perpendicular to the x-axis and has an infinite extent along the y and x axes; therefore, in a wave with wave vector **k** incident on the boundary, the boundary affects only the component of the wave vector k_x, while the components k_y and k_z remain unchanged. But $k_z = 0$, and $k_y = (\omega/c)\sin\theta$ for any wave. The frequency ω is fixed by the source, i.e., it does not depend on the boundaries. For this reason, from the condition $k = const$, we immediately obtain the general relation determining the angles of reflection and refraction θ^* for any of four reflected and refracted waves propagating with velocity c^*:

$$\sin\theta^*/\sin\theta_1 = c^*/c_1, \qquad \text{(X.20)}$$

where θ_1 and c_1 are the angle of incidence and the velocity of propagation of the incident wave.

For example, let the wave incident on the yz plane be a longitudinal wave with velocity c_{l1}. Then, for the reflected longitudinal wave,

$$\sin\theta'_{1l} = (c_{l1}/c_{l1})\sin\theta_1, \quad \theta'_{1l} = \theta_1, \qquad \text{(X.21)}$$

i.e., the angle of reflection of the longitudinal wave θ'_{1l} equals the angle of incidence θ_1.

For the reflected transverse wave, propagating with velocity c_{T1}, from the general relation (X.20) we obtain

$$\frac{\sin\theta'_{1T}}{\sin\theta_{1l}} = \frac{c_{T1}}{c_{l1}}, \quad \sin\theta'_{1T} = \frac{c_{T1}}{cl1}\sin\theta_{1l}. \qquad \text{(X.22)}$$

Since in the same medium $c_T < c_l$, $\theta_{1T} < \theta_{1l}$, i.e., the transverse wave is reflected at a smaller angle to the normal (i.e., to the x-axis). This case is illustrated in Fig. 66.

For the refracted longitudinal wave

$$\sin\theta_{2l}/\sin\theta_{1l} = c_{l2}/c_{l1}. \quad (X.23)$$

(Figure 66 illustrates the case when the longitudinal wave propagates with a higher velocity in the second medium: $c_{l2} > c_{l1}$.)

Finally, for the transverse part of the reflected wave we have

$$\sin\theta_{2T}/\sin\theta_{1T} = c_{T2}/c_{T1}. \quad (X.24)$$

Now, let the incident wave be transverse. Independent of its polarization (which is conserved in the reflected and refracted transverse waves due to the isotropy of the media), the angles of reflection θ'_{1T} and refraction θ_{2T} are determined by the following relations:

$$\theta'_{1T} = \theta_{1T}, \qquad \sin\theta_{2T}/\sin\theta_{1T} = c_{2T}/c_{T1}. \quad (X.25)$$

If the transverse incident wave is transformed into longitudinal waves, then from the relation (X.20), for the angles of reflection θ'_{1l} and refraction θ_{2l} of these longitudinal waves, we obtain

$$\sin\theta'_{1l}/\sin\theta_{1T} = c_{l1}/c_{T1} > 1 \quad (X.26)$$

for the reflected wave, and

$$\sin\theta_{2l}/\sin\theta_{1T} = c_{l2}/c_{T1} \quad (X.27)$$

for the refracted wave; in addition, the angle of refraction θ_{2l} can be both greater and less than the angle of incidence θ_{1T}, depending on the ration of the velocities c_{l2} and c_{l1}.

The expressions (X.21)–(X.27) obtained above admit a larger number of variants of total internal reflection than the case of reflection of purely longitudinal waves studied in Chap. VII. Total internal reflection occurs when the angle of refraction exceeds the angle of incidence, i.e., the velocity of the refracted wave is higher than that of the incident wave. If the velocity of the incident wave is lower than that of both refracted waves (longitudinal and shear) in medium 2, then in this case total internal reflection appears at two angles of incidence θ_1 satisfying the conditions

$$\sin(\theta_{1cr})_1 = c_1/c_{l2}, \quad \sin(\theta_{1cr})_2 = c_1/c_{T2}, \qquad (X.28)$$

where c_1 is the velocity of the incident wave. As a rule, such a situation is realized, for example, when a longitudinal wave is incident from a liquid on the boundary with a solid. In this case, by rotating the reflecting surface of the solid relative to the incident ultrasonic beam one can observe two sequential increases in the intensity of the reflected beam at angles satisfying the conditions (X.28), i.e., at angles of incidence for which first the longitudinal wave and then the transverse refracted wave vanish in the samples under study. This method can be used to measure the velocity of propagation of longitudinal and shear waves in isotropic solids. This method cannot, of course, provide high accuracy because of the errors introduced into the measurements of the angles; but, on the other hand, it can be used to perform measurements in strongly absorbing materials (for example, in solid polymers) and does not require transducers for exciting the shear waves.

§ 3. Reflection coefficient at the boundary of a solid at oblique incidence

To calculate the reflection coefficient at the boundary of a solid, i.e., the ratio of the energy of the reflected wave to the energy of the incident wave, it is necessary to formulate equations, as we have already done in Chap. VII, for all waves taking into account their directions of propagation and to use the boundary conditions expressing the fact that at the boundary the displacements must be continuous (the condition that there be no discontinuities) and the stresses, in this case, both the normal and tangential stresses, must also be continuous (the condition that action equals reaction). Here, generally speaking, from the viewpoint of practical realization, the case of the incidence of an ultrasonic wave on the boundary between a solid and a liquid or gas is of greatest interest. Direct, full acoustic contact between two solids is very rarely realized at ultrasonic frequencies; it is usually achieved with the help of transitional layers, primarily liquids. At a boundary with a liquid a shear ultrasonic wave is practically completely reflected at all angles of incidence; at a boundary with a gas (vacuum) both the shear and the longitudinal waves are completely reflected. In this respect, the case of total reflection of an elastic wave in a solid from its free surface is the most general case. We shall study this case first, after which we shall consider the possibility that the longitudinal component emerges into the contact liquid in the form of a refracted longitudinal wave.

In its turn, a shear wave polarized arbitrarily relative to the plane of incidence* can always be represented as a sum of two shear waves with mutually perpendicular displacements, one of which lies in the plane of incidence and the other lies in a plane perpendicular to it, i.e., parallel to the reflecting surface. The last part of the shear wave, as noted in the preceding section, does not create longitudinal displacements at the boundary and is reflected from it without singularities. Therefore, we need study only the reflection of a shear wave polarized in the plane of incidence.

So, let a **plane ultrasonic shear wave, polarized in the plane of incidence**, i.e., in the xy plane, be incident on the free boundary of an isotropic solid,** located at $x = 0$, at an angle θ to the x-axis (Fig. 67a). Since we are studying only one medium, we shall omit the index 1 for the incident and reflected waves, but we shall retain the remaining notation: we shall use the index l to denote quantities referring to longitudinal waves and the index τ to denote quantities referring to shear waves, and we shall use a prime to denote the characteristics of the reflected waves. Then, the wave vectors of the incident and reflected waves assume the following form: $\mathbf{k}_\tau = \mathbf{n}_\tau \omega / c_\tau$ for the incident shear wave, $\mathbf{k}'_\tau = \mathbf{n}'_\tau \omega / c_\tau$ for the reflected shear wave, and $\mathbf{k}'_l = \mathbf{n}'_l \omega / c_l$ for the reflected longitudinal wave. The corresponding unit vectors can be written (see Fig. 67a and the analysis in Chap. VII) as

---

*We recall that the plane of incidence of a wave on a surface is the plane formed by the wave vector and the normal to the surface. In acoustics the plane of polarization of a transverse wave is the plane containing the displacement vector.

**For shear waves a "free" boundary is the boundary with the vacuum (gas) or a liquid.

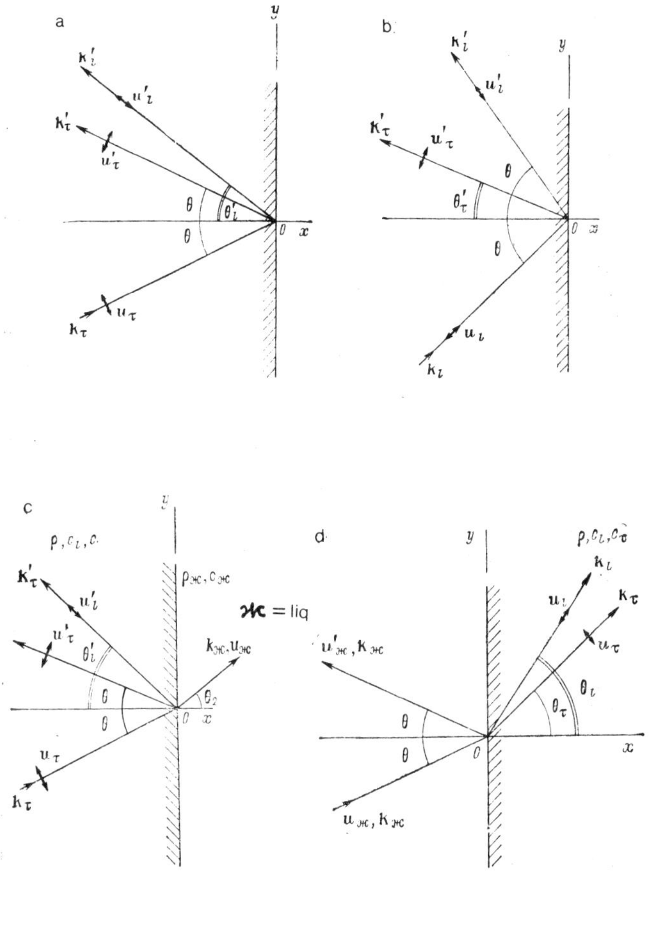

Fig. 67.

K

follows: $\mathbf{n}_T = \cos\theta + \sin\theta$; $\mathbf{n}'_T = -\cos\theta'_T + \sin\theta'_T = \cos\theta + \sin\theta$; $\mathbf{n}'_l = -\cos\theta'_l + \sin\theta'_l$, where θ is the angle of incidence (equal to the angle of reflection of the shear part of the wave θ'_T) and θ'_l is the angle of reflection of the longitudinal part of the wave, related to the angle of incidence θ by the relation (X.26). We shall likewise denote the displacements in the incident and reflected waves by $\mathbf{u}'_T$, $\mathbf{u}'_T$ and $\mathbf{u}'_l$, respectively, and their amplitudes by A_T, A'_T and A'_l. Then, the equations for the three waves can be written as follows: for the incident transverse wave

$$\mathbf{u}_T = [-\mathbf{n}_0 \times \mathbf{n}_T]A_T \exp\{i[\omega t - \mathbf{k}_T \cdot \mathbf{r}]\} =$$

$$= A_T(\cos\theta - \sin\theta)\exp\{i\omega[t - (x\cos\theta + y\sin\theta)/c_T]\}, \quad \text{(X.29)}$$

where $\mathbf{n}_0$ is the unit vector of the reflecting surface; for the reflected transverse wave, which, of course, remains polarized in the plane of incidence,

$$\mathbf{u}'_T = A'_T(\cos\theta + \sin\theta)\exp\{i\omega[t - (-x\cos\theta + y\sin\theta)/c_T]\}, \quad \text{(X.30)}$$

and, for the reflected longitudinal wave

$$\mathbf{u}'_l = A'_l(-\cos\theta'_l + \sin\theta'_l)\exp\{i\omega[t - (-x\cos\theta'_l + y\sin\theta'_l)/c_l]\}, \quad \text{(X.31)}$$

Adding all three waves, we obtain the total displacement field at all points of the body:

$$\mathbf{u} = \mathbf{u}_T + \mathbf{u}'_T + \mathbf{u}'_l. \quad \text{(X.32)}$$

We are interested in the reflection coefficient of the boundary, i.e., the ratios $\rho_{AT} = A'_T/A_T$ and $\rho_{Al} = A'_l/A_T$. To find these ratios we must use the boundary conditions, which in this case express the fact that at the boundary, i.e., at $x = 0$, both components — normal and tangential — of the stress tensor must vanish:

$$\sigma_{xx}\big|_{x=0} = 0, \qquad \sigma_{xy}\big|_{x=0} = 0. \quad \text{(X.33)}$$

Stresses in an isotropic solid are related to the deformations ϵ_{ik} by the relation (X.2), in which it is convenient to express the Lamé constants λ and μ in terms of the density of the medium and the velocity of the shear and longitudinal waves, which, substituting the relations (X.10) and (X.11), gives

$$\sigma_{ik} = \rho(c_l^2 - 2c_T^2)\Theta\delta_{ik} + 2\rho c_T^2 \epsilon_{ik} . \qquad (X.34)$$

Applying the boundary conditions (X.33) to this equation, we obtain (at $x = 0$)

$$\left.\begin{aligned} (\rho c_l^2 - 2\rho c_T^2)\left[\frac{\partial \xi}{\partial x} + \frac{\partial \eta}{\partial y}\right] + 2\rho c_T^2 \frac{\partial \xi}{\partial x} &= 0 \\ \rho c_T^2 \left[\frac{\partial \xi}{\partial y} + \frac{\partial \eta}{\partial x}\right] &= 0 \end{aligned}\right|_{x=0} , \qquad (X.35)$$

where ξ and η are the projections of the variable displacements $\mathbf{u}$ on the coordinate axes: $\xi = |\mathbf{u}|_x$ and $\eta = |\mathbf{u}|_y$. Now, performing the differentiation in Eq. (X.32), using expressions (X.29)–(X.31), required to find the deformations, and setting $x = 0$ in them and dropping the time-dependent factor $\exp(i\omega t)$, with the help of the boundary conditions (X.35) we obtain two equations, whence, using the scheme in Chap. VII, we find the ratios A_T'/A_T and A_l'/A_T sought, i.e., the amplitude reflection coefficients:

$$\rho_{AT} = \frac{A_T'}{A_T} = \frac{c_T^2 \sin 2\theta \sin 2\theta_l' - c_l^2 \cos^2 2\theta}{c_T^2 \sin 2\theta \sin 2\theta_l' + c_l^2 \cos^2 2\theta}, \qquad (X.36)$$

$$\rho_{Al} = \frac{A_l'}{A_T} = \frac{2c_l c_T \sin 2\theta \cos 2\theta}{c_T^2 \sin 2\theta \sin 2\theta_l' + c_l^2 \cos^2 2\theta}, \qquad (X.37)$$

where the angles θ and θ_l' are related by the relation $\sin\theta_l'/\sin\theta = c_l/c_T$.

To obtain the energy reflection coefficients, the energy flux density vector (the ultrasonic intensity) can be represented as two components, parallel and perpendicular to the boundary. The parallel component does not change on reflection, while the normal component decomposes into the intensities of the longitudinal and transverse waves. The transverse wave is reflected at an angle equal to the angle of incidence. For this reason, the ratio of the normal component of the intensity of the reflected transverse wave to the normal component of the intensity of the incident wave, i.e., the intensity reflection coefficient of the transverse wave ρ_{IT}, is

$$\rho_{IT} = (A_T'/A_T)^2 . \qquad (X.38)$$

The analogous relation for the longitudinal reflected wave, which is reflected at an angle $\theta_l' \neq 0$, assumes the form

$$\rho_{ll} = \frac{c_l \cos \theta_l'}{c_T \cos \theta} \left[\frac{A_l'}{A_T}\right]^2. \qquad (X.39)$$

Of course, the energy balance is maintained here, i.e., $\rho_{lT} + \rho_{ll} = 1$, which is easily verified by substituting Eqs. (X.36) and (X.37).

Analysis of these equations shows that at some angle of incidence θ, satisfying the condition

$$c_T^2 \sin 2\theta \sin 2\theta_l' - c_l^2 \cos^2 2\theta = 0, \qquad (X.40)$$

the transverse reflected wave vanishes, i.e., the incident transverse wave is completely transformed into a longitudinal wave. It is not difficult to verify, however, that this condition holds only for media in which the ratio of the velocities $c_l/c_T \geqslant \sqrt{3}$. Since, however, according to (X.16) $c_l/c_T = [(2 - 2\nu_0)/(1 - 2\nu_0)]^{1/2}$, the condition (X.40) can be satisfied only in media whose Poisson's ratio does not exceed the value $\nu_0 \leqslant 0.25$.

Furthermore, it is evident from Eq. (X.36) and (X.37) that the incident shear wave is not transformed (i.e., $\rho_{Al} = 0$ and $\rho_{AT} = 0$) at normal incidence ($\theta = 0$) and at incidence at an angle $\theta = 45°$. However, the angle $\theta = 45°$ exceeds the critical angle of incidence at which the angle of reflection of the longitudinal wave equals $\pi/2$. This critical angle is defined by the relation

$$\sin \theta_{cr} = c_T/c_l. \qquad (X.41)$$

It follows from Eq. (X.16), however, that the velocity ratio c_T/c_l cannot exceed $1/\sqrt{2}$. Thus, for any medium, $\theta_{cr} = \sin^{-1}(c_T/c_l) \leqslant 45°$ and equals 45° only for liquids, for which $\nu_0 = 0.5$. A shear wave, however, cannot propagate in a liquid, so that actually the inequality $\theta_{cr} < 45°$ always holds. Thus the values of the critical angles for different media, in accordance with the relations (X.41) and (X.16), are determined by Poisson's ratio. For $\nu_0 \simeq 0.3$, which often happens for simple solids, the critical angle is $\simeq$ 30–35°. For angles of incidence $\theta > \theta_{cr}$ total internal reflection of the transverse wave occurs:

there is no reflected longitudinal wave. At $\theta = \theta_{cr}$ the angle $\theta_l' = 90°$, i.e., the longitudinal wave propagates parallel to the reflecting boundary. In this case, according to relations (X.36) and (X.37), $\rho_{AT} = -1$ and $\rho_{Al} = (2 \sin \theta \sin 2\theta / \cos 2\theta)$. The intensity reflection coefficient of this wave, of course, equals zero, which follows from expressions (X.39); the minus sign in the amplitude reflection coefficient of the transverse wave means (see Chap. VII) that it is reflected with a change in phase by 180°, i.e., with a loss of a half-wave. For angles of incidence $\theta > \theta_{cr}$ the situation is analogous to the one that we already studied in Chap. VII for refracted longitudinal waves in liquids under the same condition. Namely, for $\theta > \theta_{cr}$, $\sin \theta_l'$ is greater than 1, i.e., the angle θ_l' is imaginary and the part of the displacement field corresponding to longitudinal deformation degenerates into a nonuniform, surface, longitudinal wave, whose amplitude decreases exponentially away from the reflecting surface (i.e., in the $-x$ direction) with an exponent proportional to $\cos \theta_l'$. In this case, the reflection coefficient of the transverse wave ρ_{AT} will be a complex number with unit modulus, while the phase will depend on $\cos \theta_l'$, i.e., on the angle of incidence θ. Thus an incident transverse wave can be transformed into a longitudinal wave only in a limited range of angles of incidence in the interval $0° < \theta < \sin^{-1}[(1 - 2\nu_0)/(2 - 2\nu_0)]^{-1/2}$, i.e., practically up to the values $\theta = 30$–$35°$.

Now, let a **purely longitudinal wave** with displacement $\mathbf{u}_l$ and wave vector $\mathbf{k}_l = k_l \mathbf{n}_l = \mathbf{n}_l \omega / c_l$ be incident on the free plane boundary of a solid at an angle θ to the x-axis (see Fig. 67b). The equation of the incident wave with displacement amplitude A_l is

$$\begin{aligned} \mathbf{u}_l &= A_l \mathbf{n}_l \exp\{i[\omega t - \mathbf{k}_l \mathbf{n}_l \cdot \mathbf{r}]\} = \\ &= A_l (\cos\theta + \sin\theta) \exp\{i\omega[t - (x\cos\theta + y\sin\theta)/c_l]\}, \end{aligned} \tag{X.42}$$

The equations of the reflected longitudinal wave with displacement amplitude A_l' and the transverse wave with amplitude A_T' will have the previous form (X.31) and (X.30), in which it is only necessary to change the notation for the angles of reflection $\theta \to \theta_l'$ and $\theta_l' \to \theta$, because the **longitudinal** wave is now reflected at the angle of incidence. The reflected shear wave, of course, will be polarized in the plane of incidence, because the longitudinal wave incident on the boundary does not create displacements along the z-axis. Thus the equations of the reflected longitudinal and shear reflected waves have the form

$$\mathbf{u}'_T = A'_T(\cos\theta'_T - \sin\theta'_T)\exp\{i\omega[t-(-x\cos\theta'_T + y\sin\theta'_T)/c_T]\}, \quad (X.43)$$

$$\mathbf{u}'_l = A'_l(-\cos\theta + \sin\theta)\exp\{i\omega[t-(-x\cos\theta + y\sin\theta)/c_l]\}. \quad (X.44)$$

The total displacement field in the medium is

$$\mathbf{u} = \mathbf{u}_l + \mathbf{u}_l + \mathbf{u}'_T. \quad (X.45)$$

The boundary conditions also remain unchanged and have the form of Eqs. (X.35). Performing in these equations the corresponding differentiation and solving the equations obtained at $x = 0$ for A'_l/A_l and A'_T/A_T, we find the amplitude reflection coefficients for the longitudinal and shear waves:

$$\rho_{Al} = \frac{A'_l}{A_l} = \frac{c_T^2 \sin 2\theta \sin 2\theta'_T - c_l^2 \cos^2 2\theta'_T}{c_T^2 \sin 2\theta \sin 2\theta'_T + c_l^2 \cos^2 2\theta'_T},$$
$$\rho_{AT} = \frac{2c_l c_T \sin 2\theta \cos 2\theta'_T}{c_T^2 \sin 2\theta \sin 2\theta'_T + c_l^2 \cos^2 2\theta'_T}, \quad (X.46)$$

where $\sin\theta'_T/\sin\theta = c_T/c_l$. For the intensity reflection coefficients we obtain, respectively, $\rho_{ll} = (A'_l/A_l)^2$ and $\rho_{lT} = c_T \cos\theta'_T/(c_l \cos\theta) \times (A_T/A_l)^2$. These expressions also show that at some angle of incidence θ, corresponding to the condition

$$c_T^2 \sin 2\theta \sin 2\theta'_T - c_l^2 \cos^2 2\theta'_T = 0, \quad (X.47)$$

the incident longitudinal wave is completely transformed into a transverse wave. In this case, the condition (X.47), like condition (X.40), holds only for a limited number of media, for which $c_l/c_T \geqslant \sqrt{3}$, i.e., $\nu_0 \leqslant 0.25$. According to Eq. (X.46), the incident longitudinal wave is not transformed at $\theta = 0$ (normal incidence) and at some angle of incidence θ corresponding to reflection of the shear wave at an angle $\theta'_T = 45°$. This situation cannot, however, be realized, because according to the foregoing discussion it corresponds to an angle of incidence $\theta > 90°$. On the other hand, since $c_T < c_l$ total internal reflection of the incident wave is impossible in this case also.

We shall now study the case of contact between **a solid and a liquid**. Let a **shear wave,** polarized in the plane of incidence (for a wave polarized in the perpendicular direction $\rho_{AT} = 1$), be incident on the plane boundary with a liquid from a solid at an angle θ to the x-axis (see Fig. 67c). We retain all of the previous notation for waves propagating in the solid. The total displacement field in the solid will have the same form (X.32) as in the case of a free surface, i.e., $\mathbf{u} = \mathbf{u}_T + \mathbf{u}'_T + \mathbf{u}_l$, where $\mathbf{u}_T$, $\mathbf{u}'_T$ and $\mathbf{u}'_l$ are the displacement vectors in the incident transverse, reflected transverse, and reflected longitudinal waves described by Eq. (X.29)–(X.31), respectively. Only the longitudinal wave, whose characteristics we shall label with the subscript liq, can exist in the liquid. The equation for the refracted longitudinal wave, in the case corresponding to Fig. 67c, can be written in the form*

$$\mathbf{u}_{liq} = A_{liq}(\cos\theta_2 + \sin\theta_2)\exp\{i\omega[t - (x\cos\theta_2 + y\sin\theta_2)/c_{liq}]\}, \quad (X.48)$$

where θ_2 is the angle of refraction, satisfying the general condition (X.20), i.e., $\sin\theta_2/\sin\theta = c_{liq}/c_T$.

Under the condition of full acoustic contact, the normal components of the stress and displacements must be continuous at the boundary between the solid and the liquid. As far as the tangential component of the stress tensor is concerned, it also must be continuous, but since there are no shear stresses in a liquid, the condition for the tangential component of the stress at the boundary remains unchanged, i.e., the tangential component vanishes at $x = 0$. The stress components are represented in terms of the true deformation and the velocity of sound by Eq. (X.34). For liquids, $c_T = 0$ and the normal component of the stress ("negative pressure") $\sigma_{xx} = (-p) = \rho_{liq}c_{liq}^2\partial\xi/\partial x$. Thus the equality of the stress components at the boundary of a solid with a liquid yields the following boundary conditions in the plane $x = 0$:

---

*In Chap. VII we wrote down the equation for a longitudinal wave in a liquid in terms of the scalar potential φ, which is related to the displacement velocity vector (or the displacement vector itself, whence the particle velocity is determined by differentiation with respect to time) by the relation $\mathbf{u} = -\text{grad}\,\varphi$. A vector potential can be introduced for shear waves in an analogous manner, but for greater clarity we study the displacement field directly.

$$(\rho c_l^2 - 2\rho c_T^2)\left(\frac{\partial \xi}{\partial x} + \frac{\partial \eta}{\partial y}\right) + 2\rho c_T^2 \frac{\partial \xi}{\partial x} = \rho_{\text{liq}} c_{\text{liq}}^2 \left(\frac{\partial \xi}{\partial x}\right)_{\text{liq}},$$

$$\rho c_T^2 \left(\frac{\partial \xi}{\partial y} + \frac{\partial \eta}{\partial x}\right) = 0. \tag{X.49}$$

In addition, the normal components of the displacements must be continuous at the boundary, which gives

$$|\mathbf{u}|_x = \xi_{\text{liq}} = \xi_T + \xi_T' + \xi_l', \tag{X.50}$$

where ξ_T, ξ_T', and ξ_l' are the x-components of the displacement vectors in the incident and reflected transverse waves in the solid, described by Eqs. (X.29)–(X.31), and ξ_{liq} is the normal component of the longitudinal displacement wave in the liquid (X.48).

The boundary conditions (X.49) and (X.50) give three equations for the unknown amplitudes of the reflected waves (A_T' and A_l') and of the refracted wave (A_{liq}). Solving these equations for A_T'/A_T, A_l'/A_T and A_{liq}/A_T, we obtain equations for the reflection coefficients (ρ_{AT} and ρ_{Al}) and the transmission coefficient (d_A) of the shear wave incident on the solid–liquid boundary:

$$\rho_{AT} = \frac{A_T'}{A_T} = \tag{X.51}$$

$$= -\frac{\rho_{\text{liq}} c_{\text{liq}}/\cos\theta_2 + (\rho c_l \cos^2 2\theta)/\cos\theta_l' - (\rho c_T \sin^2 2\theta)/\cos\theta}{\rho_{\text{liq}} c_{\text{liq}}/\cos\theta_2 + (\rho c_l \cos^2 2\theta)/\cos\theta_l' + (\rho c_T \sin^2 2\theta)/\cos\theta},$$

$$\rho_{Al} = \frac{A_l'}{A_T} = -\frac{c_l}{c_T}\,\frac{\cos 2\theta}{\sin 2\theta_l'}(1 + \rho_{AT}), \tag{X.52}$$

$$d_A = \frac{A_{\text{liq}}}{A_T} = \frac{c_T \operatorname{ctg}\theta_2}{c_{\text{liq}}\, 2\sin^2\theta}(1 + \rho_{AT}), \tag{X.53}$$

and the angles θ, θ_l', and θ_2 are related by the relations $\sin\theta_l'/\sin\theta = c_l/c_T$ and $\sin\theta_2/\sin\theta = c_{\text{liq}}/c_T$. The quantities $\rho c = z$ in expression (X.51) represent the specific acoustic impedances of the media under study relative to the longitudinal or shear wave, while the quantities $\rho c/\cos\theta$ represent their ratios to the direction cosines of the corresponding waves. If we introduce the "normal" acoustic impedances using the notation $\rho_{\rm liq}c_{\rm liq}/\cos\theta_2 \equiv z^n_{\rm liq}$, $\rho c_l/\cos\theta_l' \equiv z^n_l$, and $\rho c_T/\cos\theta \equiv z^n_T$, then Eq. (X.51) can be written in the more compact form:

$$\rho_{AT} = \frac{A_T'}{A_T} = -\frac{z^n_{\rm liq} + z^n_l\cos^2 2\theta - z^n_T\sin^2 2\theta}{z^n_{\rm liq} + z^n_l\cos^2 2\theta + z^n_T\sin^2 2\theta}.$$

Now, let a **longitudinal ultrasonic wave** be incident from the solid on the boundary with a liquid at an angle θ. The displacement field in the solid will be described by expression (X.45) together with the equations for the incident and reflected waves (X.42)–(X.44). The equation of the refracted wave in the liquid has the previous form (X.48). The boundary conditions remain unchanged, and we obtain

$$\rho_{Al} = \frac{A_l'}{A_l}\frac{z^n_{\rm liq} + z^n_T\sin^2 2\theta_T' - z^n_l\cos^2 2\theta_T'}{z^n_{\rm liq} + z^n_T\sin^2 2\theta_T' + z^n_l\cos^2 2\theta_T'}, \qquad \text{(X.54)}$$

$$\rho_{AT} = \frac{A_T'}{A_l} = \frac{c_T}{c_l}\,\frac{\sin 2\theta}{\cos 2\theta_T'}(1 - \rho_{Al}), \qquad \text{(X.55)}$$

$$d_A = \frac{A_{\rm liq}}{A_l} = \frac{\cos\theta}{\cos\theta_2\cos^2 2\theta_T'}(1 - \rho_{Al}), \qquad \text{(X.56)}$$

and $\sin\theta_T'/\sin\theta = c_T/c_l$ and $\sin\theta_2/\sin\theta = c_{\rm liq}/c_l$.

Taking into account the relations between the angles, it is evident that for $z_{\rm liq} = 0$ these equations transform into the equations obtained previously for the case of a free boundary of an isotropic solid. For $\theta = 0$ (normal incidence), in the case of an incident shear wave Eqs. (X.51)–(X.53) give $\rho_{AT} = -1$ and $\rho_{A\,l} = d = 0$; if, on the other hand, a longitudinal wave is incident on the boundary between a solid and a liquid, then, according to the relations (X.54)–(X.56), $\rho_{Al} = (z_{\rm liq} - z_l)/(z_{\rm liq} + z_l)$, $\rho_{AT} = 0$, and $d_A = 2z_l(z_{\rm liq} + z_l)$, which is the same as the result obtained in Chap. VII for the case of normal incidence of a plane longitudinal wave on the boundary between two media.

In conclusion, we shall examine the inverse problem: refraction of a

longitudinal wave incident from a liquid on the plane boundary with a solid. We solved this problem for two liquids in Chap. VII. The result, which in this case is obtained for the reflection and refraction coefficients in the form of relations (VII.39) and (VII.40), follows directly from Eqs. (X.54)–(X.56), if we set in them $c_T = 0$ (and $z_T = 0$). If, however, the longitudinal wave is incident at some angle θ from the liquid on the surface of a solid, then it excites both longitudinal and shear displacements in the solid, as a result of which two refracted waves propagating with velocities c_l and c_T at angles θ_l and θ_T arise in the solid (Fig. 67d). We shall find the reflection and refraction coefficients of these waves.

The displacement field in the liquid is $\mathbf{u} = \mathbf{u}_{\text{liq}} + \mathbf{u}'_{\text{liq}}$ and the displacement field in the solid is $\mathbf{u}_{\text{sol}} = \mathbf{u}_l + \mathbf{u}_T$. The condition of continuity of the normal components of the displacements in this case is

$$u_{\text{liq}x} + u'_{\text{liq}x} = u_{lx} + u_{Tx}\big|_{x=0}. \tag{X.57}$$

The condition of continuity of the normal and tangential components of the stresses at the boundary, however, remains in its previous form, i.e., Eq. (X.49), valid at $x = 0$. The three boundary conditions (X.49) and (X.50) yield three equations, from which the reflection and transmission coefficients ρ_A and d_A of the longitudinal wave incident from the liquid can be found. Carrying out the corresponding calculations, we obtain

$$\rho_A = \frac{A'_{\text{liq}}}{A_{\text{liq}}} = -\frac{z_l^n \cos^2 2\theta_T + {}_T^n \sin 2\theta_T - z^n{}_{\text{liq}}}{z_l^n \cos^2 2\theta_T + z_T^n \sin 2\theta_T + z_{\text{liq}}^n}, \tag{X.58}$$

$$d_{Al} = \frac{A_l}{A_{\text{liq}}} = -\frac{2z_{\text{liq}}^n \cos 2\theta_T}{z_l^n \cos^2 2\theta_T + z_T^n \sin 2\theta_T + z_{\text{liq}}^n}\,\frac{\cos\theta}{\cos\theta_l}, \tag{X.59}$$

$$d_{AT} = \frac{A_T}{A_{\text{liq}}} = -\frac{2z_{\text{liq}}^n \sin 2\theta_T}{z_l^n \cos^2 2\theta_T + z_T^n \sin 2\theta_T + z_{\text{liq}}^n}\,\frac{\cos\theta}{\cos\theta_T}, \tag{X.60}$$

and, in addition, $\sin\theta_l/\sin\theta = c_l/c_{\text{liq}}$ and $\sin\theta_T/\sin\theta = c_T/c_{\text{liq}}$.

Thus a longitudinal ultrasonic wave incident at an arbitrary angle from a liquid on the boundary with a solid decomposes into a longitudinal and a shear wave. As a result, the coefficient of reflection of the incident wave from the surface of the solid is less than the coefficient of reflection from the boundary with a liquid, which would have the value $\rho_{liq}c_{liq} = z_l$, i.e., it would have an equivalent acoustic impedance equal to the acoustic impedance of the given solid with respect to purely longitudinal waves. Indeed, for liquids $c_T = 0$ and $z_T = 0$ and the equation for the reflection coefficient (X.58) transforms, to within a phase factor (—1), into the expression for the coefficient of reflection (VII.39) from a medium which has only volume elasticity. It is not difficult to verify that the specific acoustic impedance of the boundary of the solid $z_{sol,T} = z_l^n\cos^2 2\theta_T + z_T^n\sin^2 2\theta_T$, due to the appearance of longitudinal and shear waves in it and determining the reflection coefficient from Eq. (X.58), is less than the specific impedance $z_{liq} = z_l^n = z_l/\cos\theta_l$, which a liquid with the same value of z_l as a solid would have and which determines the reflection coefficient of the liquid. Actually,

$$\frac{z_{solid}}{z_{liq}} = \cos^2 2\theta_T + \frac{z_T^n}{z_l^n}\sin^2 2\theta_T = 1 - (1 - \frac{c_T}{c_l}\,\frac{\cos\theta_l}{\cos\theta_T})\sin^2 2\theta_T.$$

Since, however, $c_l > c_T$ and $\theta_l > \theta_T$ always hold, i.e., $\cos\theta_l < \cos\theta_T$, we have $z_{sol}/z_{liq} < 1$, i.e., the coefficient of reflection from the surface of a solid is less than the coefficient of reflection from a liquid. Therefore, the shear elasticity of the reflecting medium decreases the acoustic stiffness of its boundary. The same thing can be said about longitudinal waves which are incident **from a solid body on the boundary of a liquid** and decompose into reflected longitudinal and shear waves. This is evident, in particular, from Eq. (X.54): if we set $z_T = 0$ there, then with the same value of z_l the reflection coefficient of the longitudinal wave increases. This result also follows from energy considerations: if the medium in which the incident longitudinal wave propagates has shear elasticity, then part of the energy in the reflected wave is transferred into the shear wave.

When a longitudinal wave is incident obliquely on the boundary between a liquid and a solid, the transformation of waves in the solid leads to results that differ considerably from those obtained in Chap. VII for the boundary between two media exhibiting only volume elasticity. At some angle of incidence θ of the longitudinal wave from the liquid on the boundary with a solid, this transformation can reach 100%. As is evident from Eqs. (X.59) and (X.60), such a situation occurs when the angle of refraction of the shear wave $\theta_T = 45°$. In this case, Eqs. (X.58)–(X.60) assume the following form:

$$\rho_A = \frac{z_T\cos\theta - z_{\text{liq}}\cos\theta_T}{z_T\cos\theta + z_{\text{liq}}\cos\theta_T}, \qquad d_{Al} = 0,$$
$$d_{AT} = -\frac{2z_{\text{liq}}\cos\theta}{z_T\cos\theta + z_{\text{liq}}\cos\theta_T}. \tag{X.61}$$

The angle of refraction of the shear wave $\theta_T = 45°$ corresponds to the angle of incidence

$$\theta = \sin^{-1}[(c_{\text{liq}}/c_t)\sqrt{1/2}] \tag{X.62}$$

Since $\theta_T = 45°$ almost always, in contrast to the previously studied total transformation of the wave incident from a solid, the condition (X.62) holds for most combinations of liquids and solids. For example, at a water–ice boundary at a temperature of 0 °C total transformation of a longitudinal wave into a shear wave occurs at an angle of incidence $\theta = 30°$; at a water–aluminum boundary at 20°C total transformation occurs at $\theta = 20.5°$, etc. This circumstance can be used to excite pure shear waves in a solid with the help of a transducer emitting longitudinal waves into a liquid. The reflection coefficient of the longitudinal wave is, however, always large, because in the first of Eqs. (X.61) the inequalities $\theta_T > \theta$ and $z_T > z_{\text{liq}}$ almost always hold, so that $\rho_A > 0$ (ρ_A usually equals 0.4–0.6).

Total transformation of longitudinal waves into shear waves in a solid also occurs at angles of incidence from the liquid $\theta \gg \theta_{cr} = \sin^{-1}(c_{\text{liq}}/c_l)$, i.e., in the case of total internal reflection of the longitudinal wave, and this case is also realized for almost all combinations of liquids and solids, because $c_l > c_{\text{liq}}$ and $\theta_l > \theta$ almost always (see Fig. 67d). At $\theta = (\theta_{cr})_1$ the refracted longitudinal wave propagates in the solid parallel to its boundary and for angles of incidence $\theta > (\theta_{cr})_1$, the angle θ_l becomes complex which, as is well known, corresponds to a nonuniform longitudinal wave (in the solid) decaying exponentially away from the boundary of the solid. Finally, for $\theta \geqslant (\theta_{cr})_2 = \tan^{-1}(c_{liq}/c_T)$ the same thing also occurs with a shear wave, after which the absolute magnitude of the reflection coefficient of a longitudinal wave incident on a solid for all angles of incidence equals 1, and the reflection of an ultrasonic wave from a solid no longer differs from reflection from the boundary between two liquids.

All the results obtained in Chap. VII are also valid in the case of normal incidence of a plane ultrasonic wave from a liquid on a solid boundary. In this case $(\theta = \theta_l = \theta_T = 0)$, Eqs. (X.58)–(X.60) give

$$\rho_A = \frac{z_l - z_{liq}}{z_l + z_{liq}}, \qquad d_A = \frac{2z_{liq}}{z_l + z_{liq}},$$

which are the asme as the corresponding expressions (VII.8) and (VII.9), obtained for normal incidence of a plane wave from a flexible medium on the boundary with a stiff medium. Evidently, the case of normal incidence of a plane ultrasonic wave from a liquid on a plane-parallel solid layer also gives nothing new: all equations obtained in § VII.1 will be valid for this case also.

§ 4. Rayleigh waves

In the preceding section, as in Chap. VII, we studied the effect of boundaries on the propagation of volume waves in the interior of a medium. We shall now clarify the nature of the disturbances and the propagation of these disturbances in the immediate vicinity of the free boundary of an isotropic solid. It is evident at the outset that since for any deformations the stress vanishes on a free boundary while away from the boundary it increases to some value determined by Hooke's law (X.34), the effective stiffness of the boundary layer will differ from the bulk stiffness of the elastic medium and, therefore, the nature of the elastic disturbances in this layer and the velocity of propagation of disturbances near the free boundary will differ. The propagation of such surface disturbances can evidently be described quantitatively by starting from the general wave equation, which is valid in the entire volume of the elastic medium, and finding its solution at points near the free boundary of the medium.

Let the boundary of the homogeneous isotropic solid lie, as before, in the yz plane and let the x-axis be its outer normal, i.e., the medium under study occupies the half-space $x < 0$. The general wave equation for this medium can be written in the form (X.17), i.e.,

$$\Delta \mathbf{u} = \frac{1}{c^2}\frac{\partial^2 \mathbf{u}}{\partial t^2}, \qquad (X.63)$$

where $\mathbf{u}$ denotes any displacements $\mathbf{u}_l$ or $\mathbf{u}_T$, and c is the corresponding velocity of propagation c_l or c_T. Anticipating a singularity along the x-axis related to the presence of a free boundary, we write the solution of Eq. (X.63) in the form $\mathbf{u} = f(x)\exp[i(\omega t - ky)]$, separating out the part depending on x and examining, as before, a two-dimensional pattern of displacements. Substituting this solution into Eq. (X.63), we obtain an equation for the function $f(x)$:

$$\partial^2 f(x)/\partial x^2 - (k^2 - \frac{\omega^2}{c^2}) f(x) = 0, \qquad (X.64)$$

where c, by definition, is the velocity of propagation of the longitudinal or shear volume wave and k is the wave number of the wave under study. If $\omega^2/c^2 > k^2$, then the solution of Eq. (X.64) will be a sinusoidal function of x and, therefore, Eq. (X.64) gives an ordinary plane wave (longitudinal or shear), which exists for all values of x. This means that the interesting solution of Eq. (X.64) is the case when $\omega^2/c^2 < k^2$. Then, this solution will represent an exponential function with the exponent $\pm x\sqrt{(k^2 - \omega^2/c^2)}$. Furthermore, the solution with the minus sign indicates that $\mathbf{u}$ increase without limit in a direction into the body ($x < 0$), i.e., it is physically meaningless. The real solution is therefore the function $f(x) = f_0 \exp[x\sqrt{(k^2 - \omega^2/c^2)}]$, and we arrive at the following solution of the wave equation (X.63):

$$\mathbf{u} = f_0 \exp(\chi x)\exp[i(\omega t - k_R y)], \qquad (X.65)$$

where f_0 is a constant which is independent of coordinates and time and $\chi \equiv \sqrt{(k_R^2 - \omega^2/c^2)}$. This solution corresponds to a wave propagating along the y-axis and decaying exponentially in the $-x$ direction (into the body), i.e., a wave which exists in a thin surface layer of the solid. Such waves are called *surface* or *Rayleigh waves*, because the first calculations of these waves were performed by Rayleigh.[1]

We recall that, according to the meaning of the solution obtained (X.65), the velocity c appearing in it through the parameter χ is different for different components of the displacement: the component $\mathbf{u}_l$ corresponds to the velocity c_l and the shear component $\mathbf{u}_T$ corresponds to the velocity c_T. These components can propagate independently in the volume of a solid body with corresponding velocities, i.e., volume waves can be purely longitudinal as well as purely shear waves. In a surface wave, however, due to the presence of a free boundary the displacement $\mathbf{u}$ is always mixed: it includes different components which, generally speaking, are no longer "longitudinal" or "transverse." The corresponding calculation, using the boundary conditions, shows that the displacement trajectory of particles in the surface wave is an ellipse whose long axis is perpendicular to the surface and whose short axis is parallel to the surface and oriented along the direction of propagation of the surface wave, i.e., along the y-axis in this case. The ration of the axes depends on the ratio of the velocities c_l/c_T, i.e., on Poisson's ratio, and for $\nu_0 = 0.3$, ≈ 1.5 for particles on the surface ($x = 0$). The velocity of propagation of a Rayleigh surface wave $c_R = \omega/k_R$ also depends on c_l/c_T, i.e., on ν_0, and does not depend on the frequency ω.

It is interesting to note that these results can be formally obtained based on the relations derived in the preceding section by interpreting a Rayleigh wave as a degenerate case of reflection of plane waves, in which the coefficient of reflection of the incident wave from the free boundary becomes infinite. Since the physical origin of the reflection and refraction of waves at boundaries lies in the radiation by the oscillating boundary, the indicated condition ($\rho_A = \infty$) corresponds to a wave process propagating along the boundary without the incident wave, i.e., a free surface wave. Its propagation velocity c_R can be determined from the velocity of the track of the reflected wave with an infinite reflection coefficient. For example, for a reflected shear wave $c_R = c_T/\sin\theta_T^{\infty}$ at $\rho_{AT} = \infty$. Setting $z_{\text{liq}} = 0$ in Eq. (X.51) for the reflection coefficient of the shear wave and equating its denominator to zero, we obtain the equation $(c_l/\cos\theta_l)\cos^2\theta_T + (c_T/\cos\theta_T)\sin^2 2\theta_T = 0$, whence, taking into account the relation between the angles θ_l and θ_T $(\sin\theta_T/\sin\theta_l = c_T/c_l)$, it is not difficult to calculate the values of $\sin\theta_T^{\infty}$ that determine the velocity of the Rayleigh wave as a function of the ratio c_T/c_l for a given medium, i.e., as a function of Poisson's ratio, since according to (X.16) $c_T/c_l = \{(1-2\nu_0)/[2(1-\nu_0)]\}^{1/2}$. The results of such a calculation are presented in Fig. 68, whence it is evident that as ν_0 varies between the two limiting values 0 and 1/2, the velocity of the Rayleigh wave for different media varies from $0.874\ c_T$ to $0.955\ c_T$, i.e., it differs insignificantly from the velocity of volume shear waves.

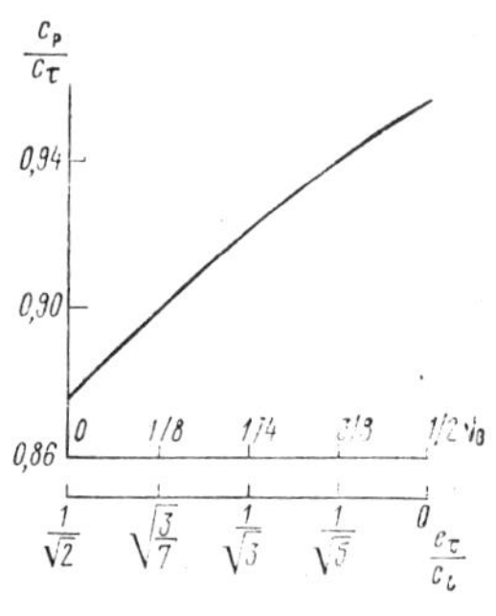

Fig. 68.

Surface ultrasonic waves of various types, including Rayleigh waves, play an important role in modern engineering and physical ultrasonics. There are a number of special reviews and monographs devoted to such waves, which can better acquaint the reader with these problems (see, e.g., Refs. 74–76).

§ 5. Love waves

The Rayleigh waves studied above can propagate along a free surface of a solid. There is another case of interest in physical acoustics. This is the situation in which a thin layer of another solid material lies on the surface of a solid with different acoustic properties. Special types of elastic waves can propagate in such a layer under certain conditions. Amongst these waves, the so-called *Love waves*, i.e., shear waves with displacement parallel to the interface, are of greatest interest. As an analog of this case, we shall examine a plane-parallel layer with thickness d lying on the surface of a solid half-space, whose characteristics we shall label with the subscript 1; the characteristics of the layer will remain unlabelled. We shall orient the axis perpendicular to the layer into the half-space and the y-axis along the boundary, and we shall seek the solution of the wave equation (X.4) for both media with nonzero components of the displacements $u_z = \zeta$, which do not depend on z, making the assumption that the displacements vanish in the limit $x \to \infty$. For such displacements, we obtain from Eq. (X.4) for the layer and the half-space, respectively:

$$\frac{\partial^2 \zeta}{\partial x^2} + \frac{\partial^2 \zeta}{\partial y^2} = \frac{1}{c_T^2}\frac{\partial^2 \zeta}{\partial t^2}, \quad \frac{\partial^2 \zeta_1}{\partial x^2} + \frac{\partial^2 \zeta_1}{\partial y^2} = \frac{1}{c_{1T}^2}\frac{\partial^2 \zeta_1}{\partial t^2}, \qquad (X.66)$$

where c_T and c_{1T} are the velocities of propagation of shear waves in the layer and in the half-space. For the boundary conditions, we shall choose the absence of stress on the free boundary of the layer, i.e., at $x = -d$, and the continuity of the displacements and stresses at the boundary between the layer and the half-space, i.e., at $x = 0$. These conditions give:

$$\partial \zeta / \partial x = 0 \text{ at } x = -d,$$

$$\zeta = \zeta_1, \ \rho c_T^2 \frac{\partial \zeta}{\partial x} = \rho_1 c_{1T}^2 \frac{\partial \zeta_1}{\partial x} \text{ at } x = 0, \qquad (X.67)$$

where ρ and ρ_1 are the densities of the material in the layer and in the half-space, respectively. As before, we shall seek the solutions of Eq. (X.66) in the form of sinusoidal plane waves with frequency ω, separating out factors that depend on x,

$$\zeta = f(x)\exp[i(\omega t - ky)]; \quad \zeta_1 = f_1(x)\exp[i(\omega t - ky)], \qquad (X.68)$$

where $k = \omega/c_L$ is the wave number of the wave sought and c_L is the velocity of the wave. Substituting (X.68) into Eq. (X.67), we obtain

$$\frac{\partial^2 f(x)}{\partial t^2} + k^2 \alpha f(x) = 0, \quad \frac{\partial^2 f_1(x)}{\partial t^2} - k^2 \beta^2 f_1(x) = 0, \qquad (X.69)$$

where $\alpha \equiv \surd(c_L^2/c_T^2 - 1)$ and $\beta \equiv \surd(1 - c_L^2/c_{1T}$. It is easy to see that Eq. (X.69) has real roots when the inequalities $c_T < c_L < c_{1T}$ are satisfied, which includes the case $c_T < c_{1T}$, i.e., the velocity of shear waves in the layer can be lower than in the substrate. From Eq. (X.69), we have for the functions $f(x)$ and $f_1(x)$:

$$f(x) = A\sin(\alpha kx) + B\cos(\alpha kx),$$
$$f(x) = C\exp(-\beta kx) + D\exp(\beta kx). \tag{X.70}$$

In order for the solution $f_1(x)$ to remain finite we must set $D = 0$, so that $f_1(x) = c\exp(-\beta kx)$. From the boundary conditions (X.67) it follows that

$$B = C, \quad A = -C\rho_1 c_{1T}^2/(\rho c_T^2)(\beta/\alpha). \tag{X.71}$$

Substituting these results into the general solution of the wave equation (X.68), we obtain finally

$$u_z = \zeta = C\left[\cos(\alpha kx) - \frac{\rho_1 c_{1T}^2}{\rho c_T^2}\frac{\beta}{\alpha}\sin(\alpha kx)\right]\exp[i(\omega t - ky)];$$

$$\zeta_1 = C\exp[-\beta kx + i(\omega t - ky)].$$

This solution describes a Love wave propagating along the y-axis with velocity c_L and with displacement parallel to the boundary of the layer and perpendicular to the direction of propagation. The velocity of this wave is easily found from Eq. (X.70) taking into account the first boundary conditions (X.67) and relations (X.71), which give:

$$-C\rho_1 c_{1T}^2/(\rho c_T^2)(\beta/\alpha)\cos(\alpha kd) + C\sin(\alpha kh) = 0,$$

whence

$$\tan(\alpha kd) = \rho_1 c_{1T}^2\beta/(\rho c_T^2\alpha). \tag{X.72}$$

Since $k = \omega/c_L$, this expression determines the velocity of Love waves as a function of the thickness of the layer and the relations between the densities and velocities of propagation of the usual shear waves in the materials of the layer and substrate. Since the energy of Love waves is concentrated near the surface of the substrate, these waves, like Rayleigh waves, are weakly damped waves and can propagate over large distances. However, their velocity of propagation, according to relation (X.72), depends on the frequency, i.e., unlike Rayleigh waves, Love waves are dispersive. Another difference is that Love waves are purely transverse, and they do not have longitudinal displacements. For this reason, a liquid at the free boundary of the layer, unlike the case of Rayleigh waves, should have no effect on the propagation of Love waves (if the liquid is assumed to be ideal). In a real liquid, as we know, however, in the presence of shear displacements viscous stresses arise in the boundary layer, which should change the boundary conditions at a free boundary. Because Love waves are very sensitive to the boundary conditions, however, the presence of contact with a liquid should change their propagation velocity. For this reason, Love waves can be used to study the shear properties of liquids, which is an important problem in molecular acoustics.

Love waves can also propagate in a free layer (plate).[64] Other types of waves with different polarization, used for different purposes in engineering ultrasonics, can also exist in layers and plates. An analysis of these waves, referred to in general as *Lamb waves*, is given in Refs. 64, 74, 76, and 77.

§ 6. Geometric dispersion of sound in rods

We shall now study the propagation of volume ultrasonic waves in an isotropic solid, bounded along the ultrasonic beam, i.e., in a homogeneous rod. The rigorous analysis of longitudinal oscillations in rods in the theory of elasticity is performed based on Hamilton's principle[73] and involves quite cumbersome calculations. In the low-frequency approximation, however, when the wavelength of the sound wave greatly exceeds the transverse dimensions of the rod, the tensile and compressive elasticity of the rod can be described by Young's modulus E and the problem of propagation of uniform longitudinal waves in a thin (relative to the wavelength) rod can be solved in a simple manner.

We orient the x-axis along a rod with cross-sectional area S, and we separate in the rod a volume element with thickness dx. The deformation at the point x is $\partial\xi/\partial x$; according to Hooke's law, $\partial\xi/\partial x = F_x(ES)$, where $F_x = ES(\partial\xi/\partial x)$ is the force applied to the cross section with coordinate x. A force F_{x+dx}, which in the linear approximation can be represented by $F_{x+dx} = F_x + (\partial F_x/\partial x)dx$, acts on the opposite face with coordinate $x + dx$. Thus the resulting force, due to the deformation, acting on the element singled out is

$$F = F_{x+dx} - F_x = \frac{\partial F_x}{\partial x} dx = \frac{\partial}{\partial x} (ES \frac{\partial \xi}{\partial x}) dx. \qquad (X.73)$$

This force imparts an acceleration $m\partial^2\xi/\partial t^2$ to the element with mass m. Equating this acceleration to the right side of expression (X.73) and assuming $S = const$, we obtain a wave equation for an infinitely long thin rod with density $\rho = m/(Sdx)$ (as before, we drop the zero index): $E\partial^2\xi/\partial x^2 = \rho\partial^2\xi/\partial t^2$ or

$$\partial^2\xi/\partial x^2 = c_{rod}^{-2}\, \partial^2\xi/\partial t^2, \qquad (X.74)$$

where

$$c_{rod} = \sqrt{E/\rho}\ . \qquad (X.75)$$

Equation (X.74) describes the propagation of a one-dimensional tension wave along the rod with velocity c_{rod}, determined by Eq. (X.75), in which the role of the effective stiffness in this case is played by Young's modulus E (dynamic, of course). We note, in passing, that since we are examining a one-dimensional problem, Eq. (X.75) is equally valid for a rod cut out of an anisotropic material in a direction in which the elasticity of the material is characterized by the given value of Young's modulus E.

Let us compare the result obtained (X.75) for the velocity of sound in a rod with the velocity of propagation of longitudinal waves in an unbounded solid, expressed in terms of Young's modulus. According to Table 11, this velocity equals

$$c_l = \left[\frac{E}{\rho} \frac{(1 - \nu_0)}{(1 + \nu_0)(1 - 2\nu_0)}\right]^{1/2}, \qquad (X.76)$$

where ν_0 is Poisson's ratio. Since $\nu_0 < 1/2$, the effective stiffness in this case is higher by an amount determined by the factor $(1 - \nu_0)/[(1 + \nu_0)(1 - 2\nu_0)] > 1$, so that the velocity of sound is higher in an unbounded medium than in a thin rod. Since usually $\nu_0 = 0.25$–0.30, this excess can be significant: $\approx 10\%$. For real bodies, the condition of unboundedness means that the wavelength of ultrasound Λ is much less than the transverse dimensions of the body. Therefore, the velocity of sound c_l determined by expression (X.76) refers to the high-frequency limit ($\omega \to \infty$) and the quantity c_{rod} in Eq. (X.75) refers to the low-frequency limit ($\omega \to 0$), when the wavelength is much longer than the transverse dimensions of the rod. Thus in rods the velocity of sound is dispersive, which in this case is called *geometric dispersion*, because it is related to the dimensions of the body.

The physical reason for the difference in the limiting values of c_{rod} and c_l is easily understood taking into account the fact that this difference is related to Poisson's ratio, which determines the reduction in the transverse dimensions of the rod as the length of the rod increases. In the case of a thin rod, the change in its transverse dimensions accompanying longitudinal deformations is not resisted by the external medium, which is equivalent to a lower effective stiffness than for an unbounded body with $\nu_0 \neq 0$. In its turn, the presence of transverse pulsations accompanying the propagation of longitudinal waves in a thin rod indicates that the transverse dimensions of the rod, i.e., the area S, depend on the coordinate x, which was neglected in the derivation of Eq. (X.74). When this circumstance is taken into account, which was done by Rayleigh[1] for a round rod with radius R, the velocity c_{rod} decreases with increasing frequency for $R < \Lambda$. The physical reason for this phenomenon is that when longitudinal deformations of the rod are accompanied by the excitation of radial oscillations the kinetic energy of the oscillating particles is higher than for purely longitudinal oscillations, which is equivalent to a larger oscillating mass, i.e., lower effective stiffness for longitudinal waves. When the wavelength Λ is comparable to the diameter of the rod, the transverse effect gives rise to resonant radial oscillations. In the resonance region, an anomalous dispersion is observed: the velocity of longitudinal waves drops to zero and then, as the frequency increases further, it returns rapidly from infinity, approaching a new, high-frequency limit $c(\infty) = c_l$

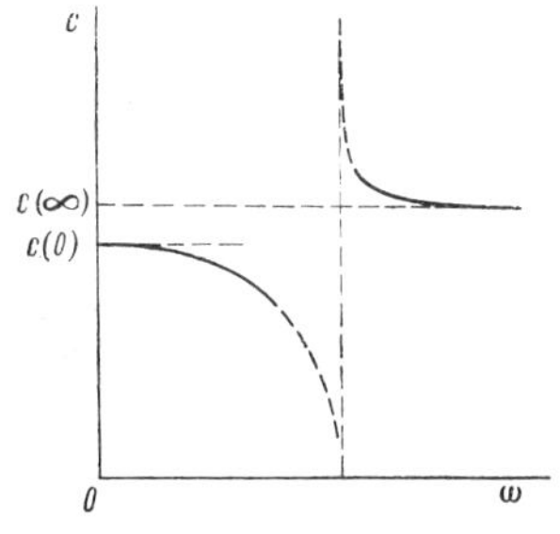

Fig. 69.

determined by Eq. (X.76). The general picture of geometric dispersion is qualitatively illustrated in Fig. 69, which is in good agreement with the experimental data.[12] The entire region of high dispersion in this picture lies in a small frequency band, corresponding to a change in the wavelength Λ by $\pm$(30–40)% relative to the radius of the rod. As experiments show, however, in exact measurements of the velocity of propagation of ultrasonic waves in rods, geometric dispersion is observed even when the geometric dimensions of the rod are tens and hundreds of times greater than the wavelength of the ultrasonic wave.[78]

One way or another, far from the region of dispersion the velocity of longitudinal waves in rods is determined by Eq. (X.76) for $\Lambda \ll R$ or Eq. (X.75) for $\Lambda \gg R$. The values of c measured, with higher or lower accuracy, at ultrasonic frequencies always correspond to the limiting case of an unbounded medium; these are the values presented in this book in all tables referring to solids. The values of c_{rod} can be easily calculated from them using Eq. (X.76) when the value of Poisson's ratio is known.

In conclusion, we note that the characteristic frequencies of oscillations of a plane-parallel layer consisting of a solid material, which are related to the finiteness of the thickness of the layer d, will evidently be determined by the previous equation (VIII.24), obtained for a transversely infinite medium, i.e., $\omega_n = n\pi c/d$ or $\nu_n = nc/(2cd)$, where $n = 1, 2, 3, \ldots$. However, for longitudinal waves, when the transverse dimensions of the layer are finite, the velocities of sound appearing in these equations will depend, owing to geometric dispersion, on the wavelength corresponding to the given frequency ν_n. For the fundamental frequency ($n = 1$), the wavelength of the traveling wave equals twice the thickness of the layer $2d$; the dispersion jump in the velocity of sound, however, in a cylindrical specimen with radius R occurs near $\Lambda \simeq R$. Therefore, a cylindrical layer whose longitudinal dimensions are greater than its transverse dimensions will behave relative to the fundamental characteristic frequency of longitudinal oscillations like a "rod" and a layer whose thickness is less than the transverse dimensions will behave like a "plate" in the acoustic–geometric sense of these words, i.e., in the first case the frequency of the fundamental tone will be determined by the value of the velocity

c_{rod} and in the second case it will be determined by the high-frequency value $c(\infty) = c_l$. Of course, for excitations at the overtone frequencies the reverse relationship will hold.

§ 7. Nonlinear elasticity and the origin of the nonlinear acoustics of solids

If in the problems involving ultrasonic propagation in solids the deformations are not assumed to be infinitesimal, then the complete form of the strain tensor ϵ_{ik} (I.5) must be used, i.e.,

$$\epsilon_{ik} = \frac{1}{2}\left[\frac{\partial u_i}{\partial x_k} + \frac{\partial u_k}{\partial x_i} + \frac{\partial u_l}{\partial x_i}\,\frac{\partial u_l}{\partial x_k}\right].$$

The stress tensor σ_{ik}, as shown in Chap. I, for adiabatic processes can be expressed in terms of the internal energy U by the general relation (I.20): $\sigma_{ik} = \partial U/\partial(\partial u_i/\partial x_k)$. The internal energy of the isotropic solid is invariant under transformations of the coordinates. On the other hand, it is a function only of the deformation of the body (dissipative processes are neglected and deformations are assumed to be ideally elastic). Therefore, the internal energy must depend only on the invariants of the strain tensor. In this case these invariants will be the quantities [6, 19, 29]: $J_1 = \Theta = \epsilon_{ll}$, $J_2 = (\epsilon_{ll}^2 - \epsilon_{ik}^2)/2$, and $J_3 = [\epsilon_{ik}\epsilon_{il}\epsilon_{kl} - (3/2)\,\epsilon_{ik}^2\epsilon_{ll} + (1/2)\,\epsilon_{ll}^2]/3$. Since the deformations are always small, the internal energy U can again be expanded in a series about the energy of the undeformed state. Assuming that this is the equilibrium state and retaining third-order infinitesimals, we now obtain:

$$U = \frac{\lambda}{2}\,\epsilon_{ll}^2 + \mu\epsilon_{ik}^2 + \frac{A}{3}\,\epsilon_{ik}\epsilon_{il}\epsilon_{kl} + B\epsilon_{ik}^2\epsilon_{ll} + \frac{C}{3}\,\epsilon_{ll}^3. \qquad (X.77)$$

In contrast to the previous expression (I.29), which was obtained for the case of linear elasticity, which is characterized by two moduli λ and μ, three additional constants A, B, and C appear here. Since these constants appear in the expansion of the energy in front of the cubic terms, they are called *third-order elastic moduli* or *nonlinear moduli*. Thus the elasticity of an isotropic solid to a first approximation is described on the whole by five constants and, for this reason, the nonlinear theory of elasticity based on this approximation is called a five-constant theory. The next approximation would require the introduction of four more fourth-order moduli, five more fifth-order moduli, etc. In what follows, we shall base our analysis on the five-constant theory.

Substituting into Eq. (X.77) the components of the strain tensor (I.5) and introducing instead of the Lamé constants the bulk modulus of elasticity $K = \lambda + (2/3)\mu$ and the shear modulus $G = \mu$, we obtain up to third-order infinitesimals:

$$U = \frac{G}{4}\left[\frac{\partial u_i}{\partial x_k} + \frac{\partial u_k}{\partial x_i}\right]^2 + \left[\frac{K}{2} - \frac{G}{3}\right]\left[\frac{\partial u_l}{\partial x_l}\right]^2 +$$
$$+ \left(G + \frac{A}{4}\right)\frac{\partial u_i}{\partial x_k}\frac{\partial u_l}{\partial x_i}\frac{\partial u_l}{\partial x_k} + \left(\frac{B+K}{2} - \frac{G}{3}\right)\frac{\partial u_l}{\partial x_l}\left[\frac{\partial u_i}{\partial x_k}\right]^2 +$$
$$+ \frac{A}{12}\frac{\partial u_i}{\partial x_k}\frac{\partial u_k}{\partial x_l}\frac{\partial u_l}{\partial x_k} + \frac{B}{2}\frac{\partial u_i}{\partial x_k}\frac{\partial u_k}{\partial x_i}\frac{\partial u_l}{\partial x_l} + \frac{C}{3}\left[\frac{\partial u_l}{\partial x_l}\right]^3 . \tag{X.78}$$

Using this expression, we write for the components of the stress tensor from their definition (I.19):

$$\sigma_{ik} = G\left[\frac{\partial u_i}{\partial x_k} + \frac{\partial u_k}{\partial x_i}\right] + \left(K - \frac{2}{3}G\right)\frac{\partial u_l}{\partial x_l}\delta_{ik} +$$
$$+ \left(G + \frac{A}{4}\right)\left[\frac{\partial u_l}{\partial x_i}\frac{\partial u_l}{\partial x_k} + \frac{\partial u_k}{\partial x_l}\frac{\partial u_i}{\partial x_l} + \frac{\partial u_l}{\partial x_k}\frac{\partial u_i}{\partial x_l}\right] +$$
$$+ \left(\frac{K}{2} - \frac{G}{3} + \frac{B}{2}\right)\left[\left[\frac{\partial u_l}{\partial x_i}\right]^2\delta_{ik} + 2\frac{\partial u_i}{\partial x_k}\frac{\partial u_l}{\partial x_l}\right] + \frac{A}{4}\frac{\partial u_k}{\partial x_l}\frac{\partial u_l}{\partial x_i} +$$
$$+ \frac{B}{2}\left[\frac{\partial u_l}{\partial x_j}\frac{\partial u_j}{\partial x_l}\delta_{ik} + 2\frac{\partial u_k}{\partial x_i}\frac{\partial u_l}{\partial x_l}\right] + C\left[\frac{\partial u_l}{\partial x_l}\right]^2\delta_{ik}, \tag{X.79}$$

where δ_{ik} is the Kronecker delta. It is not difficult to see that when the quadratic terms are neglected, expression (X.79) transforms into the linear Hooke's law for an isotropic solid (X.2). In the first nonlinear approximation the relation between the stresses and strains becomes considerably more complicated, even for isotropic bodies.

Substituting now the expressions for the stresses (X.79) into the equation of motion for an isotropic solid (X.1), we obtain the equation of motion in the form

$$\rho_0 \frac{\partial^2 u_i}{\partial t^2} - G \frac{\partial^2 u_i}{\partial x_k^2} - (K + \frac{G}{3}) \frac{\partial^2 u_l}{\partial x_i \partial x_l} =$$

$$= (G + \frac{A}{4}) \left[\frac{\partial^2 u_l}{\partial x_k^2} \frac{\partial u_l}{\partial x_i} + \frac{\partial^2 u_l}{\partial x_k^2} \frac{\partial u_i}{\partial x_l} + 2 \frac{\partial^2 u_i}{\partial x_k^2} \frac{\partial u_l}{\partial x_k} \right] +$$

$$+ (K + \frac{G}{3} + \frac{A}{4} + B) \left[\frac{\partial^2 u_l}{\partial x_i \partial x_k} \frac{\partial u_l}{\partial x_k} + \frac{\partial^2 u_k}{\partial x_l \partial x_k} \frac{\partial u_i}{\partial x_l} \right] +$$

$$+ (K - \frac{2}{3} G + B) \frac{\partial^2 u_i}{\partial x_k^2} \frac{\partial u_l}{\partial x_l} + (\frac{A}{4} + B) \left[\frac{\partial^2 u_k}{\partial x_l \partial x_k} \frac{\partial u_l}{\partial x_i} + \right.$$

$$\left. + \frac{\partial^2 u_l}{\partial x_i \partial x_k} \frac{\partial u_k}{\partial x_l} \right] + (B + 2C) \frac{\partial^2 u_k}{\partial x_i \partial x_k} \frac{\partial u_l}{\partial x_l} . \qquad (X.80)$$

Equation (X.80) together with the boundary and initial conditions is the basic equation of the nonlinear (five-constant) theory of elasticity. Its nonlinearity is a result of two factors. The first one is the purely geometric nonlinearity related to the nonlinearity of the strain tensor (X.5). The second one is the nonlinearity related to the deviation of the elasticity of the body from the linear Hooke's law, i.e., a "physical" nonlinearity. The physical nonlinearity is characterized by the third-order moduli A, B, and C, which can be determined from measurements of the dependence of the velocity of propagation of ultrasonic waves with different polarization on the static stresses.[19, 80] Such measurements of the third-order moduli are important in modern solid-state physics, because the physical nonlinearity of solids is related to their structural characteristics. Thus far, however, such measurements have been performed only for a very limited number of isotropic solids and some highly symmetric crystals.[80]

In conclusion, we note that since the five-constant theory of elasticity takes into account only second-order infinitesimals, in order to solve the equation of motion (X.80) it is natural to use the small-parameter method (see § IV.8), representing the displacement vector $\mathbf{u}$ in the form $\mathbf{u} = \mathbf{u}' + \mathbf{u}'' + \ldots$, where $\mathbf{u}'$ is the displacement vector in the first (linear) approximation and $\mathbf{u}''$ is the displacement vector in the second approximation, which is small compared to $\mathbf{u}'$. Then, from Eq. (X.80) we obtain the equation

$$\rho_0 \frac{\partial^2 u_i'}{\partial t^2} - G \frac{\partial^2 u_i'}{\partial x_k^2} - (K + \frac{G}{3}) \frac{\partial^2 u_l'}{\partial x_l \partial x_i} = 0 \qquad \text{(X.81)}$$

in the first approximation and the equation

$$\rho_0 \frac{\partial^2 u_i''}{\partial t^2} - G \frac{\partial^2 u_i''}{\partial x_k^2} - (K + \frac{G}{3}) \frac{\partial^2 u_l''}{\partial x_l \partial x_i} = 0 \qquad \text{(X.82)}$$

in the second approximation, where $f_i' = f_i'(\mathbf{u}')$ is the entire right side of Eq. (X.80) as a function of the linear part of the displacement vector $\mathbf{u}'$. From the form of Eq. (X.82) it follows that second-order infinitesimals, as in the case of liquids, arise under the action of forces created by displacements in the first (linear) approximation.

Differentiating Eq. (X.82) with respect to x_1 with $i = 1$, with respect to x_2 with $i = 2$, and with respect to x_3 with $i = 3$, and adding the resulting equaitons, we obtain

$$\rho_0^2 \frac{\partial^2 (\nabla \cdot \mathbf{u}'')}{\partial t^2} - (K + \frac{4}{3} G) \Delta (\nabla \cdot \mathbf{u}'') = \nabla \cdot \mathbf{f}'.$$

Analogously

$$\rho_0 \frac{\partial^2 (\nabla \times \mathbf{u}'')}{\partial t^2} - G \Delta (\nabla \times \mathrm{u}'') = \nabla \times \mathbf{f}'.$$

In these equations the longitudinal component in the second approximation, for which $\nabla \cdot \mathbf{u}'' \neq 0$, is separated from the transverse component, for which $\nabla \times \mathbf{u}'' \neq 0$. We thus arrive at two nonlinear wave equations, describing to a second approximation the propagation of ultrasonic finite-amplitude waves in an isotropic solid and corresponding to the longitudinal and transverse components, respectively, of the displacements in this approximation. This is the main difference between the nonlinear acoustics of solids and the propagation of finite-amplitude waves in liquids and gases, in which only longitudinal waves are possible and which we studied in detail in Chap. IV.

XI. Propagation of Ultrasound in Crystals

§ 1. General acoustic equations for crystals

A crystal is an anisotropic medium, so that its equation of motion in the linear approximation must be expressed in the general form (I.11). The stress σ_{ik} can be expressed in the same approximation in terms of the strain with the help of relation (I.13c):

$$\sigma_{ik} = c_{iklj}\epsilon_{lj}, \qquad \text{(XI.1)}$$

in which the elastic moduli c_{iklj} must be written with four indices (i, k, l, j = 1, 2, 3). Since, according to the definition (I.2), ϵ_{lj} = (1/2) $(\delta u/\delta x_l + \delta u_l/\delta x_j)$ for small deformations, the substitution of relation (XI.1) into the equation of motion (I.11) reduces the latter to an equation for one unknown – the displacement vector:

$$\rho_0 \frac{\partial^2 u_i}{\partial t^2} = -c_{iklj}\left(\frac{\partial^2 u_j}{\partial x_k \partial x_l} + \frac{\partial^2 u_l}{\partial x_k \partial x_j}\right). \qquad \text{(XI.2)}$$

If in the expression $c_{iklj}\delta^2 u_l/\delta x_k \delta x_j$ we interchange the dummy indices l and j, we obtain $c_{ikjl}\delta^2 u_j/\delta x_k \delta x_l$. The tensor c_{ikjl}, however, is symmetric relative to the second pair of indices; both terms in parentheses in Eq. (XI.2) are therefore equal and the equations of motion for the components of the displacement vector **u** assume the form

$$\rho_0 \frac{\partial^2 u_i}{\partial t^2} = c_{iklj}\frac{\partial^2 u_j}{\partial x_k \partial x_l}. \qquad \text{(XI.3)}$$

This expression, as before, contains three equations for the components of the displacements $u_j = u_1, u_2, u_3$ (i.e., ξ, η, and ζ); the indices k, l, and j are summed over in each equation.

For plane monochromatic waves, the displacement vector can be written in the form

$$\mathbf{u} = \mathbf{u}_{max} \exp\{i[\omega t - \mathbf{k}\cdot\mathbf{r}]\}, \qquad \text{(XI.4)}$$

where $\mathbf{u}_{max}$ is the vector amplitude of the displacement and is independent of the coordinates and time, $\mathbf{r}(x_1, x_2, x_3)$ is the position vector, and $\mathbf{k}$ is the wave vector. By definition $\mathbf{k} = k\mathbf{n} = \mathbf{n}\omega/c_0$, where $\mathbf{n}$ is a unit vector normal to the wave front with the components along the axes of a rectangular coordinate system (direction cosines) $n_1 = n_x$, $n_2 = n_y$, and $n_3 = n_z$. Therefore, the wave vector $\mathbf{k}$ has the components $k_j = k_1, k_2, k_3 = kn_1, kn_2, kn_3 = n_j\omega/c_0$. Taking this into account, we perform the double differentiation in Eq. (XI.3). Keeping in mind that $\mathbf{k}\cdot\mathbf{r} = \Sigma_j k_j x_j$ so that differentiation of Eq. (XI.4) with respect to x_j is therefore equivalent to multiplication by $-ik_j = -in_j\omega/c_0$, while differentiation with respect to time is equivalent to multiplication by $i\omega$, we obtain for the i-th component of the displacement

$$\rho_0 c_0^2 u_i = c_{iklj} n_k n_l u_j. \qquad \text{(XI.5)}$$

Using the Kronecker delta

$$\delta_{ij} = 1 \text{ if } i = j$$

$$= 0 \text{ if } i \neq j$$

and writing u_i as $u_i = \delta_{ij} u_j$, Eq. (XI.5) can be put into the following form, again dropping the zero indices corresponding to the linear approximation:

$$(\rho c^2 \delta_{ij} - c_{iklj} n_k n_l) u_j = 0 \quad (j = 1, 2, 3). \qquad \text{(XI.6)}$$

This very general relation is known in the theory of elasticity as *Christoffel's equation*. It is most conveniently written in terms of another tensor Γ_{ij} in the form

$$(\rho c^2 \delta_{ij} - \Gamma_{ij}) u_j = 0, \qquad \text{(XI.7)}$$

where

$$\Gamma_{ij} = c_{iklj} n_k n_l. \qquad \text{(XI.8)}$$

Equations (XI.5)--(XI.7) form a system of three homogeneous linear equations for the unknown quantities $u_i = u_x, u_y, u_z$. This system has nontrivial simultaneous solutions if the determinant formed from the coefficients multiplying u_j vanishes, i.e.,

$$|\Gamma_{ij} - \rho c^2 \delta_{ij}| = 0. \qquad \text{(XI.9)}$$

This is a cubic equation for c^2, which, in general, has three roots, depending on the direction of propagation of the plane wave. The quantity c^2 in Eqs. (XI.5)--(XI.9) is, by definition, the square of the velocity of sound, i.e., the velocity of propagation of the given displacement u_i. Therefore, in a plane wave propagating in a crystal in an arbitrary direction the resulting displacement can be written as a sum of three components u_i (which are, by definition, mutually orthogonal), each of which is characterized by a different velocity of propagation. In other words, we can say that for each direction in the crystal there are "three independent waves" with different phase velocities and mutually orthogonal displacements. In addition, as follows from Eqs. (XI.5)--(XI.9), the components of the displacements u_i in general are not necessarily normal or parallel to the wave front, as in an isotropic solid. This means that in an elastically anisotropic medium the wave vector in general is not oriented normal to the wave front, i.e., a plane wave propagates in the medium at some angle to the direction of the beam. In addition, such a wave is not, in general, either a purely longitudinal (in which the displacement coincides with the wave normal) or purely transverse wave. It is often possible to single out in this wave one component u_i which makes a small angle with the normal to the wave front $\mathbf{n}$; the two other components then make a small angle with the plane of the wave front. Waves corresponding to such displacements are called *quasilongitudinal* and *quasitransverse*.

Analysis of Eqs. (XI.5)–(XI.9) shows, however, that in crystals it is possible to single out orientations of **n** along which one of the components of the displacement vector completely coincides with the wave vector, i.e., it corresponds to a purely longitudinal wave. Since the three components of the displacement are perpendicular to one another, in this case the two other components will lie in the plane of the wave front and will correspond to shear waves. Thus, in crystals, it is possible to single out directions along which purely longitudinal and purely transverse waves can propagate (with a velocity that depends on the polarization). These directions are called "isonormal" directions; several such directions can exist in a given crystal. They are usually associated with axes of high symmetry. There also exist directions along which only one purely shear wave with definite polarization can propagate. In general, any direction along which one pure ultrasonic wave can propagate is called a *characteristic direction*.[81-87] Evidently, the laws of propagation of a wave in a particular characteristic direction in the crystal will be the same as those for a wave with the same polarization in an isotropic body, and the corresponding equations for it can be written in a scalar form. By analogy with optics the concept of *acoustic axes*, defined as the directions along which two transverse waves have the same phase velocities, is sometimes used in the literature.[83,84] Unlike optical axes, however, several acoustic axes can exist in crystals.

The velocities of propagation of longitudinal and shear waves in different directions are related to different dynamic elastic moduli c_{iklj}. The latter, in their turn, can be determined by measuring the velocity of ultrasound in some "sections" of the crystal, i.e., in crystal samples whose faces are cut perpendicular to a chosen direction.[88,89] Such measurements comprise one of the important problems of ultrasonics. In this connection, in the following sections we shall study the relation between the velocities of sound and the elastic moduli in the most convenient sections for the different crystallographic groups presented in Table 1.

§ 2. Relationship between the elastic moduli and the velocities of propagation of ultrasound in crystals

For convenience, we shall write Eqs. (XI.5)--(XI.8) in the expanded form. We replace the indices i, k, l, j by x, y, z and the four-valued tensor indices for the moduli by two-valued matrix indices $n = 1, 2, \ldots, 6$ and $m = 1, 2, \ldots, 6$, used in Chap. I and Table 1, and we shall employ the previous notation ξ, η, ζ for the components of the displacement u_x, u_y, u_z. We note that Eqs. (XI.5)--(XI.7) are equally valid for the variable displacements and their amplitudes, because the variable quantities differ from the amplitudes only by the phase factor $\exp\{i[\omega t - \mathbf{k}\cdot\mathbf{r}]\}$, which in Eqs. (XI.5) - (XI.7) can be dropped. As a result, the system of equations will assume the following compact form

$$
\begin{aligned}
\rho c^2 \xi_{max} &= \xi_{max}\Gamma_{11} + \eta_{max}\Gamma_{12} + \zeta_{max}\Gamma_{13},\\
\rho c^2 \eta_{max} &= \xi_{max}\Gamma_{21} + \eta_{max}\Gamma_{22} + \zeta_{max}\Gamma_{23},\\
\rho c^2 \zeta_{max} &= \xi_{max}\Gamma_{31} + \eta_{max}\Gamma_{32} + \zeta_{max}\Gamma_{33},
\end{aligned}
\qquad \text{(XI.10a)}
$$

where $\xi_{max} (= u_{xmax})$, $\eta_{max} (= u_{ymax})$, and $\zeta_{max} (= u_{zmax})$ are the amplitudes of the displacements u_i, and the components of the symmetric tensor Γ_{ij}, by definition (XI.8), can be written in an expanded form as follows:

$$
\begin{aligned}
\Gamma_{11} &= n_x^2 c_{11} + n_y^2 c_{66} + n_z^2 c_{55} + 2n_y n_z c_{56} + 2n_x n_z c_{15} + 2n_x n_y c_{16},\\
\Gamma_{22} &= n_x^2 c_{66} + n_y^2 c_{22} + n_z^2 c_{44} + 2n_y n_z c_{24} + 2n_x n_z c_{46} + 2n_x n_y c_{26},\\
\Gamma_{33} &= n_x^2 c_{55} + n_y^2 c_{44} + n_z^2 c_{33} + 2n_y n_z c_{34} + 2n_x n_z c_{35} + 2n_x n_y c_{45},\\
\Gamma_{13} &= \Gamma_{31} = n_x^2 c_{15} + n_y^2 c_{46} + n_z^2 c_{35} + n_y n_z (c_{45} + c_{46}) +\\
&\quad + n_x n_z (c_{13} + c_{55}) + n_x n_y (c_{56} + c_{14}),\\
\Gamma_{23} &= \Gamma_{32} = n_x^2 c_{56} + n_y^2 c_{24} + n_z^2 c_{34} + n_y n_z (c_{23} + c_{44}) +\\
&\quad + n_x n_z (c_{45} + c_{36}) + n_x n_y (c_{46} + c_{25}),\\
\Gamma_{12} &= \Gamma_{21} = n_x^2 c_{16} + n_y^2 c_{26} + n_z^2 c_{45} + n_y n_z (c_{46} + c_{25}) +\\
&\quad + n_x n_z (c_{56} + c_{14}) + n_x n_y (c_{12} + c_{56}).
\end{aligned}
\qquad \text{(XI.10b)}
$$

Equations (XI.10a) enable the relative magnitudes of three mutually perpendicular displacements in a plane wave propagating in the direction of the normal **n** to the wave front to be determined. In order that one of these directions correspond to a purely longitudinal wave the total vector displacement **u** must be parallel to **n**. The remaining two waves in this case must be transverse. The condition of collinearity of two vectors **u** and **n** is expressed mathematically as $\mathbf{u} \times \mathbf{n} = 0$, which gives

$$\begin{aligned} \xi_{max} n_y - \eta_{max} n_x &= 0; \\ \eta_{max} n_z - \zeta_{max} n_y &= 0; \\ \zeta_{max} n_x - \xi_{max} n_z &= 0. \end{aligned} \qquad \text{(XI.11)}$$

From here we find that the components of the displacement vector along the coordinate axes (in particular, their amplitudes ξ_{max}, η_{max} and ζ_{max}) will be related in the same manner as are the components of the vector normal **n**, i.e., the condition of collinearity of the vectors **u** and **n** leads to the relation

$$n_x : n_y : n_z = \xi_{max} : \eta_{max} : \zeta_{max}. \qquad \text{(XI.12)}$$

This relation enables the components of the displacements u_i in Eq. (XI.10) to be replaced by the corresponding components of the vector normal n_i proportional to them, thereby reducing these equations to the equations for n_i for the isonormal directions. An additional condition for the propagation of purely transverse waves is the condition

$$\mathbf{u} \cdot \mathbf{n} = 0, \qquad \text{(XI.13)}$$

i.e.,

$$\xi_{max} n_x + \eta_{max} n_y + \zeta_{max} n_z = 0. \qquad \text{(XI.14)}$$

The relations (XI.11)–(XI.14) thus make it possible to select from the various solutions of the system (XI.10a) the particular ones that satisfy these conditions. The latter conditions, in their turn, determine the directions in a given crystal along which purely longitudinal and purely transverse ultrasonic waves can propagate. In addition, the existence of symmetry elements reduces the number of independent, nonzero elastic moduli c_{nm}, thereby simplifying Eqs. (XI.10a), i.e., making it easier to solve them and to determine the characteristic directions. Crystals with cubic symmetry have the simplest table of elastic moduli. We shall perform detailed calculations for these crystals. For crystals with lower symmetry we shall present relations that relate the velocity of sound to the elastic moduli in the optimal sections.

§ 3. Cubic crystals

According to Table 1, for cubic crystals (group IX) $c_{11} = c_{22} = c_{33}$, $c_{12} = c_{21} = c_{13} = c_{31} = c_{23} = c_{32}$, $c_{44} = c_{55} = c_{66}$, i.e., there are three independent moduli and 12 nonzero moduli. Equations (XI.10a), after combining similar terms, assume in this case the form

$$
\begin{aligned}
(\rho c^2 - c_{44})\,\xi_{max} &= (c_{11} - c_{44})\,\xi_{max} n_x^2 + (c_{12} + c_{44})\,\eta_{max} n_x n_y + \\
&\quad + (c_{12} + c_{44})\,\zeta_{max} n_x n_z, \\
(\rho c^2 - c_{44})\,\eta_{max} &= (c_{12} + c_{44})\,\xi_{max} n_x n_y + (c_{11} - c_{44})\,\eta_{max} n_y^2 + \\
&\quad + (c_{12} + c_{44})\,\zeta_{max} n_y n_z, \\
(\rho c^2 - c_{44})\,\zeta_{max} &= (c_{12} + c_{44})\,\xi_{max} n_x n_z + (c_{12} + c_{44})\,\eta_{max} n_y n_z + \\
&\quad + (c_{11} - c_{44})\,\zeta_{max} n_z^2 .
\end{aligned}
\qquad \text{(XI.15)}
$$

Solving this system of equations for ξ_{max}, η_{max} and ζ_{max}, we obtain

$$
\xi_{max} = \frac{n_x}{A + Bn_x^2}, \qquad \eta_{max} = \frac{n_y}{A + Bn_y^2}, \qquad \zeta_{max} = \frac{n_z}{A + Bn_z^2},
$$

where $A \equiv (\rho c^2 - c_{44})/c_{44}$ and $B \equiv (c_{12} - c_{11} + 2c_{44})/c_{44}$. Applying now the condition of isonormality (XI.12) to each of the displacement amplitudes found, we obtain the relations for the components of the unit normal vector:

$$
n_x : n_y : n_z = \frac{n_x}{A + Bn_z^2} : \frac{n_y}{A + Bn_y^2} : \frac{n_z}{A + Bn_z^2} . \qquad \text{(XI.16)}
$$

These relations determine all possible directions in the crystal along which purely longitudinal (and purely transverse) waves can propagate.

It is not difficult to see that the relation (XI.16) holds for any n with $B = 0$, i.e., if $c_{44} = (c_{11} - c_{12})/2$, which is the condition for isotropy. As a result, the quantity

$$b \equiv 2c_{44}/(c_{11} - c_{12}) \qquad \text{(XI.17)}$$

is a measure of the anisotropy of a cubic crystal and can be called the *anisotropy factor*. Then $B = 2(1 - 1/b)$; for an isotropic body, in accordance with the expression (I.14), $b = 1$ and $B = 0$. Thus in an isotropic medium all directions are "isonormal"; a longitudinal wave can propagate in any of them, while in the presence of shear elasticity a transverse wave whose velocity does not depend on the polarization can also propagate.

If $B \neq 0$, the relation (XI.16) is satisfied under the following conditions:

$$n_x = 1,\ n_y = n_z = 0;\ n_y = 1,\ n_x = n_z = 0;\ n_z = 1,\ n_x = n_y = 0;$$

$$n_x = n_y,\ n_z = 0;\ n_x = n_z,\ n_y = 0;\ n_y = n_z,\ n_z = 0;$$

$$n_x = n_y = n_z. \qquad \text{(XI.18)}$$

To relate these directions to the crystallographic axes we must decide how to choose these axes. The principal crystallographic axes a, b, and c are customarily oriented along the edges of the unit cell of the crystal, marking off on them the unit segments a_0, b_0, and c_0, which are determined by the period of the crystal lattice.[90-93] The angles between these axes α, β, and γ can differ from 90°. In addition, a rectangular coordinate system X, Y, Z (we shall denote the axes fixed in the crystal by capital letters), chosen in such a way that the Z-axis ($\parallel c$) coincides, when possible, with the axis of symmetry of highest order and the other two coordinate axes coincide with two or at least with one of the axes a and b, is used. The customary choice of orientations of the axes a, b, and c and the rectangular "crystallographic" axes X, Y, and Z are presented for different crystallographic systems in Figs. 70 (a–f), which also show the symmetry elements corresponding to these axes and the unit-cell parameters. The symbols for the symmetry elements of the crystal, relative to which the corresponding axis is oriented, are shown in the broken circles. Arbitrary directions in the crystal relative to the crystallographic axes are indicated in brackets, using Miller's notation, in terms of the relative magnitudes of the projections of the elementary segments.

For crystals with cubic symmetry, the crystallographic axes coincide with the edges of a cube, i.e., with the fourth-order symmetry axes C^4, while the lengths of the unit segments a_0, b_0, and c_0 are equal to one another (see Fig. 70a). Figure 71 shows the crystallographic directions corresponding to the conditions (XI.18). The first condition corresponds to three equivalent directions along the edges of the cube: [100], [010], or [001]; the second condition corresponds to equivalent directions along the diagonals of the faces of the cubic cell, i.e., the crystallographic directions [110], [101], or [011]; finally, the third condition in (XI.18) corresponds to three equivalent directions along the body diagonals of the cube ([111], etc.). All indicated directions coincide with the symmetry axes of cubic crystals and are isonormal. It is not difficult to verify that purely shear waves can propagate in these directions. Indeed, the condition for the existence of a purely transverse wave is that the vectors **n** and **u** must be orthogonal, i.e., the condition (XI.14) must hold. Applying this condition to the solutions found for the displacement amplitudes (XI.15), we find the same directions as those determined by the relations (XI.18).

To calculate the velocities of propagation of sound along isonormal directions in terms of the corresponding elastic moduli the combinations of components n_i obtained must be substituted into Eqs. (XI.15) and these equations must be solved for ρc^2. The relations obtained will give the velocity for each specific direction. For this reason, we shall denote by a subscript to c the direction of propagation of a given wave, and we shall use a supercript to denote the polarization of the wave. For

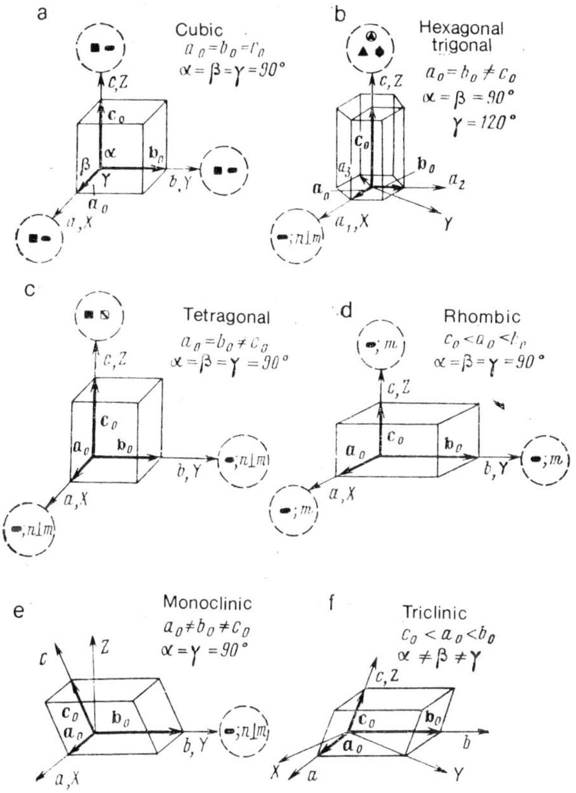

■ – C^4 axis ▬ C^2 axis ▲ C^3 axis, ⬢ – C^6 axis

● – inversion axis C^{6i} , ◪ – inversion axis C^{4i}

m – axis lies in the symmetry plane

$n \perp m$ – normal to symmetry plane.

Fig. 70.

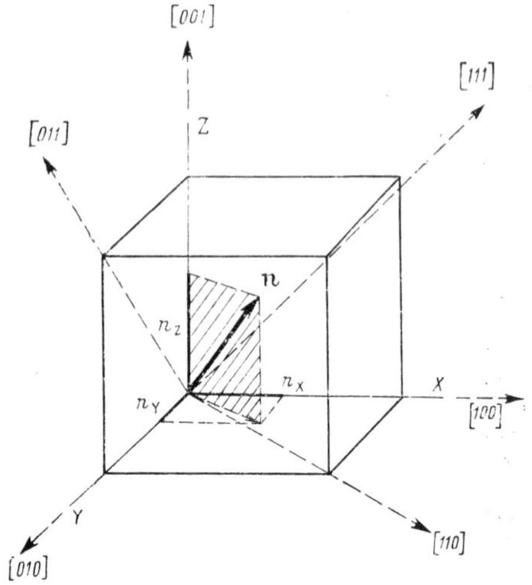

Fig. 71.

example, the symbol $c^{[100]}_{[010]}$ will denote the velocity of propagation of a transverse wave along the axis [010] = Y with displacement along the axis [100] = X, which is equivalent to the notation c^X_Y. In order to distinguish more easily longitudinal waves from transverse waves, for longitudinal waves we shall replace the superscript (repeating, in this case, the subscript) by the index l. Thus, using the first condition (XI.18) in the general equation (XI.15) and taking into account the fact that the vector **n** is a unit normal vector, i.e., $n^2 = 1$, we find that $c_{11} - c_{44} = \rho c^2 - c_{44}$, whence

$$\rho\,(c^l_{[100]})^2 = c_{11}. \qquad \text{(XI.19)}$$

Of course, this result is valid for any other cubic axis: [010] and [001].

The second condition in (XI.18), corresponding to the propagation of a longitudinal wave along the axes [110], [101], or [011], is: $n_x = n_y$ and $n_z = 0$. Since the components of the unit normal vector are direction cosines for the given direction of propagation, i.e., $n_x = \cos\psi$ and $n_y = \sin\psi$, where ψ is the angle between the vector **n** (lying in the xy

plane) and the X axis, $n_x^2 + n_y^2 = 1$, so that the second condition gives $n_x = n_y = 1/\sqrt{2}$ and $n_z = 0$. Substituting these values of n_i into Eq. (XI.15), we obtain

$$\frac{1}{2}(c_{11} - c_{44})\xi_{max} + \frac{1}{2}(c_{12} + c_{44})\eta_{max} = (\rho c^2 - c_{44})\xi_{max},$$

$$\frac{1}{2}(c_{12} + c_{44})\xi_{max} + \frac{1}{2}(c_{11} - c_{44})\eta_{max} = (\rho c^2 - c_{44})\eta_{max}.$$

Equating the determinant formed from the coefficients in front of ξ_{max} and η_{max} in these equations to zero

$$\begin{vmatrix} \frac{1}{2}(c_{11} + c_{44}) - \rho c^2 & \frac{1}{2}(c_{12} + c_{44}) \\ \frac{1}{2}(c_{12} + c_{44}) & \frac{1}{2}(c_{11} + c_{44}) - \rho c^2 \end{vmatrix} = 0, \quad \text{(XI.20)}$$

we find the solution sought for ρc^2

$$\rho (c'_{[110]})^2 = (c_{11} + c_{12} + 2c_{44})/2. \quad \text{(XI.21)}$$

Of course, the quadratic equation (XI.20) has a second solution, but this solution refers to the transverse wave.

The third condition (XI.18) corresponds to the [111] direction. Since $n_x^2 + n_y^2 + n_z^2 = 1$, it follows from the equality of the components n_i that

$$n_x = n_y = n_z = 1/\sqrt{3}. \quad \text{(XI.22)}$$

Substituting these values of n_i into Eq. (XI.15) and solving them for ρc^2, we obtain analogously

$$\rho (c'_{[111]})^2 = (c_{11} + 2c_{12} + 4c_{44})/3. \quad \text{(XI.23)}$$

For purely shear waves, for which the propagation conditions are determined by the same relations (XI.18), it is necessary to include also the different directions of the displacements (polarization), which must be perpendicular to the chosen direction of propagation, i.e., to the vector **n**. Thus, in the first condition (XI.18), corresponding to propagation along one of the fourth-order axes with $n_x = 1$ and $n_y = n_z = 0$, the component of the displacement $\xi_{max} = 0$ and the component η_{max} or ζ_{max} can be different from zero, etc. In any case, these conditions make all terms on the right side of Eq. (XI.15) vanish, while a term with nonvanishing displacement remains on the left. This gives, for the shear wave propagating along one of the cubic axes,

$$\rho(c^{T}_{[100]})^2 = c_{44} \tag{XI.24}$$

for any direction of the displacement. Thus the velocity of propagation of an ultrasonic shear wave along the fourth-order axes in a cubic crystal does not depend on the polarization, i.e., the conditions for the propagation of transverse waves in these directions are the same as those for an isotropic solid. These directions refer to the aforementioned "acoustic" directions. The latter, however, also include the directions along which quasitransverse waves propagate with the same phase velocities.[84] For this reason, we shall call these directions *transversely isotropic*, and we shall denote the velocity of propagation of shear waves in these directions (denoted by the subscript) by the superscript T, as for an isotropic medium. The [111] direction in cubic crystals is also the transversely isotropic direction, which can be easily verified from the equalities (XI.22) and the condition that the displacements are transverse (XI.14). Using these relations in Eq. (XI.15), we obtain for a shear wave propagating along the [111] direction with arbitrary polarization

$$\rho(c^{T}_{[111]})^2 = (c_{11} - c_{12} + c_{44})/3. \tag{XI.25}$$

Finally, we shall study the propagation of purely transverse waves in the [110] direction, characterized by the direction cosines $n_x = n_y = 1/\sqrt{2}$ and $n_z = 0$. The displacement vector **u** for these waves lies in the (110) plane and can be oriented in different directions. For a displacement, for example, along $[001] \parallel Z$, we have $\xi_{max} = \eta_{max} = 0$ and $\zeta_{max} \neq 0$. Equations (XI.15) in this case give $(\rho c^2 - c_{44})\zeta_{max} = 0$, i.e.,

$$\rho(c^{[001]}_{[110]})^2 = c_{44}, \tag{XI.26}$$

as in the case of the propagation of a transverse wave along the [001] direction. For the case of displacements in the xy plane (i.e., along the [110] direction), however, $\zeta_{max} = 0$ and $\eta_{max} = \xi_{max}$, which gives*

$$\rho (c_{[110]}^{[1\bar{1}0]})^2 = (c_{11} - c_{12})/2. \qquad \text{(XI.27)}$$

As we can see, the greatest difference in the velocities of propagation of transverse waves in a given direction in a cubic crystal is determined by the previously introduced anisotropy factor b (XI. 17). For the most highly anisotropic crystals of a cubic system, the anisotropy factor can have a value of $b = 2$--3, which corresponds to a difference of up to 100% for the velocities indicated. For example, for a KBr single crystal

$$c_{[110]}^{[1\bar{1}0]} = 2300\ m/s \text{ and } c_{[110]}^{[001]}\ (= c_{[001]}^{T}) = 1360\ m/s.$$

The velocity anisotropy, which is related to the difference in the directions of propagation, can also be quite large. Thus in a KBr crystal $c_{[100]}^{l} = 3550\ m/s$, $c_{[110]}^{l} = 3020\ m/s$, and $c_{[111]}^{l} = 2840\ m/s$.

The expressions (XI.19), (XI.21), and (XI.23)–(XI.27) contain different combinations of the elastic moduli. The elastic properties of the cubic crystal are completely characterized by three independent moduli: c_{11}, c_{12}, and c_{44}. To determine them it is necessary and sufficient to measure the velocity of propagation of the three types of ultrasonic waves, propagating in any of the directions found. To this end certain sections of the crystal are selected. We shall denote these sections by the symbols for the axes or directions that are perpendicular to the sections in which the crystals are cut: for example, the X-section is the section perpendicular to the X-axis, etc. The choice of optimal sections for measurements of the elastic moduli may be dictated by different considerations: the simplicity of the relations between the velocity of sound and the corresponding moduli, the accuracy with which the section can be oriented in practice relative to the crystallographic axes, the presence of natural facets or convenient cleavage planes, the number of independent measurements that can

*This result is obtained in the same manner as the second root of the quadratic equation (XI.20).

be performed on a given sample, etc. The most general consideration is taken into account in selecting the optimal sections for crystals of different symmetry studied below.

As follows from the relations obtained, the elastic moduli c_{11} and c_{44} for a cubic crystal are determined directly from the measurements of the velocity of propagation of longitudinal and arbitrarily polarized transverse waves along the fourth-order axis C^4. The relation between these velocities and the moduli c_{11} and c_{44} is given, respectively, by Eqs. (XI.19) and (XI.24). To find the third independent modulus of elasticity c_{12}, we can use the 45°-section (cut at an angle of 45° to the X-axis), measuring in this section, for example, the velocity of propagation of a longitudinal wave along the [110] direction (Fig. 72). This velocity, according to Eq. (XI.21), is related to the effective stiffness, determined by a combination of all three moduli, two of which are already known. It is possible, however, to determine all independent elastic moduli of a cubic crystal using only the 45°-section, measuring in this section the velocity of propagation of the longitudinal wave in the [110] direction and the velocities of the two transverse waves with displacements in mutually perpendicular directions: [001] and [110]. According to the relations (XI.21), (XI.26), and (XI.27), these velocities are given by

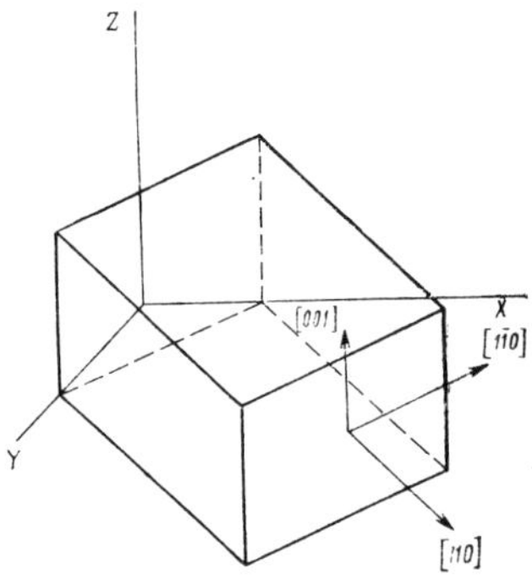

Fig. 72.

$$c^{l}_{[110]} = \left[\frac{c_{11} + c_{12} + 2c_{44}}{2\rho}\right]^{1/2},$$

$$c^{[001]}_{[110]} = \left[\frac{c_{44}}{\rho}\right]^{1/2}, \quad c^{[1\bar{1}0]}_{[110]} = \left[\frac{c_{11} - c_{12}}{2\rho}\right]^{1/2}.$$

Table 12

Relationship Between the Propagation Velocities of Ultrasonic Waves and the Elastic Moduli for Crystals in the Cubic System

Section symbol	Direction of propagation	Type of wave and polarization	Symbol for sound velocity	ρc^2
X; (100)	X; [100]	Longitudinal, [100]	$c^l_{[100]}$	c_{11}
(110)	[110]	Longitudinal, [110]	$c^l_{[110]}$	$\frac{c_{11}+c_{12}}{2}+c_{44}$
		Transverse, [110]	$c^{[1\bar{1}0]}_{[110]}$	$\frac{1}{2}(c_{11}-c_{12})$
X; (100)	[100]	Transverse, arbitrary	$c^T_{[100]}$	c_{44}
(110)	[110]	Transverse, [001]	$c^{[001]}_{[110]}$	c_{44}

From these relations, we find three moduli: c_{11}, c_{12}, and c_{44}. To reduce the errors related to the uncertainty in the orientation of the [110] faces and the polarization of the shear waves, two more control measurements of the velocity of the longitudinal and arbitrarily polarized transverse waves in the [001] direction can be made in the same sample. For these velocities, according to Eq. (XI.19) and (XI.24), we have $c^l_{[001]} = (c_{11}/\rho)^{1/2}$ and $c^T_{[001]} = (c_{44}/\rho)^{1/2}$. All chosen directions in the 45°-section of the cubic crystal are shown in Fig. 72. The equations presented above are summarized in Table 12. Table 13 presents the results of measurements of the dynamic elastic moduli by ultrasonic methods in cubic crystals together with the corresponding values of the velocities of sound. Detailed data on the elastic moduli of cubic (and other) crystals, studied since 1960, can also be found in the review by K. S. Aleksandrov and T. R. Ryzhova.[89]

Table 13.

Elastic Moduli and Propagation Velocities of Ultrasonic

Crystal	Chemical formula	$\rho \cdot 10^{-3}$, kg/m^3	T, °C	$c_{nm} \cdot 10^{-10}$	
				c_{11}	c_{12}
1	2	3	4	5	6
Diamond	C	3.51	20	107.6	12.5
Aluminum	Al	2.70	20	10.69	6.26
Antimonide					
gallium	GaSb	5.619	20	8.887	4.033
indium	InSb	5.789	20	6.72	3.67
Arsenide					
gallium	GaAs	5.31	20	11.88	5.38
indium	InAs	5.655	20	8.34	4.54
Sodium bromate	$NaBrO_3$	3.339	25	5.57	1.70
Bromide					
ammonium	NH_4Br	2.436	20	3.38	0.91
potassium	KBr	2.75	20	3.46	0.58
sodium	NaBr	3.20	20	3.87	0.97
silver	AgBr	6.47	20	5.63	3.38
thallium	TlBr	7.57	20	3.78	1.48
cesium	CsBr	4.45	20	3.10	0.84
Vanadium	V	6.022	20	22.79	11.87
Tungsten	W	19.3	20	51.26	20.58
Germanium	Ge	5.32	20	12.92	4.79
α-Iron	Fe	7.86	20	24.2	14.65
Gold	Au	19.32	20	18.6	15.7
Iodide					
potassium	KI	3.13	20	2.67	0.43
sodium	NaI	3.6714	20	2.931	0.782
cesium	CsI	4.52	20	2.45	0.71
Potassium	K	0.91	−83	0.457	0.374
Silicon	Si	2.33	20	16.57	6.39
Lithium	Li	0.55	−98	1.342	1.125
Lithium-Indium	LiIn	5.158	20	5.589	4.169
Copper	Cu	8.94	20	16.84	12.14
Molybdenum	Mo	10.19	20	46.0	17.6
Sodium	Na	1.01	−63	0.615	0.496
Nickel	Ni	8.90	20	24.65	14.73
Oxide					
barium	BaO	5.72	20	12.57	4.81
magnesium	MgO	3.58	20	28.6	8.70
Palladium	Pd	12.132	20	22.213	17.710

Continuation of Table 13

Waves in Crystals in the Cubic System

N/m^2	$c \cdot 10^{-3}$, m/s					
c_{44}	$c^l_{[100]}$	$c^T_{[100]}$	$c^l_{[110]}$	$c^{[1\bar{1}0]}_{[110]}$	$c^l_{[111]}$	$c^T_{[111]}$
7	8	9	10	11	12	13
57.6	17.5	12.8	18.3	11.6	18.6	12.0
2.85	6.29	3.26	6.27	2.86	6.53	3.00
4.324	3.96	2.77	4.381	2.078	4.51	2.33
3.02	3.39	2.29	3.80	1.61	3.88	1.84
5.94	4.71	3.34	5.24	2.47	5.40	2.79
3.95	3.84	2.64	4.29	1.83	4.42	2.14
1.51	4.08	2.13	3.93	2.41	3.87	2.32
0.685	3.72	1.68	3.41	2.25	3.30	2.08
0.505	3.55	1.36	3.02	2.29	2.84	2.19
0.97	3.48	1.74	3.26	2.13	3.18	2.01
0.72	2.95	1.05	2.83	1.34	2.79	1.25
0.756	2.23	0.999	2.12	1.23	2.07	1.16
0.75	2.64	1.30	2.47	1.59	2.41	1.50
4.25	6.15	2.66	5.99	3.011	5.93	2.90
15.27	6.15	2.81	5.15	2.82	5.15	2.82
6.70	4.92	3.55	5.41	2.75	5.56	3.04
11.2	5.55	3.77	6.24	2.46	6.46	2.97
4.20	3.10	1.47	3.33	0.87	3.39	1.10
0.421	2.92	1.16	2.51	1.89	2.33	1.68
0.737	2.83	1.42	2.66	1.73	2.60	1.62
0.62	2.33	1.17	2.21	1.39	2.16	1.32
0.263	2.24	1.70	2.73	0.68	2.88	1.13
7.96	8.43	5.85	9.13	4.67	9.35	5.09
0.960	5.19	4.43	6.67	1.45	7.09	2.82
2.666	3.29	2.27	3.82	1.17	3.76	1.62
7.54	4.34	2.90	4.96	1.62	5.16	2.14
11.0	6.72	3.29	6.48	3.73	8.19	2.39
0.592	2.44	2.41	3.31	0.84	3.58	1.55
12.47	5.26	3.74	6.01	2.36	6.24	2.90
3.55	4.6921	2.4422	4.5326	2.60	4.61	2.57
14.8	8.94	6 43	9.66	5.27	9.89	5.68
7.137	4.28	2.43	4.73	1.42	5.06	1.86

Continuation of Table 13

1	2	3	4	5	6
Lead	Pb	11.34	20	4.66	3.92
Zinc selenide	ZnSe	5.264	27	8.95	5.39
Silver	Ag	10.49	20	12.20	9.15
Iron silicate	Fe_3Si	7.191	20	23.2	15.6
Strontium nitrate	$Sr(NO_3)_2$	2.986	20	4.73	2.18
Zinc sulfide	ZnS	4.088	27	9.81	6.27
Telluride					
mercury	HgTe	8.081	17	5.386	3.676
cadium	CdTe	5.854	20	5.33	3.65
Strontium titanate	$SrTiO_3$	5.116	20	31.81	10.25
Cobalt-zinc ferrite	$Co_{0.32}Zn_{0.22}F_{2.2}O_4$	5.43	27	26.6	15.3
Fluorite	CaF_2	3.18	20	16.44	5.02
Fluoride					
lithium	LiF	2.60	20	11.44	4.26
magnesium	MgF_2	3.98	20	17.54	—
Sodium chlorate	$NaClO_3$	2.49	20	4.99	1.41
Chloride					
potassium	KCl	1.984	20	3.98	0.62
sodium	NaCl	2.168	20	4.87	1.24
silver	AgCl	5.56	20	6.01	3.62
cesium	CsCl	3.99	20	3.64	0.92
Chromite	$FeO \cdot Cr_2O_3$	4.32–4.57	20	32.25	14.37
Spinel	MgO·3.5	3.63	20	30.05	15.37

§ 4. Crystals with lower symmetry

We shall not repeat the detailed analysis of all directions in which pure waves can propagate for crystals with other types of symmetry. We shall only summarize the useful relations, which enable the independent elastic moduli to be determined together with an indication of the corresponding directions relative to the crystallographic axes. For a more detailed derivation of these relations, the reader is referred to the original literature.[94-104]

Hexagonal and trigonal systems. For crystals with hexagonal and trigonal symmetry, four crystallographic axes are usually chosen: the axis $c \parallel Z$, coinciding with the axis of highest symmetry C^3, C^6, or C^{6i},

Continuation of Table 13

7	8	9	10	11	12	13
1.44	2.03	1.13	2.25	0.57	2.32	0.80
3.984	4.12	2.75	4.61	1.84	4.75	2.19
4.48	3.41	2.07	3.79	1.20	3.92	1.55
13.5	5.68	4.33	6.76	2.30	7.09	3.13
1.46	3.98	2.21	4.05	2.07	3.79	2.12
4.483	4.90	3.31	5.54	2.09	5.74	2.56
2.116	2.58	1.62	2.87	1.03	2.96	1.26
2.044	3.02	1.87	3.34	1.20	3.44	1.46
12.36	7.876	4.910	8.098	4.918	8.141	4.703
7.8	7.00	3.80	7.28	3.23	7.37	3.42
3.47	7.19	3.30	6.68	4.24	6.50	3.95
6.28	6.63	4.91	7.37	3.71	7.60	4.16
5.52	6.64	3.72	—	—	—	—
1.17	4.47	2.17	4.18	2.68	4.09	2.52
0.625	4.48	1.78	3.85	2.91	3.61	2.59
1.26	4.74	2.41	4.72	2.90	4.73	2.45
0.625	3.29	1.06	3.13	1.47	3.07	1.35
0.80	3.02	1.42	2.78	1.85	2.69	1.71
11.67	8.51	5.12	8.87	4.48	8.98	4.70
15.86	9.09	6.61	10.30	4.50	10.68	5.30

and the axes a_i (a_1, a_2, and a_3) in three symmetric directions lying in a plane perpendicular to the principal axis (Fig. 70b). These directions can be second-order axes or normals to the three symmetry planes or straight lines parallel to the possible edges of the crystal. The X-axis of the rectangular coordinate system coincides with the a_1 axis, and the Y axis is chosen so that it is perpendicular to the X and Z axes and forms a right-handed system.

In crystals belonging to the trigonal system, the unit cell is a rhombohedron. In many cases, however, the trigonal lattice is described by hexagonal axes. In this variant, the rhombohedron is replaced by a hexagonal cell with three times the volume; the choice of coordinate axes then corresponds to the case of a hexagonal crystal.

The table of elastic moduli of crystals belonging to the hexagonal and trigonal systems, corresponding to this choice of coordinate axes, is presented in Chap. I (groups VIII and VI in Table 1). Using this table in Eqs. (XI.10a) and repeating the preceding procedure, it is not difficult to find all directions along which purely longitudinal or transverse ultrasonic waves can propagate, and then to calculate the corresponding effective stiffnesses, i.e., the quantities ρc^2, expressed in terms of some combination of elastic moduli.

Table 14

Relationship Between the Propagation Velocities of Ultrasonic Waves and the Elastic Moduli for Hexagonal Crystals in Optimal Directions Relative to the Crystallographic Axies

Section symbol	Direction of propagation	Type of wave and polarization	Symbol for sound velocity	ρc^2
Z (00 · 1)	[00 · 1]	Longitudinal, [00 · 1]	$c^l_{[00\cdot1]}$	c_{33}
		Transverse, arbitrary in the (00 · 1) plane	$c^T_{[00\cdot1]}$	c_{44}
X (12 · 0)	[10 · 0]	Longitudinal, [10 · 0]	$c^l_{[10\cdot0]}$	c_{11}
		Transverse [12 · 0] (Y axis)	$c^{[12\cdot0]}_{[10\cdot0]}$	$\frac{1}{2}(c_{11}-c_{12})$
X′	45° relative to X and Z axes in the XZ plane	Quasilongitudinal	$c^l_{X'}$	$-(c_{11}+c_{33}+2c_{44})+ + \left\{\left[\frac{1}{2}(c_{11}-c_{33})\right]^2 + +(c_{13}+c_{44})^2\right\}^{1/2}$

It is evident from Table 1 that the elastic properties of **hexagonal crystals** are characterized by five independent moduli: $c_{11} = c_{22}$, $c_{13} = c_{23}$, c_{12}, c_{33}, and $c_{44} = c_{55}$; in addition, $c_{66} = (c_{11} - c_{12})/2$. To determine them it is necessary and sufficient to perform five measurements of the velocity of sound in the most convenient sections, which are shown in Fig. 73 and presented in Table 14. This table shows the corresponding sections relative to the axes shown in Fig. 70b, the Miller indices of the direction of wave propagation, as well as the wave type and the direction of the displacements for transverse waves (their "polarization").

Table 15 presents the measurements of the elastic moduli of hexagonal crystals performed by ultrasonic methods, as well as the velocities of propagation of longitudinal and shear waves in the [001] direction (i.e., along the axis $Z \parallel c$).

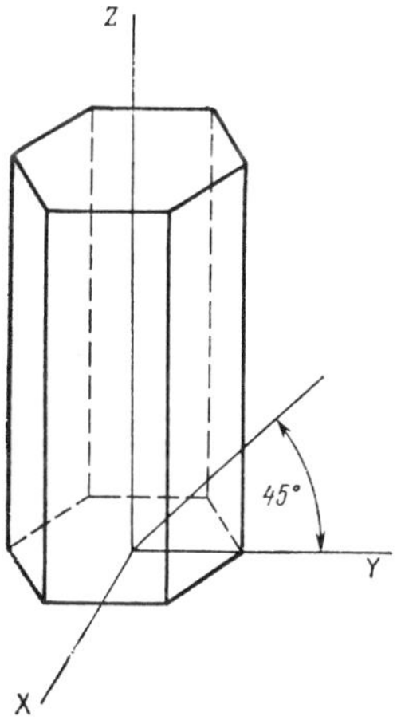

Fig. 73.

The table of elastic moduli for crystals in the **trigonal system,** belonging to the classes D_3, C_{3v}, and D_{3d} (see Table 1, group VII), contains 18 nonvanishing moduli of which six are independent: $c_{11} = c_{22}$, c_{12}, $c_{13} = c_{23}$, $c_{14} = c_{56} = -c_{24}$ and $c_{44} = c_{55}$. Their relation to the velocities of propagation of ultrasonic waves along the optimum directions is shown in Table 16.

The table of elastic moduli for crystals in the trigonal system, belonging to the classes C_3 and C_{3i}, with the same orientation of the axes, contains an additional (seventh) independent modulus c_{25} (see Table 1, group VI). This modulus can be found with the help of the more complicated relations presented in Refs. 95, 96, 99, and 104. Table 17 presents the elastic moduli of a number of trigonal crystals, measured at room temperature by ultrasonic methods.

Tetragonal system. All crystals in the tetragonal system, belonging to the classes D_4, D_{2d}, C_{4v}, and D_{4h}, have a simple fourth-order axis C^4 or

Table 15

The Elastic Moduli and the Velocities of Propagation of

Crystal	Chemical formula	$\rho \cdot 10^{-3}$ kg/m^3
Beryllium	Be	1.87
Lead vanadate-germanate	$Pb_5(GeO_4)(VO_4)_2$	7.15
Lead vanadate-silicate	$Pb_5(SiO_4)(VO_4)_2$	7.02
Lithium iodate	α-$LiIO_3$	4.5
Yttrium	Y	4.477
Cadmium	Cd	8.64
Cancrinite	$(Na_2Ga)_4(AlSiO_4)_6CO_3(H_2O)_{0\text{-}3}$	2.42–2.48
β-Quartz (580°C)	SiO_2	2.533
Cobalt	Co	8.836
Ice (−5°C)	H_2O	0.94
Magnesium	Mg	1.79
Rhenium	Re	20.53
Ruthenium	Ru	12.1
Zinc selenide	ZnSe	5.558
Sulfide		
Cadmium	CdS	4.83
Zinc	NzS	4.091
Barium titanate	$BaTio_3$	5.5
Zinc	Zn	7.18
Zincite	ZnO	5.7036

an inversion axis C^{4i}. These axes are used as the crystallographic c axis, with which the Z-axis coincides. The axes a and b lie in a plane perpendicular to the c-axis, forming a right angle. These axes coincide either with the second-order symmetry axes or with the normals to the

Continuation of Table 15

Ultrasonic Waves in Hexagonal Crystals ($T = 20°C$)

$c_{nm} \cdot 10^{-10}$, N/m²					$c \cdot 10^{-3}$, m/s	
c_{11}	c_{33}	c_{44}	c_{12}	c_{13}	$c^{l}_{[001]}$	$c^{T}_{[001]}$
29.23	33.64	16.25	2.67	1.4	13.41	9.32
7.1	8.4	1.7	2.1	3.3	3.427	1.54
7.7	9.2	2.1	2.5	3.6	3.62	1.73
8.3	5.7	1.8	3.9	—	4.01	1.99
7.79	7.69	2.431	2.85	2.1	4.14	2.33
12.1	5.13	1.85	4.81	4.42	3.74	1.46
5.2	8.26	2.38	0.86	1.24	5.81	3.12
11.66	11.04	3.606	1.67	3.28	6.61	3.78
30.7	35.81	7.55	16.50	10.30	5.91	2.93
1.38	1.50	0.319	0.707	0.581	3.83	1.84
5.97	6.17	1.64	2.62	2.17	5.84	3.06
64.45	71.70	16.85	27.70	19.59	5.92	2.87
57.63	64.05	18.92	18.74	16.74	7.28	3.95
11.81	3.82	1.17	4.75	3.2	2.62	1.45
7.78	8.81	1.47	4.47	4.79	4.270	1.75
13.12	14.08	2.86	6.63	5.09	5.818	2.627
16.6	16.2	4.29	7.66	7.75	5.50	2.79
16.1	6.10	3.83	3.42	5.01	4.75	2.31
20.70	22.1	4.61	11.17	10.13	6.30	2.84

symmetry planes, or they are parallel to the possible edges of the crystal. The X-axis can be oriented parallel to the a or b axis. Thus in tetragonal crystals there are two possible orientations for the X and Y axes (Fig. 70c). The table of elastic moduli, referred to these axes (see

Table 1, group V), for the indicated classes contains six independent moduli: $c_{11} = c_{22}$, c_{12}, $c_{13} = c_{23}$, c_{33}, $c_{44} = c_{55}$ and c_{66}. They can be determined with the help of the directions and relations presented in Table 18. For the classes C_4, S_4, and C_{4h} of tetragonal crystals one more nondiagonal modulus c_{16} is added to the table of elastic moduli referred to the same coordinate axes (see Table 1, group IV). The results following from the corresponding relations for this more complicated case can be found in Refs. 95 and 104. Table 19 presents measurements of the elastic moduli of several tetragonal crystals, performed at room temperature.

Rhombic (orthorhombic system) (classes D_2, C_{2v}, D_{2h}). The symmetry elements of crystals in this system are three mutually perpendicular second-order axes (class D_2), through which three mutually perpendicular symmetry planes (class D_{2h}) can pass, or one second-order axis and two symmetry planes (class C_{2v}) perpendicular to one another and intersecting on this axis, i.e., the symmetry elements in these crystals always form three mutually orthogonal directions, along which the coordinate axes a, b, and c and, correspondingly, X, Y, and Z are drawn (Fig. 70d). All these crystals have the same table of elastic moduli (see Table 1, group III), referred to the given coordinate axes. It contains nine independent moduli: c_{11}, c_{12}, c_{13}, c_{22}, c_{23}, c_{33}, c_{44}, c_{55}, and c_{66}. To determine them by acoustic methods it is necessary to perform a series of measurements of the velocity of

Table 16

Relationship Between the Velocities of Propagation of Ultrasonic Waves and Elastic Moduli of Trigonal Crystals (Classes D_3, C_{3v}, D_{3d})

Section symbol	Direction of propagation	Type of wave and polarization	Velocity of sound	ρc^2
X: (12·0)	X: [10·0]	Longitudinal, [10·0]	$c^{l}_{[10\cdot0]}$	c_{11}
Z: (00·1)	Z: [00·1]	Longitudinal, [00·1]	$c^{l}_{[00\cdot1]}$	c_{33}
		Transverse, arbitrary polarization	$c^{T}_{[00\cdot1]}$	c_{44}
Y: (01·0)	Y: [12·0)	Transverse, [10·0]	$c^{[10\cdot0]}_{[12\cdot0]}$	$\frac{1}{2}(c_{11}-c_{12})$
Y'	45° to the Y and Z axes in the YZ plane	Longitudinal	$c^{l}_{Y'}$	$\frac{1}{4}(c_{11}+c_{33}+2c_{44}-2c_{14})+\left[\frac{1}{4}(c_{11}-c_{33}-2c_{14})^2+(c_{13}+c_{44}-c_{14})^2\right]^{1/2}$
		Transverse, [10·0]	$c^{[10\cdot0]}_{Y'}$	$-(4c_{14}+2c_{44}+c_{11}-c_{12})$

propagation of ultrasonic waves in the directions indicated in Table 20. Data from such measurements for several rhombic crystals are presented in Table 21.

Monoclinic system. Monoclinic crystals have one second-order symmetry axis (class C_2) or one symmetry plane (class C_S) or both at

Table 17

Elastic Moduli of Crystals in

Crystal	Chemical formula	$\rho \cdot 10^{-3}$, kg/m^3
Bismuth	Bi	9.80
Quartz	α-SiO_2	2.6487
Corundum	Al_2O_3	3.97
Lithium niobate	$LiNbO_3$	4.644
Vanadium oxide	V_2O_3	4.87
Proustite	Ag_3AsS_3	5.6
Sapphire		
man-made	Al_2O_3	4.00
natural	. . .	3.4—3.6
Antimony	Sb	6.684
Lithium tantalate	$LiTaO_3$	7.451
Tellurium	Te	
Tourmaline	. . .	2.90—3.25
Aluminum phosphate	$AlPO_4$	2.566

the same time (class C_{2h}). For all these monoclinic crystals the standard coordinate system is the rectangular system X, Y, and Z shown in Fig. 70e. The C^2 symmetry axis, or the normal to the plane of symmetry, coinciding with the second-order symmetry axis, is taken as the b-axis, along which the Y-axis is oriented, and the X-axis is chosen so as to coincide with the crystallographic axis a. The a-and c-axes are chosen in a plane perpendicular to the b-axis. The table of elastic moduli, referred to such axes, for all three classes of monoclinic crystals contains 13 independent moduli (see Table 1, group II), c_{11}, c_{12}, c_{22}, c_{13}, c_{23}, c_{33}, c_{44}, c_{46}, c_{15}, c_{25}, c_{35}, c_{55}, and c_{66}. To determine them it is necessary to measure the velocities of propagation of ultrasonic waves in six nonequivalent crystallographic directions: [100], [010], [001], [110], [101], and [011] (see Ref. 10). All three elastic waves propagating along the [010] direction in a monoclinic crystal are pure waves. In addition, along the [001], [101], and [100] directions, one of the three waves is a purely shear wave polarized along the [010] axis. The effective stiffness for these three types of waves directly determines the moduli c_{22}, c_{66}, and c_{44}. A derivation of the relations

Continuation of Table 17

Trigonal Systems *(T = 20°C)*

$c_{nm} \cdot 10^{-10}$ N/m^2					
c_{11}	c_{33}	c_{44}	c_{12}	c_{13}	c_{14}
6.35	3.81	1.13	2.47	2.45	0.72
8.680	10.575	5.818	0.709	1.20	–1.805
49.7	49.8	14.7	16.4	11.1	–2.4
20.16	25.17	6.01	5.68	7.50	–1.38
21.6	33.2	8.0	7.1	14.8	1.5
5.70	3.64	0.90	3.18	–	–
49.68	49.81	14.74	16.36	11.09	–2.35
46.5	56.3	23.3	12.4	11.7	10.1
7.92	4.27	2.85	2.61	1.05	–
23.78	28.27	9.43	5.23	8.0	–2.23
3.59	7.64	3.41	0.90	2.75	1.37
27.2	16.5	6.5	4.0	3.5	–0.68
10.5	13.4	2.31	2.93	6.93	–1.27

for finding all elastic moduli of monoclinic crystals can be found in Refs. 102 and 103. Table 22 presents the elastic moduli measured by ultrasonic methods of some crystals belonging to the monoclinic system.

Triclinic system. Triclinic crystals do not have any symmetry axes or planes. The rectangular X-, Y-, and Z-axes and their positive directions for each class of the triclinic system are uniquely chosen relative to the edges of the triclinic unit cell (see Fig. 70f). The positive Z-axis is parallel to the positive c-axis and, therefore, parallel to the (100) and (010) planes; the X-axis is perpendicular to the c-axis and lies in the ac plane; the Y-axis is perpendicular to the (010) plane and forms a right-handed system of coordinates with the Z- and X- axes. Both symmetry classes of the triclinic system have a complete collection of independent elastic moduli, i.e., 21 moduli $c_{nm} \neq 0$. The relations between the velocities of propagation of acoustic waves and the moduli of the triclinic crystals can be found in Ref. 96.

Table 18

Relationship Between the Velocities of Propagation of Ultrasonic Waves and Elastic Moduli of the Tetragonal System (Classes D_4, D_{2d}, C_{4v}, D_{4h})

Section symbol	Direction of propagation	Type of wave and polarization	Velocity of sound	ρc^2
Z; (001)	[001]	Longitudinal, [001]	$c^{l}_{[001]}$	c_{33}
		Transverse, any orientation in the (001) plane	$c^{T}_{[001]}$	c_{44}
X; (100)	[100]	Longitudinal, [100]	$c^{l}_{[100]}$	c_{11}
		Transverse, [010]	$c^{[010]}_{[100]}$	c_{66}
		Transverse, [001]	$c^{[001]}_{[100]}$	c_{44}
(110)	[110]	Longitudinal, [110]	$c^{l}_{[110]}$	$\frac{c_{11}+c_{12}+2c_{66}}{2}$
		Transverse, $[1\bar{1}0]$	$c^{[1\bar{1}0]}_{[110]}$	$\frac{c_{11}-c_{12}}{2}$
X'	45° between the [100] (X) and [001] (Z) axes	Quasilongitudinal	$c^{l}_{X'}$	$\frac{c_{11}+c_{33}+2c_4+[(c_{11}-c_{33})^2+4)\,c_{13}+c_{44})^2]^{1/2}}{4}$
	45° between the [100] (X) and [001] (Z) axes	Quasi-transverse, [101]	$c^{[10\bar{1}]}_{X'}$	$\frac{c_{11}+c_{33}+2c_{44}-[(c_{11}-c_{33})^2+4(c_{13}+c_{44})^2]^{1/2}}{4}$

In conclusion, we note that we have considered only the linear elasticity of crystals and, correspondingly, we have been concerned with the second-order elastic moduli, i.e., the linear moduli. To describe nonlinear elasticity, even for cubic crystals, 14 third-order elastic moduli are required; for triclinic crystals there are 56 moduli.[80] For this reason, the equations of the nonlinear acoustics of crystals are usually constructed for special crystallographic directions, for which they have the form of the nonlinear equations of elasticity of an isotropic solid, studied above, with the appropriate set of nonlinear parameters. These parameters, i.e., the third-order elastic moduli, are also determined from ultrasonic measurements.[80] Few such measurements have been performed, even though nonlinear acoustic effects play an important role in quantum acoustics in the description of processes such as phonon-phonon interactions as well as spin-phonon, photon-phonon, and other types of interactions.[87] These interesting problems, however, fall outside the scope of this book.

Table 19

Elastic Moduli of Crystals with

Crystal	Chemical formula	$\rho \cdot 10^{-3}$, kg/m^3
Calcium tungstenate	$CaWO_4$	6.120
Iron digermanide	$FeGe_2$	–
Potassium dihydrogen arsenate	KH_2AsO_4	2.867
Ammonium dihydrogen phosphate (ADP)	$(NH)_4H_2PO_4$	1.803
deuterated	$(ND)_4D_2PO_4$	–
Potassium dihydrogen phosphate	KH_2PO_4	2.340
deuterated	KD_2PO_4	–
Indium	In	7.31
Calomel	$HgCl_2$	7.19
Molybdate		
calcium	$CaMoO_4$	4.5
lead	$PbMoO_4$	6.92
Barium-sodium niobate	$Ba_2NaNb_5O_{15}$	5.3
Strontium potassium lithium niobate	$SrKLiNb_{10}O_{30}$	–
Niobium oxide	NbO_2	5.90
Tin	Sn	7.30
Paratelluride	TeO_2	6.0
Rutile	TiO_2	4.264
Barium titanate (150°*C*)	$BaTiO_3$	5.5
Zirconium	Zr	6.49
Nickel zirconide	$NiZr_2$	7.234

Continuation of Table 19

Tetragonal Symmetry *(T = 20°C)*

$c_{nm} \cdot 10^{-10}$ N/m^2					
c_{11}	c_{33}	c_{44}	c_{66}	c_{12}	c_{13}
14.3	12.8	3.40	4.49	5.54	5.04 (C_{16}=2.21)
24.44	24.94	5.70	8.87	6.70	—
5.31	3.7	1.2	0.7	–0.6	–0.2
6.89	3.35	0.856	0.595	0.40	1.89
6.2	3.0	0.91	0.61	–0.5	1.4
7.14	5.62	1.27	0.68	–0.49	1.29
7.04	—	—	0.607	0.46	—
4.45	4.44	0.655	1.22	3.95	4.05
1.89	8.04	0.846	1.23	1.72	1.56
14.4	12.6	3.69	4.61	6.48	4.48
10.8	9.52	2.64	3.54	6.32	5.07 (c_{16}=1.58)
23.9	13.5	6.5	7.6	10.4	5.0
24.4	19.4	6.2	6.7	11.0	7.5
43.3	38.8	9.4	5.7	9.3	17.1
7.35	8.7	2.2	2.27	2.34	2.8
5.6	10.51	2.70	6.68	5.16	2.72
27.3	48.4	12.5	19.4	17.6	14.9
27.5	17.81	5.43	11.3	18.65	14.16
7.35	4.60	1.38	1.60	0.90	–0.54
15.477	14.480	2.399	0.966	12.82	8.57

Table 20

Relations Between the Elastic Moduli and the Velocities of Propagation

Section symbol	Direction of propagation	Type of wave and polarization	Symbol for velocity of sound
X; (100)	*X*; [100]	Longitudinal, [100]	$c^{l}_{[100]}$
Y; (010)	*Y*; [010]	Longitudinal, [010]	$c^{l}_{[010]}$
Z; (001)	*Z*; [001]	Longitudinal, [001]	$c^{l}_{[001]}$
Y; (010)	*Y*; [010]	Transverse, [001]	$c^{[001]}_{[010]}$
Z; (001)	*Z*; [001]	Transverse, [010]	$c^{[010]}_{[001]}$
X; (100)	*X*; [100]	Transverse, [001]	$c^{[001]}_{[100]}$
Z; (001)	*Z*; [001]	Transverse, [100]	$c^{[100]}_{[001]}$
X; (100)	*X*; [100]	Transverse, [100]	$c^{[010]}_{[100]}$
Y; (010)	*Y*; [010]	Transverse, [100]	$c^{[100]}_{[010]}$
X'	45° to the *X* and *Y* axes in (*XY*) plane	Quasilongitudinal, in (001) plane	$c^{l}_{X'}$
Y'	45° to the *Y* and *Z* axes in (*XZ*) plane		$c^{l}_{Y'}$
Z'	45° to the *Y* and *Z* axes in (*YZ*) plane	Quasilongitudinal, in (100) plane	$c^{l}_{Z'}$

Continuation of Table 20

of Ultrasonic Waves for Crystals in the Rhombic System

Relationship between (ρc^2) and the elastic moduli
c_{11}
c_{22}
c_{33}
c_{44}
c_{44}
c_{55}
c_{55}
c_{66}
c_{66}
$c_{12} = 2\sqrt{[(c_{66}+c_{22})/2 - \rho(c'_{X'})^2][(c_{66}+c_{11})/2 - \rho(c'_{X'})^2]} - 2c_{66}$
$c_{13} = 2\sqrt{[(c_{55}+c_{11})/2 - \rho(c'_{Y'})^2][(c_{55}+c_{33})/2 - \rho(c'_{Y'})^2]} - 2c_{55}$
$c_{23} = 2\sqrt{[(c_{44}+c_{22})/2 - \rho(c'_{Z'})^2][(c_{44}+c_{33})/2 - \rho(c'_{Z'})^2]} - 2c_{44}$

Table 21

Elastic Moduli of Crystals in

Crystal	Chemical formula	$\rho \cdot 10^{-3}$ kg/m^3	c_{11}
Benzophenone	$(C_6H_5)CO$	1.219	10.70
Lithium germanate	Li_2GeO_3	3.5	13
Iodic acid	HIO_3	4.63	3.03
Potassium pentaborate	$KB_5H_8 \cdot 4H_2O$	—	5.82
Lithium-ammonium-tartrate	$LiNH_4C_4H_4O_6$	1.71	3.86
Magnesium-sulfate-heptahydrate	$MgSO_4 \cdot 7H_2O$	1.687	6.98
Sodium-ammonium-tartrate	...	1.587	3.68
Sodium-tartrate	$Na_2C_4H_4O_6 \cdot H_2O$	1.818	4.61
Sodium-ammonium selenate dihydride	$NaNH_4SeO_4$	2.025	2.863
Resorcinal	$C_6H_4(OH)_2$	1.272–1.289	1.03
Sulfur	S	2.07	2.40
Strontium formate	...	2.25	4.39
Terpine monohydride	$C_{10}H_{18}(OH)_2 \cdot H_2O$	1.11	1.25
Topaz	Al_2SiO_3	28.2	34.9
α-Uranium	U	19.0	21.5
Celestine	...	3.955	10.44
Zinc-sulfate-hepta-hydride	$ZnSO_4 \cdot 7H_2O$	1.974	4.00
Rochelle salt	$NaKC_4H_6O_6 \cdot 4H_2O$	1.775	2.55
Olivine	...	3.324	32.4

The problem of the reflection and refraction of ultrasonic waves at the boundaries of anisotropic media is also just as complicated. Since in crystals three waves can propagate in any direction, the general equations for the reflection and refraction coefficients, even for a specific crystal, are very complicated. For this reason, problems of this type have been solved only for the simplest particular cases, for which the reader is referred to the review by L. K. Zarembo and V. V. Shklovskaya-Kordi.[83]

Finally, the analysis presented above neglected the influence of the

Continuation of Table 21

the Rhombic System *(T = 20°C)*

$c_{nm}\cdot 10^{-10}$ N/m²							
c_{22}	c_{33}	c_{44}	c_{55}	c_{66}	c_{12}	c_{13}	c_{23}
10.00	7.10	2.03	1.55	3.58	5.5	1.69	3.21
12	15	5.9	5	3.6	3.6	4.2	4.9
5.45	4.36	1.84	2.19	1.74	1.19	1.17	0.55
3.59	2.55	1.64	0.463	0.57	2.29	1.74	2.31
5.39	3.63	1.19	0.67	2.33	1.65	0.87	2.01
5.29	8.22	1.07	2.33	2.22	3.90	2.82	2.83
5.09	5.54	1.06	0.303	0.87	2.72	3.08	3.47
5.47	6.65	1.24	0.31	0.98	2.86	3.20	3.52
3.379	2.074	0.536	0.506	0.523	0.826	1.11	1.01
1.44	1.29	0.33	0.44	0.40	0.62	0.74	0.69
2.05	4.83	0.43	0.87	0.76	1.33	1.71	1.59
3.48	3.74	1.54	1.07	1.72	1.04	−1.49	−0.14
0.99	1.53	0.243	0.223	0.346	0.38	0.62	0.410
29.5	10.8	13.3	13.10	12.6	12.6	8.5	8.50
19.9	26.7	12.4	7.3	7.4	4.6	2.2	10.7
10.61	12.86	1.35	2.79	2.66	7.73	6.05	6.19
3.22	5.45	0.50	1.70	1.81	1.32	1.08	1.19
3.81	3.71	1.34	0.321	0.979	1.41	1.16	1.46
19.8	24.9	6.67	8.10	7.93	5.9	7.9	7.8

piezoelectric properties of crystals, which consists of the fact that the elastic deformation wave in such crystals may be accompanied by an electric-field wave, and the latter, in its turn, gives rise to additional mechanical stresses, which can affect the effective stiffness for the corresponding "piezoactive" wave, i.e., its velocity of propagation. The piezoelectric effect occurs in crystals which do not have a center of symmetry, i.e., most crystals.[105, 106] Since, however, the piezoelectric effect affects the measurements of elastic moduli of crystals performed by ultrasonic methods, it is worthwhile to consider this

Table 22

Elastic Moduli of Crystals in

Crystal	Chemical formula	$\rho\cdot 10^{-3}$ kg/m^3	c_{11}	c_{22}	c_{33}
Tartaric acid	$C_4H_6O_6$	1.760	9.3	1.93	4.65
Dibenzyl	$C_6H_5CH_2{=}CH_2C_6H_5$	0.995	0.945	0.680	0.720
Potassium tartrate (DKT)	$K_2C_4H_4O_6\cdot\frac{1}{2}(H_2O)$	1.988	3.11	3.90	5.54
Lithium-sulfate-monohydrate	$Li_2SO_4\cdot H_2O$	2.06	5.25	5.06	5.4
Sodium-thiosulfate	$Na_2S_2O_3$	1.667	3.31	3.02	4.57
Naphthalene	$C_{10}H_8$	1.168	0.78	0.99	1.19
l-Rhamnose-monohydrate	...	1.471	3.82	2.19	1.98
Stilbene	$C_6H_5CH{=}CHC_6H_5$	1.164	0.930	0.920	0.790
Tolan	$C_6H_5C{=}CC_6H_5$	0.996	0.785	0.855	0.645
Triglycine-sulfate (TGS)	$(NH_2CH_2COOH)_3\times$ $\times H_2SO_4$	1.68	4.55	3.21	2.63
Ethylene diamine tartrate (EDT)	$CHH_{14}N_2O_6$	1.538	5.7	3.29	2.01

question briefly in a separate concluding section, which can be regarded as an appendix to the last chapter of this book.

§ 5. Influence of the piezoelectric effect on the elastic properties of crystals

The influence of the piezoelectric effect on the velocity of propagation of ultrasonic waves in crystals can be determined by taking into account the additional mechanical stress that arises under the action of the sound-induced electric field **E**. For this, we shall use the *equation of the inverse piezoelectric effect*

$$\sigma_{ik} = c^E_{iklj}\,\epsilon_{lj} - f_{lik}E_l, \qquad \text{(XI.28)}$$

Continuation of Table 22

the Monoclinic System (T = 20°C)

$c_{nm} \cdot 10^{-10}$, N/m^2									
c_{44}	c_{55}	c_{66}	c_{12}	c_{13}	c_{23}	c_{15}	c_{25}	c_{35}	c_{46}
0.81	0.82	1.06	2.03	3.67	1.4	-1.2	-0.398	-0.0388	0.138
0.310	0.255	0.260	0.395	0.415	0.335	-0.24	2.08	0.07	0.08
0.87	1.040	0.826	1.72	1.69	1.33	0.287	0.182	0.71	0.072
1.4	1.565	2.77	1.715	1.73	0.368	-0.196	0.571	-0.254	-0.054
0.57	1.11	0.60	1.83	1.84	1.68	0.25	1.04	-0.69	-0.27
0.33	0.21	0.415	0.230	0.340	0.445	-0.06	-0.27	0.29	-0.05
0.537	0.502	0.911	1.60	1.66	0.888	-0.03	0.122	-0.118	0.022
0.325	0.640	0.245	0.570	0.570	0.485	-0.03	-0.05	-0.05	0.05
0.290	0.545	0.185	0.350	0.115	0.350	0.03	0.25	0.09	0.01
0.95	1.11	0.62	1.72	1.98	2.08	-0.30	-0.036	-0.5	-0.026
0.52	1.185	0.523	1.07	2.25	0.901	1.2	-0.064	0.668	-0.01

in which the coefficients f_{lik}, which are called the *piezoelectric constants* and form a tensor of rank three, determine precisely the magnitude of the additional mechanical stress sought. Expression (XI.28) can be called the equations of the mechanical state of a piezoelectric material. It is not difficult to see that in the absence of the piezoelectric effect, i.e., for $f_{lik} = 0$, Eq. (XI.28) transforms into the expression for the generalized Hooke's law (XI.1). In their turn, the components E_l of the electric intensity vector are related to the components of the electric induction vector **D** by the well-known equations of the electric state for a piezoelectric crystal, which take into account the direct piezoelectric effect:

$$D_i = \epsilon^u_{il} E_l + f_{ilj} \epsilon_{lj}, \qquad \text{(XI.29)}$$

where ϵ^u_{il} is the dielectric-constant tensor and ϵ_{ij} is the strain tensor.* In a piezoelectric material the dielectric and elastic constants depend on the conditions under which they are measured. For this reason the quantities c_{iklj} in Eq. (XI.28) and ϵ_{il} in Eq. (XI.29) contain superscripts indicating that the first equation refers to elastic moduli measured with a constant electric intensity ($\mathbf{E}$ = const), while the second equation refers to the dielectric constant measured with constant strain ($\mathbf{u}$ = const).

Solving now the equations of motion for anisotropic media (I.11), i.e., $\delta\sigma_{ik}/\delta x_k = \rho\delta^2 u_i/\delta t^2$, together with Eqs. (XI.28) and (XI.29) and Maxwell's equations for a nonconducting crystal div $\mathbf{D} = 0$ and $\nabla \times \mathbf{E} = 0$, we obtain

$$\rho\frac{\partial^2 u_i}{\partial t^2} = c^E_{iklj}\frac{\partial \epsilon_{lj}}{\partial x_k} - f_{lik}\frac{\partial E_l}{\partial x_k}, \tag{XI.30}$$

$$\frac{\partial D_i}{\partial x_i} = 0 = f_{ilj}\frac{\partial \epsilon_{lj}}{\partial x_i} + \epsilon^u_{il}\frac{\partial E_l}{\partial x_i}, \tag{XI.31}$$

$$\mathbf{E} = -\,\mathrm{grad}\,\varphi_e,$$

where φ_e is the electric potential. We shall seek the solution of these equations, as before, in the form of a monochromatic plane wave with frequency ω: $u_i = u_{i\max}\exp\{i[\omega t - \mathbf{k}\cdot\mathbf{r}]\}$. Substituting this solution into Eqs. (XI.30) and (XI.31) and eliminating from them the potential φ_e, we find the system of equations for the components of the elastic displacement vector $\mathbf{u}$:

$$u_j\left[c^E_{iklj}k_k k_l + \frac{(f_{lik}k_k k_l)(f_{ilj}k_i k_l)}{\epsilon^u_{il}k_i k_l} - \rho\omega^2\delta_{ij}\right] = 0.$$

*Unfortunately, because of State standard specifications (see A. G. Chertov "Units of Physical Quantities" (Vysshaya Shkola, Moscow (1977)), the same notation is used for two different physical quantities in this equation. The dielectric constant is distinguished from the strain by the superscript u (ϵ^u).

Equating the determinant of this system to zero and taking into account the fact that the wave number $k = \omega/c$, where c is the phase velocity of sound, we obtain

$$|\rho c^2 \delta_{ij} - \Gamma^*_{ij}| = 0, \tag{XI.32}$$

where

$$\Gamma^*_{ij} \equiv c_{iklj} k_k k_l + \frac{(f_{lik} k_k k_l)(f_{ilj} k_i k_l)}{\epsilon^u_{il} k_i k_l}. \tag{XI.33}$$

Comparing this result to Christoffel's equation (XI.7) we see that it differs only by the presence of an additional term, proportional to the square of the piezoelectric constant, in the tensor Γ_{ij}. In crystals with a weak piezoelectric effect this additional term is usually small and can be neglected. However, for crystals with a strong piezoelectric effect, such as rock salt, lithium niobate and iodate, etc., the additional term in (XI.33) can have a considerable magnitude. Since, however, Eq. (XI.32) determines the velocity of propagation of ultrasonic waves in crystals, this means that the piezoelectric effect can substantially affect the effective stiffness for those elastic waves that are accompanied by a longitudinal electric-field wave generated by the piezoelectric effect.[85,97,107]

Ultrasonic waves of this type are said to be *piezoelectrically active*. Their velocity of propagation, according to the relation (XI.3), will be determined by the effective modulus of elasticity, corresponding to the condition under which it is measured $\mathbf{D}$ = const and including the correction proportional to f^2, i.e., by the modulus c^D_{ijkl}. If, on the other hand, the piezoelectric field created by the wave is perpendicular to the wave vector, then the effective elastic constant for such a wave will be the modulus c^E_{iklj} measured under the condition $\mathbf{E}$ = const, i.e., as in the case with no piezoelectric effect.

As an example, we shall consider the hexagonal crystal lithium iodate (α-$LiIO_3$), which exhibits a strong piezoelectric effect.[108] In this crystal there are two transverse ultrasonic waves: one propagating along the Z-axis with displacement along the X-axis and the other propagating along the X-axis with displacement along the Z-axis, creating the same strain ϵ_{xz}. According to Table 1 (group VIII), they are characterized by the same elastic constant c_{44}. The measured velocities of propagation of these waves are, however, $2.0 \cdot 10^3$ and $2.5 \cdot 10^3$ m/s, respectively, i.e., they differ by tens of percent. This difference is attributable to the piezoelectric effect: the first wave is not piezoelectrically active, while the second wave is piezoelectrically active. Correspondingly, the velocity of propagation of the first wave gives the elastic modulus c^E_{44}, as in the case of no piezoelectric effect, and the velocity of the second wave gives the constant c^D_{44}. The difference between these moduli determines the magnitude of the so-called *electromechanical coupling constant*. We note that this important characteristic of a piezoelectric crystal, and together with it its piezoelectric constants, can be determined by purely ultrasonic measurements. In piezoelectric crystals, of course, amongst the different directions, including the characteristic directions, there are directions along which there is no coupling between the piezoelectric fields and the ultrasonic wave. For example, in the α-$LiIO_3$ crystal the transverse wave propagating along the X-axis with displacement along the Y-axis does not create a piezoelectric field, because, according to the symmetry of the piezoelectric properties, the corresponding piezoelectric constant vanishes for this wave.[105]

The piezoelectric effect also strongly affects the conditions of propagation of surface waves along the free boundary of a piezoelectric crystal.[74,75] It turns out that special, purely shear surface waves, called *Gulyaev–Blyustein waves*,[109] which play an important role in acoustoelectronics, can propagate along the surface of a piezoelectric crystal in certain directions. This problem, however, also falls outside the scope of this book.

Problems

CHAPTER 1

1.1. Prove that the stress tensor is a symmetrical tensor, i.e. $\tau_{ik} = \sigma_{ki}$.

1.2. Prove that the second-order tensor of the moduli of elasticity is symmetrical to permutations of the pairs of indices, i.e. $c_{ijkl} = c_{klij}$.

1.3. Find the relative change in the volume of a crystal of arbitrary symmetry under the action of a hydrostatic pressure *p*.

1.4. Prove that if the transition to the Voigt "matrix" notation in the stress tensor and the second-order tensor of the moduli of elasticity is carried out by simply changing the indices, the components of the deformation tensor must be introduced in accordance with the following rule: $\varepsilon_1 = \varepsilon_{xx}$, $\varepsilon_2 = \varepsilon_{yy}$, $\varepsilon_3 = \varepsilon_{zz}$, $\varepsilon_4 = 2\varepsilon_{zy}$, $\varepsilon_5 = 2\varepsilon_{xz}$, $\varepsilon_6 = 2\varepsilon_{xy}$.

1.5. Find the relative expansion of a copper rod $\Delta l / l$ when a tensile stress $\sigma = 10^5$ N/m^2 is applied to it.

CHAPTER 2

2.1. Estimate the local changes in the temperature of the air when an ultrasonic wave passes through it.

2.2. Estimate the wavelength of an ultrasonic wave of frequency $\nu = 1$ MHz in water and in air under normal conditions.

2.3. Estimate the degree of applicability of the calculation of the velocity of ultrasonic waves in a liquid from data for the isothermal moduli of elasticity using the example of glycerine under normal conditions.

CHAPTER 3

3.1 Analyse the phase difference between the pressure and the oscillatory velocity in a field of superimposed forward and backward plane acoustic waves with different amplitude.

3.2. Show that the instantaneous energy density in an acoustic wave satisfies the wave equation.

3.3. Derive a relation between the instantaneous acoustic energy density and the instantaneous acoustic-energy flux density, i.e. the so-called differential law of conservation of acoustic energy in a medium.

3.4. Show that the viscosity of a medium leads to considerable dispersion of the velocity of ultrasound in the frequency range where the absorption of the ultrasound is high.

CHAPTER 4

4.1. Write an expression for the acoustic energy density in the case of an ideal gas or liquid, which obeys Tate's equation.

4.2. Using the solution of the previous problem obtain an expression for the acoustic energy density of an ideal gas or a Tate liquid to a second approximation.

4.3. Show that for an arbitrary but constant phase shift of the harmonics with respect to the phase of the first harmonic, the summation of two distorted waves of the same amplitude, one of which is shifted with respect to the other by 180° in degrees of the first harmonics, leads to isolation of the even harmonics.

4.4. How does relaxation of the viscosity (the dependence of the viscosity on frequency) affect the distortion of the form of a wave of finite amplitude ?

4.5. Estimate the relative amplitude of the second harmonic generated in water at a distance $x = 10$ m from the source when an ultrasonic wave of frequency $\nu = 1$ MHz and intensity 1 W/cm^2 propagates.

CHAPTER 5

5.1. Calculate the radiation pressure on a plane-parallel partially absorbing plate, oriented perpendicular to a sonic beam, if the reflection coefficient ρ_I and the absorption coefficient of the plate $a_I = Q/(c_0\overline{w}_1)$ are known, where Q is the energy absorbed per unit surface of the plate in unit time.

5.2. Obtain the value of the maximum intensity of ultrasound of frequency $\nu = 1$ MHz in benzine in the region of a radiator for which, due to nonlinear effects, the error in calculating the velocity of ultrasonic flow using Eq. (5.38) and tabulated data for α_0 does not exceed 10%.

5.3. Give a schematic sketch of the direction of the gas flows occurring in a gas pipe under the condition that in pipe a) an ultrasonic travelling wave regime has been created, and b) a stationary wave regime has been created.

5.4. J.E. Piercy and J. Lamb [111] have used the following method for the determination of the coefficient of absorption of ultrasound: they inserted in the fluid filling the measuring chamber, depicted in the diagram, an outlet tube in which the speed of the acoustic flow was measured. Explain how to calculate the coefficient of absorption of ultrasound.

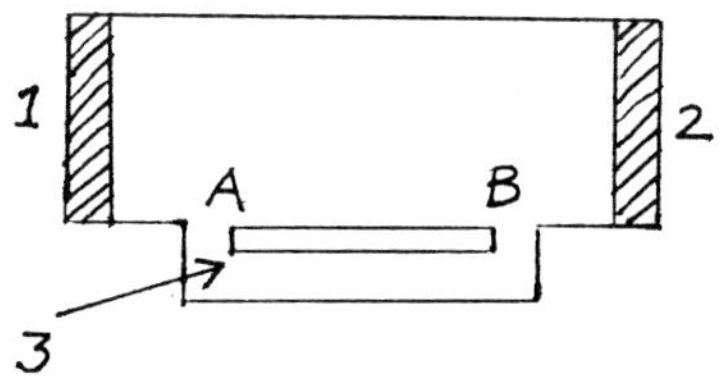

1 is an ultrasonic radiator, 2 is a sound-absorbing material, and 3 is an outlet tube.

5.5. Estimate the resonant size of an air bubble in water under normal conditions for an ultrasonic frequency of $\nu = 20$ kHz.

CHAPTER 6

6.1. Estimate, from thermodynamic considerations, the probability of the fluctuational occurrence of a gas bubble in a liquid.

6.2. To simplify the numerical analysis of the equations describing the pulsation of a cavitational cavity, it has been proposed that they can be converted to a form that does not include an explicit dependence on the frequency. To carry out this transformation of the Noiting-Nepaires equation and indicate under what conditions this transformation is possible.

6.3. Estimate the intensity cavitation threshold for water, containing bubbles with dimensions of the order of 10^{-4} cm ($P_H = 2.6\times10^{-8}$ at, and $\sigma = 80$ dynes/cm). Indicate the applicability of the method of calculation for frequencies of 20 kHz and 10 MHz.

CHAPTER 7

7.1. How is the intensity I_n of Rayleigh scattering of ultrasound at an angle $\theta = \pi/2$ in water by an air bubble in the preresonance region related to the scattering intensity I_H for scattering by an absolutely incompressible sphere of the same radius ?

7.2. In polycrystalline materials the attenuation of ultrasound is governed to a considerable extent by Rayleigh scattering at the grain boundaries. In this case how will the attenuation factor depend on the grain size ?

7.3. Determine the pressure exerted by a sound wave for normal incidence on the boundary of separation between two liquids.

7.4. An ultrasonic radiator operating at a frequency $\nu = 20$ kHz is placed in the lower part of a tube filled with water. The top of the tube is closed with a thin metal cover. A standing ultrasonic wave is set up in the tube. At what distance from the cover will an air bubble of radius $R = 10$ μm be in equilibrium ?

CHAPTER 8

8.1. Obtain an expression for the reflection coefficient of an ultrasonic wave from a plane-parallel layer situated between two similar media, assuming that the angle of incidence on the layer exceeds the angle of total internal reflection, and analyse the expression obtained.

8.2. Determine the natural frequencies of acoustic oscillations of a liquid in a vessel having the shape of a parallelepiped.

8.3. Determine the passband of an acoustic system in which there is no radiation into the surroundings.

CHAPTER 9

9.1. Determine the natural frequencies of spherically symmetrical acoustic oscillations in a spherical vessel.

9.2. Obtain the frequency of radial oscillations of a spherical cavity of radius R in an unbounded solid isotropic elastic medium, in which the condition $c_e \gg c_\tau$ is satisfied.

CHAPTER 10

10.1. Prove that when ultrasonic waves of finite amplitude propagate in an isotropic solid only a longitudinal second harmonic is generated.

10.2. Obtain the frequencies of natural longitudinal oscillations of a rod of length L, one end of which ($x = 0$) is clamped, and the second ($x = L$) is free.

CHAPTER 11

11.1. Show that elastic waves propagating along the second-order axis of crystals with symmetry C_2 or C_{2h} are pure modes.

11.2. Prove that an axis of symmetry of order higher than two is the acoustic axis.

11.3. Show that in a plane perpendicular to a fourth- or sixth-order axis of symmetry, a purely shear wave can propagate with a polarization parallel to the axis of symmetry, and its velocity is independent of the direction of propagation in the plane.

11.4. Obtain the reflection and transmission coefficients for a transverse ultrasonic wave, incident at an angle on the boundary of separation between two crystals of hexagonal symmetry, assuming that the wave is polarized perpendicular to the plane of incidence, and the directions of the sixth-order axes of symmetry of both crystals coincide with the displacement in the wave.

11.5. Show that for a GaAs crystal (point symmetry T_d) the velocity of a transverse wave propagating in a plane perpendicular to one of the fourth-order axes (to be specific the z axis), and polarized along this axis, depends on the direction of propagation because of the piezoelectric effect.

Solutions to Problems

CHAPTER 1

1.1. Consider the moment of the forces acting on a certain volume of the deformed body :

$$\mathbf{u} = \int_V [\mathcal{F}\times\mathbf{r}]\, dV$$

where $\mathcal{F}$ is the force acting on unit volume of the body, and $\mathcal{F}_i = \partial\sigma_{ik}/\partial x_k$. The moment of these forces, like the total force acting on any volume, must be expressed in terms of an integral over the surface of the volume V. We will convert $\mathbf{u}$ into a sum of two integrals: over the surface and over the volume. From the definition of the vector product of the i-th component of a vector, $\mathbf{u}$ can be written in the form

$$u_i = \int_V (\mathcal{F}_k x_l - \mathcal{F}_l x_k)\, dV,$$

where i, k, and l are the cyclic permutation of the Cartesian coordinates x,y,z. Substituting the expression for $\mathcal{F}_k$ in terms of the stress tensor, we obtain

$$u_i = \int_V \left(\frac{\partial\sigma_{kj}}{\partial x_j}x_l - \frac{\partial\sigma_{lj}}{\partial x_j}x_k\right)dV = \int_V \frac{\partial(\sigma_{kj}x_l - \sigma_{lj}x_k)}{\partial x_j}\,dV -$$

$$- \int_V \left(\sigma_{kj}\frac{\partial x_l}{\partial x_j} - \sigma_{lj}\frac{\partial x_k}{\partial x_j}\right)dV = \int_S (\sigma_{kj}x_l - \sigma_{lj}x_k)\,dS_j + \int_V (\sigma_{kl} - \sigma_{lk})\,dV$$

Since the integral over the volume must be zero, we have $\sigma_{kl} = \sigma_{lk}$.

1.2. The energy of elastic deformation of the crystal has the form $U =$ $= 1/2c_{ijkl}\varepsilon_{ij}\varepsilon_{kl}$ (see Section 5). It follows directly from this expression that the second-order tensor of the moduli of elasticity is symmetrical to a permutation of the pairs of indices i, j, and k,l: $c_{ijkl} = c_{klij}$.

1.3. For a hydrostatic compression $\sigma_{ik} = -p\delta_{ik}$. Hooke's law then takes the form $\varepsilon_{jl} = k_{jlik}\sigma_{ik} = -pk_{jlii}$, where k_{jlik} are the elastic susceptibilities (Section 4). Using the formula for the relative change in the volume θ (see Section 1) we obtain the expression $\theta = \varepsilon_{jj} = -pk_{jjii}$. Then the compressibility $\chi = -\theta/p$ is equal to $\chi = k_{jjii}$.

1.4. To prove this it is sufficient to consider the expression for the component σ_{xx} in the usual notation

$$\sigma_{xx} = c_{xxkl}\varepsilon_{kl} = c_{xxxx}\varepsilon_{xx} + c_{xxyy}\varepsilon_{yy} + c_{xxzz}\varepsilon_{zz} +$$

$$+ 2c_{xxxy}\varepsilon_{xy} + 2c_{xxxz}\varepsilon_{xz} + 2c_{xxyz}\varepsilon_{yz}$$

and in the "matrix" notation

$$\sigma_1 = c_{11}\varepsilon_1 + c_{12}\varepsilon_2 + c_{13}\varepsilon_3 + c_{16}\varepsilon_6 + c_{15}\varepsilon_5 + c_{14}\varepsilon_4 .$$

These expressions are identically equal only if the relations proved are satisfied. It is easy to verify that when these relations are satisfied, the other components of the stress tensor are identically equal to one another in the different notations.

1.5. According to (1.24) $\Delta l/l = \sigma/E$. From Table 1.2 for copper $E =$ $= 12.0 \times 10^{10}$ N/m^2. Hence, $\Delta l/l = 8.3 \times 10^{-7}$.

CHAPTER 2

2.1. According to (2.3), $(T_0 + \Delta T)/T_0 = (P/P_0)^{\frac{\gamma-1}{\gamma}}$. Since $P = P_0 + P$ (where p is the pressure in the acoustic wave) and $p \ll P_0$, we have $(P/P_0)^{\gamma-1} = I + \frac{\gamma-1}{\gamma} \cdot \frac{p}{P}$ and $\Delta T/T_0 = \frac{\gamma-1}{\gamma} \cdot \frac{p}{P_0}$. Under normal conditions in air $\Delta T = 8 \times 10^{-4} p$ (if p is measured in N/m^2). At the threshold of audibility $p \simeq 10^{-4}$ N/m^2 and at the threshold of pain $p \simeq 10^2$ N/m^2 (see Section 3, in Chapter 3). The corresponding amplitudes of the change in temperature are $\Delta T \simeq 8 \times 10^{-8}$ K and ΔT 8×10^{-2} K. These small changes in temperature also give rise to a 20% difference in the velocities of sound for adiabatic and isothermal processes in gases.

2.2. The wavelength is $\Lambda = c_0/\nu$. Using the data for c_0 from Table 2.2, we obtain for air $\Lambda = 3\times10^{-4}$ m = 0.3mm and for water $\Lambda = 1.4\times10^{-3}$m = 1.4mm.

2.3. The isothermal and adiabatic moduli of elasticity are related to one another by equation 2.29, whence $(K_{ad} - K_{is})/K_{is} = \alpha_T^2 T K_{is}/(\rho c_V)$. Using the data for ρ_0 given in Table 2.2, and also the data from handbooks: $\alpha_T = 0.59\times10^{-3}\ K^{-1}$ and $c_V =$ = 2.39 kJ/kg.K, and $\kappa_{is} = 0.22\times10^{-9}$ m^2/N, we obtain that at $T \simeq 300$ K, $(K_{ad} - K_{is})/K_{is} \simeq 1.5\times10^{-1}$. Hence the relative error of the calculation of the velocity of sound in glycerine from the data for the isothermal modulus of elasticity is 1.5×10^{-1} = 15%. Note that for water for $T = 4.2°K$, $\alpha_T = 0$ and $K_{ad} = K_{is}$.

CHAPTER 3

3.1. We use the formula for the acoustic impedance, obtained in Section 2:

$$\tilde{Z} = \rho_0 c_0 \frac{e^{i(\omega t - kx)} + e^{i(\omega t + kx)}}{e^{i(\omega t - kx)} - e^{i(\omega t + kx)}}$$

This formula can easily be reduced to the form $\tilde{Z} = i\rho_0 c_0 \cot(kx)$, from which it follows that the acoustic impedance is a purely imaginary quantity and the shift between the pressure and the oscillatory velocity is ±90°.

3.2. It follows from Eq. (3.20) that the energy density in an acoustic wave can be written in the form $W = W_{max}\sin^2(\omega t - kx)$. If the function $v = v_{max}\sin(\omega t - kx)$ is the solution of the wave equation, its square is also the solution of the wave equation, and, consequently, the function W is also a solution of the wave equation. In fact,

$$\frac{\partial^2 W}{\partial x^2} = W_{max}\frac{\partial^2}{\partial x^2}\sin^2(\omega t - kx) = 2k^2 W_{max}\cos(2\omega t - 2kx) =$$

$$= 2\frac{\omega^2}{c_0^2} W_{max}\cos(2\omega t - 2kx) = \frac{1}{c_0^2}\frac{\partial^2 W}{\partial t^2}$$

Hence, the energy in a travelling acoustic wave is transmitted with the velocity of sound.

3.3. We use Eqs. (2.5) and (2.11). Using relation (2.30) we can convert Eq. (2.11) to the form

$$-\frac{1}{\rho_0}\left(\frac{\partial\rho}{\partial p}\right)_0\frac{\partial p}{\partial t} = \nabla\mathbf{v}\,,$$

and then, taking into account the definition of the compression coefficient κ, we can reduce it to the form $-\kappa\dfrac{\partial p}{\partial t} = \nabla\mathbf{v}$. Multiplying the equation obtained by p and adding it, after scalar multiplication by v, to Eq. (2.5), we obtain

$$p\nabla\mathbf{v}+\rho_0\mathbf{v}\frac{\partial \mathbf{v}}{\partial t}+\kappa p\frac{\partial p}{\partial t}+\mathbf{v}\nabla p = \frac{\partial}{\partial t}\left\{\frac{1}{2}\rho_0 v^2+\frac{1}{2}\kappa p^2\right\}+\nabla(p\mathbf{v}) = 0\ .$$

The first term is the derivative with respect to time of the instantaneous acoustic energy density W. The second term is the divergence of the instantaneous acoustic energy flux density $\mathbf{I}_t = p\mathbf{v}$. Hence, the differential law of conservation of acoustic energy is given by the relation

$$\frac{\partial W}{\partial t} + \nabla\mathbf{I}_t = 0$$

3.4. Separating the complex wave number $\tilde{k}$, given by Eq. (3.33), into real and imaginary parts in the general case, without assuming that the viscous forces are much less than the elastic forces, we obtain

$$\tilde{k} = \omega\sqrt{\frac{\rho_0 k}{k^2+\omega^2\eta^2}}\left\{\frac{1+\sqrt{1+\omega^2\eta^2/k^2}}{2} - \frac{i\omega\eta/k}{\sqrt{2+2\sqrt{1+\omega^2\eta^2/k^2}}}\right\}$$

It follows from this expression that the velocity of sound in a viscous medium $\tilde{c}$ is

$$\tilde{c} = c_0\sqrt{\frac{1+\sqrt{1+\omega^2\eta^2/k^2}}{2(1+\omega^2\eta^2/k^2)}}$$

The value of $\tilde{c}$ differs from c_0 when the ratio $\omega^2\eta^2/k^2$ is close to unity. However, in this case the absorption coefficient of ultrasound, calculated from the imaginary part of the complex wave number

$$\alpha_0 = \frac{\omega^2\eta^2}{\sqrt{k^2+\omega^2\eta^2}}\sqrt{\frac{\rho_0}{k}}\,\frac{1}{\sqrt{2+2\sqrt{1+\omega^2\eta^2/k^2}}}$$

becomes equal to $2\pi/\Lambda$ in order of magnitude, which denotes physically that the ultrasound is attenuated in a path length of the order of Λ. Note that this conclusion holds when there is no frequency dependence of the viscosity [13].

CHAPTER 4

4.1. The acoustic energy density of a medium can be written in the form of the sum of the kinetic energy of the particles of the medium due to oscillatory motion, and the difference in the internal energy in the medium perturbed and unperturbed by the sound : $W = \rho v^2/2 + U - U_0$. Using for the internal energy of an ideal gas the expression $U = \frac{P_0}{\gamma-1}\left(\frac{\rho}{\rho_0}\right)^{\gamma-1}$, we obtain the following expression for W:

$$W = \frac{\rho v^2}{2} + \frac{P_0}{\gamma-1}\left[\left(\frac{\rho}{\rho_0}\right)^{\gamma} - 1\right] = \frac{\rho v^2}{2} + \frac{P-P_0}{\gamma-1} = \frac{\rho v^2}{2} + \frac{P-P_0}{2(\varepsilon_0-1)}$$

For a Tate liquid we must replace γ by n.

4.2. Substituting relation (4.34) into the expression obtained for W in the previous problem, we have

$$W = \frac{\rho_0 v^2}{2} + \frac{1}{2(\varepsilon_0-1)}\left[c_0^2\Delta\rho + (\varepsilon_0-1)\, c_0^2\Delta\rho^2/\rho_0\right].$$

4.3. The oscillatory velocities for both waves can be written in the form

$$v_1 = A\sin(\omega t-kx) + B\sin[2(\omega t-kx)+\phi_2] +$$
$$+ C\sin[3(\omega t-kx)+\phi_3] + D\sin[4(\omega t-kx)+\phi_4] + \ldots$$

$$v_2 = A\sin(\omega t-kx+\pi) + B\sin[2(\omega t-kx+\pi)+\phi_2] + C\sin[3(\omega t-kx+\pi)+\phi_3] + \ \ldots\ .$$

Then, in the overall wave, the oscillatory velocity is

$$v = v_1+v_2 = 2B\sin[2(\omega t-kx)+\phi_2] + 2D\sin[4(\omega t-kx)+\phi_4] + \ldots\ .$$

This effect is used for interference selection of even harmonics [110].

4.4. When a monochromatic wave propagates in a nonlinear medium the generation of higher acoustic harmonics is limited by the increase in the absorption coefficient of the ultrasound as the frequency increases. The presence of relaxation processes when $\omega\tau \gtrsim 1$ leads to a weakening of the frequency dependence of the absorption coefficient of the sound, as a result of which the amplitude of the higher harmonics increases and the distortion of the form of the wave is therefore increased.

4.5. We use expressions (4.43) for the estimate. According to this expression $v_{max2}/v_{max} = v_{max}\omega\varepsilon_0 x/(2c_0^2)$. For v_{max} we obtain (see Section 3 in Chapter 3) $v_{max} = 10$ cm/s, and $\varepsilon_0 = 1 + B/2K$. The values of c_0 and B/K are given in Tables 2.2 and 4.1 respectively. Substituting the numerical values into the formula for v_{max2}/v_{max} we obtain $v_{max2}/v_{max} \simeq 0.9\times10^{-4}$.

CHAPTER 5

5.1. Substituting the expression for the energy density in the reflected and transmitted waves into Eq. (5.13) and taking into account the fact that $\bar{w}_3 = (1 - \rho_I)\bar{w}_1 - a_I\bar{w}_1$, we obtain the radiation pressure π

$$\pi = \bar{w}_1 + \rho_I\bar{w}_1 - (1 - \rho_I + a_I)\bar{w}_1 = \bar{w}_1(2\rho_I + a_I)$$

5.2. According to (5.38) the velocity of ultrasonic flow is proportional to the absorption coefficient α of the ultrasound in the medium. Consequently, to obtain the required accuracy in calculating v_0 the increase in the absorption coefficient of the ultrasound due to nonlinear effects should not exceed 10%. From relation (4.83) for $\Delta\alpha/\alpha_0 \leq 0.1$ we obtain $\mathrm{Re} \leq 1/\varepsilon_0$. Using the definition of the Reynolds number $\mathrm{Re} = \rho_0 c_0 v_{max}/(b\omega)$ (see Section 8, Chapter 4) and assuming that we can take the usual viscosity η for b, we obtain $v_{max} \leq b\omega/(\rho_0 c_0 \varepsilon_0)$. Since $\eta = \rho_0 c_0^3 \alpha_0/(2\pi^2\nu^2)$ (Eq. 3.38), we have

$$v_{max} \leq \frac{c_0^2 \nu}{\varepsilon_0 \pi}\left(\frac{\alpha_0}{\nu^2}\right) \text{ and } I = \frac{\rho_0 c_0 v_{max}^2}{2} \leq \frac{\rho_0 c_0^5 \nu^2}{2\varepsilon_0^2 \pi^2}\left(\frac{\alpha_0}{\nu^2}\right)^2 .$$

The data required for the calculation are given in Tables 2.2, 3.2, and 4.1. Substituting these data and taking into account the fact that $\varepsilon_0 = I + B/2K$, we find that $I \leq 0.5$ *Bt* /cm^2. Note that the values η_c must not be used for b since the experimental value of α_0/v^2 in benzol is considerably in excess of the calculated value (see Table III.2).

5.3. The direction of the gas flow in the pipe is shown in the diagrams:

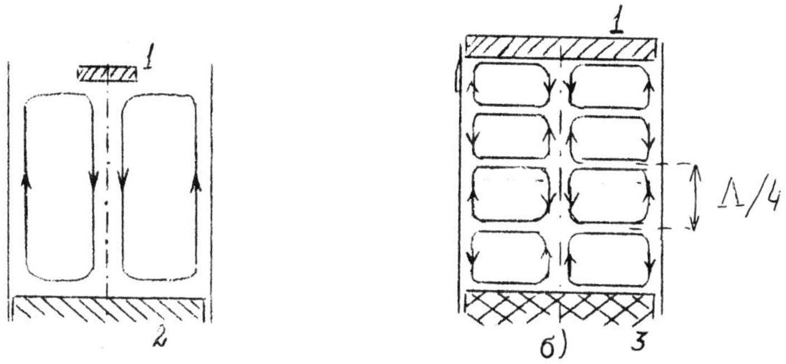

a) b)

1 is the ultrasound radiator,

2 is sound-absorbant material,

3 is an ultrasound reflector.

In case b) the direction of the flows around the walls is from the anti-node to the node of the stationary wave.

5.4. The velocity v_0 of acoustic flow in a tube is related to the pressure difference ΔP at the points A and B by Poisseuile's formula (5.36), whence we obtain $\Delta P = 4\eta l v_0/R^2$ (l is the length of the tube). On the other hand, converting (5.37) for ΔP, it is easy to obtain the expression $\Delta P = \overline{w}(0)\,[e^{-2\alpha_0 x_A} - e^{-2\alpha_0 x_B}]$, where x_A and x_B are the distances from the radiator to the points A and B. Equating the two equations for ΔP we obtain a relation between v_0 and α_0.

5.5. We use Eq. (5.27) and the values of the velocity of ultrasound and the density of water and air given in Table 2.2. We thereby obtain $R_{res} =$ $= \sqrt{3}b/k\sqrt{a} \simeq 200$ μm.

CHAPTER 6

6.1. The probability of the fluctuational occurrence of a bubble is proportional to $\exp[-A_{min}/(k_b T)]$ where A_{min} is the minimum work which must be done to form it, and k_B is Boltzmann's constant. A_{min} is equal to the change in the thermodynamic potential when the nucleating centre is being formed, which is determined by the change in the pressure in the volume of the bubble and the surface tension

$$A_{min} = -(P' - P)V_0 + \sigma S_0$$

where P' and P are the pressure in the bubble and in the surrounding liquid, and V_0 and S_0 are the volume and area of the surface of the bubble. Assuming that $V_0 = 4\pi R^3/3$, $S_0 = 4\pi R^2$ and $R = 2\sigma/(P' - P)$ (Eq. (6.2), where R is the radius of the bubble, we obtain

$$A_{min} = \frac{16\pi\sigma^3}{3(P' - P)^2}$$

6.2. Instead of R and t we introduce the new variables $a = \frac{\omega}{\omega_0} \cdot \frac{R}{R_0}$ and $\tau = \omega t$, where $\omega_0 = \omega_{res}/\sqrt{3\gamma}$, and ω_{res} is the resonance frequency of an equilibrium bubble of radius R_0. Substituting a and τ into (6.24) and using the relationship between P_0 and ω_0 given by Eq. (6.5), we obtain

$$a\frac{d^2a}{d\tau^2} + \frac{3}{2}\left(\frac{da}{d\tau}\right)^2 + \frac{1}{\rho_0\omega_0^2R_0^2}\left[\omega_0^2R_0^2\rho_0 - \frac{2\sigma}{R_0} - P_H - p_{max}\sin\tau + \right.$$

$$\left. + \frac{2\sigma}{R_0}\frac{\omega}{a\omega_0} - (\omega_0^2R_0^2\rho_0 - P_H)\left(\frac{\omega}{a\omega_0}\right)^{3n}\right] = 0$$

whence

$$a\frac{d^2a}{d\tau^2} + \frac{3}{2}\left(\frac{da}{d\tau}\right)^2 - \frac{1}{\rho_0\omega_0^2R_0^2}\left\{\frac{2\sigma}{R_0}\left(1 - \frac{\omega}{a\omega_0}\right) + \right.$$

$$\left. + P_H\left[1 + \left(\frac{\omega}{a\omega_0}\right)^{3n}\right] + p_{max}\sin\tau\right\} + \left[1 - \left(\frac{\omega}{a\omega_0}\right)^{3n}\right] = 0 .$$

The expression obtained already contains an explicit dependence on ω. Provided that $\omega/(a\omega_0) \ll 1$, i.e. $R/R_0 \gg 1$, corresponding to the region in which the

Noiting-Nepaires equation can be used, this dependence can be eliminated

$$a\frac{d^2a}{d\tau^2} + \frac{3}{2}\left(\frac{da}{d\tau}\right)^2 - \frac{1}{\rho_0\omega_0^2R_0^2}\left(\frac{2\sigma}{R_0} + P_H + p_{max}\sin\tau\right) = 0.$$

6.3. Cavitation occurs when the negative pressure in the acoustic wave reaches a critical value, calculated from Eq. (6.4)

$$p_{max} = P_{cr} = P_H - \frac{2}{3\sqrt{3}}\sqrt{\frac{(2\sigma/R_0)^3}{P_0 - P_H + 2\sigma/R_0}} = 1.3 \text{ at},$$

$$I = P_{max}^2 / (2\rho_0 c) = 0.55 \text{ W/cm}^2 .$$

The resonance frequency of a bubble with dimensions of 10^{-4} cm, as can easily be obtained from Eq. (5.28), is $\simeq 5$ MHz. Hence, one can use the calculation for 20 kHz but not for 10 MHz.

CHAPTER 7

7.1. According to Eqs. (7.50b) and (7.54) for $\theta = \pi/2$, $I_n/I_H = (1 - K_2/K_1)^2$. Since for a gas bubble $K_2 \gg K_1$, we have $I_n/I_H \simeq K_2^2/K_1^2$. It follows from Table 2.1 that for water $K_2 = 0.22\times10^{10}$ N/m^2, and for water it follows from Table 2.2 that $K_1 = \gamma P_0 = c^2/\rho_0 \simeq 1.5\times10^5$ N/m^2, whence $I_n/I_H \simeq 1.5\times10^8$.

7.2. Assuming the grain to be approximately spherical and taking into account the fact that the radius of the grain R and their concentration n_0 are related by the approximate expression $n_0 = 3/(4\pi R^3)$, we obtain, using the formula for α'_{sc} from Section 6, that in polycrystalline materials $\alpha'_{sc} = 7n_0\pi k^4R^6/9 \sim R^3$.

7.3. At the boundary of two liquids, the power reflection coefficient, according to (7.14) is

$$\rho_I = [(\rho_1c_1 - \rho_2c_2)/(\rho_1c_1 + \rho_2c_2)]^2$$

Substituting this expression for ρ_I into (5.15) we obtain

$$\pi = \overline{w}\left[2\,\frac{\rho_1^2c_1^2 + \rho_2^2c_2^2}{(\rho_1c_1 + \rho_2c_2)^2} - \frac{4\rho_1c_1\rho_2c_2}{(\rho_1c_1 + \rho_2c_2)^2}\,\frac{c_1}{c_2}\right] =$$

$$= 2\overline{w}\,\frac{\rho_1^2c_1^2+\rho_2^2c_2^2-2\rho_1\rho_2c_1^2}{(\rho_1c_1+\rho_2c_2)^2}$$

7.4. There will be a pressure node at the upper end of the tube, closed with the thin cover. At a distance $\Lambda/4$ from it there will be a pressure anti-node. The air bubble will move to a pressure node or a pressure anti-node in the standing wave depending on the ratio between its dimension and the resonance dimension of an air bubble in water. It follows from the solution of problem 5.5 that the resonance radius of an air bubble $R_{res} \simeq 200\mu m$. Since $R < R_{res}$, the air bubble will be in equilibrium at a pressure anti-node (see Section 3 in Chapter 5), i.e. at a distance $\Lambda/4 = c_0/(4\nu) \simeq 1.9$ cm from the cover.

CHAPTER 8

8.1. If the angle θ_1 exceeds the angle of total internal reflection, $\cos\theta$ is a purely imaginary quantity

$$\cos\theta = \sqrt{1-\sin^2\theta} = i\sqrt{(\frac{c}{c_1}\sin\theta_1)^2-1} \equiv ia\,,$$

where a is a positive number.

Substituting $\cos\theta$ into expression (8.3) for the pressure reflection coefficient ρ_p (converted to a convenient form)

$$\rho_p = \frac{z_1^2\cos^2\theta - z_2^2\cos^2\theta_1}{\sqrt{(z_1^2\cos^2\theta + z_2^2\cos^2\theta_1)^2 + 4z_1^2z_2^2\cos^2\theta_1\cot^2(dk\cos\theta)}}$$

we obtain for the modulus of ρ_p the expression

$$|\rho_p| = \frac{z_1^2a^2 + z_2^2\cos^2\theta_1}{\sqrt{z_2^4\cos^4\theta_1 + z_1^4a^4 + 2z_1^2z_2^2a^2\cos^2\theta_1[2\coth^2(dka)-1]}}$$

Since $\coth^2(dka) \geq 1$, we have $|\rho_p| \leq 1$ and $|\rho_p| \to 1$ as $d \to \infty$. Hence, for a finite thickness of the layer the acoustic wave will always penetrate through the layer. This phenomenon is similar to the phenomenon of tunnelling through a potential barrier in quantum mechanics.

8.2. We seek a solution of Eq. (2.32) in the form

$$\phi_0 = C\cos(k_x x)\cos(k_y y)\cos(k_z z)$$

where C is a constant.

Substituting ϕ_0 into (2.32) we obtain the relation $k_x^2 + k_y^2 + k_z^2 = \omega^2/c_0^2$. On the walls of the vessel for $x = 0$ and a, we have $v_x = \partial\phi_0/\partial x = 0$, for $y = 0$, b, we have $v_y = \partial\phi_0/\partial y = 0$, and for $z = 0$, d we have $v_z = \partial\phi_0/\partial z = 0$, where a, b, and d are the lengths of the sides of the parallelepiped. It follows from the condition

$$\partial\phi_0/\partial x = -Ck_x \sin(k_x x)\cos(k_y y)\cos(k_z z) = 0$$

when $x = 0$ and $x = a$ that $\sin(k_x a) = 0$ and $k_x = m_x\pi/a$, where m_x is any integer. In a similar way $k_y = m_y\pi/b$ and $k_z = m_z\pi/d$. Consequently, the natural frequencies are

$$\omega^2 = c_0^2(k_x^2 + k_y^2 + k_z^2) = c_0^2\pi^2\left(\frac{m_x^2}{a^2} + \frac{m_y^2}{b^2} + \frac{m_z^2}{d^2}\right).$$

8.3. The Q-factor of an acoustic system in which there is no radiation into the surroundings is determined by the internal losses (see Section 6): $Q_a = \pi/(\alpha_0\Lambda)$. Substituting this quantity for Q_a into (8.66) we obtain for the passband $\Delta\nu$

$$\Delta\nu = \nu_0\alpha_0\Lambda/\pi = \alpha_0 c_0/\pi.$$

As the frequency of the sound increases the passband of such a system increases due to the increase in α_0.

CHAPTER 9

9.1. The velocity potential in a monochromatic spherical wave, which is the solution of the wave equation (9.1), has the form (9.4). Substituting this expression for ϕ into boundary condition $\partial\phi/\partial r = 0$ with $r = r_0$, corresponding to the radial component of the velocity on the walls of the vessel being zero, we obtain the transcendental equation $\tan(kr_0) = kr_0$ or $\tan(\omega_r r_0/c_0) = \omega_r r_0/c_0$, which determines the natural frequencies ω_r. The lowest natural frequency $\omega_{r1} = 4.49\, c_0/r_0$.

9.2. At the boundary of the cavity, like on a free surface, the condition for the normal stresses to be zero must be satisfied : $\sigma_{rr}(r = R) = 0$. Using relation (1.15) in a spherical system of coordinates, we can write σ_{rr} in the form

$$\sigma_{rr} = \lambda u_{ee} + 2\mu u_{rr} = \rho[(c_e^2 - 2c_\tau^2)\, u_{ee} + 2c_\tau^2 u_{rr}]$$

Since the velocity potential, according to (9.5), can be written in the form $\phi = \phi_0 e^{i(\omega t - kr)}/r$, we have $u_{rr} = \partial^2\phi/\partial r^2$ and $\Delta\phi = \frac{1}{r^2}\frac{\partial}{\partial r}(r^2 \frac{\partial\phi}{\partial r}) = -k^2\phi$ for $k = \omega/c_e$, and

$$\sigma_{rr} = \rho\{ (c_e^2 - 2c_\tau^2)\, \Delta\phi + 2c_\tau^2 \frac{\partial^2\phi}{\partial r^2}\} = \rho\{(2c_\tau^2 - c_e^2) + 2c_\tau^2 \frac{\partial^2\phi}{\partial r^2}\}$$

Substituting this expression for σ_{rr} into the boundary condition, we obtain, after some reduction, the equation $\omega^2 R^2 - 4c_\tau^2 \left(1 + \frac{i\omega R}{c_e}\right) = 0$, whence $\omega =$ $= 2\frac{c_\tau}{R}\left(1 + i\frac{c_\tau}{c_e}\right)$, assuming that $c_e \gg c_\tau$. The real part of ω gives the natural frequency of the oscillations $\omega_0 = 2c_\tau/R$, while the imaginary part is proportional to the attenuation factor. Since $c_e \gg c_\tau$, the attenuation factor is small. It should be noted that the oscillations considered are due to the shear elasticity ($\mu \neq 0$).

CHAPTER 10

10.1. Consider a longitudinal plane wave propagating along the x axis. The boundary conditions have the form $u_x(x = 0) = u_0 \cos \omega t$, and $u_y(x = 0) =$ $= u_z(x = 0) = 0$. The solution of the equation of the first approximation (10.81) has the form $u'_x = u_0 \cos(\omega t - k_e x)$, $u'_y = u'_z = 0$. Substituting this solution into the equation of the second approximation (10.82) we obtain

$$\rho_0 \frac{\partial^2 u''_x}{\partial t^2} - (K + \tfrac{4}{3}G)\frac{\partial^2 u''_x}{\partial x^2} = \beta \frac{\partial^2 u'_x}{\partial x^2}\frac{\partial u'_x}{\partial x} = -\beta k_e^3 \sin[2(\omega t - k_e x)]\,,$$

$$\rho_0 \frac{\partial^2 u''_{y,z}}{\partial t^2} - G\frac{\partial^2 u''_{y,z}}{\partial x^2} = 0$$

where $\beta = 3K + 4G + 2A + 6B + 2C$.

Taking the boundary conditions into account, it follows from the equations for u''_y and u''_z that $u''_y = u''_z = 0$ over the whole volume of the medium. For u''_x we can obtain the solution

$$u''_x = x \frac{\beta k_e^2 u_0^2}{K + 4G/3} \cos[2(\omega t - k_e x)]$$

Hence, when a longitudinal wave of finite amplitude propagates, only a longitudinal second harmonic with a linearly increasing amplitude is generated.
Consider a transverse plane wave propagating along the x axis and polarized, to be specific, along the y axis. The boundary conditions have the form $u_y(x = 0) = u_0 \cos \omega t$, and $u_x(x = 0) = u_z(x = 0) = 0$. It is easy to see that in this case it follows from Eqs. (10.82) that $u''_y = u''_z = 0$. For u''_x we obtain

$$u''_x = \frac{[2c_\tau^2 + c_e^2 + (A/2 + B)/\rho_0]\omega u_0^2}{4c_\tau^3[(c_e/c_\tau)^2 - 1]} \sin[(k_\tau - k_e)x] \cos[2\omega t - (k_\tau + k_e)x]$$

When a transverse wave propagates, the amplitude of the second longitudinal harmonic does not increase due to the absence of spatial synchronism between the transverse wave of the linear approximation and the longitudinal wave of the second approximation.

10.2. The equation of natural longitudinal oscillations of a rod has the form (10.74). On the basis of the conditions of the problem we can write the boundary conditions for ξ : $\xi(x = 0) = 0$, and since $\sigma_{xx}(x = L) = 0, \frac{\partial \xi}{\partial x}(x = L) = 0$. We seek a solution of Eq. (10.74) in the form $\xi = \xi_0 \cos(\omega t + \alpha_0) \sin(k_x)$ (α_0 is the initial phase and $k = \omega / c_{C\tau}$). The condition $\xi(x = 0) = 0$ is satisfied for all values of k, and it follows from the condition $\frac{\partial \xi}{\partial x}(x = L) = 0$ that $\cos(kL) = 0$, whence we obtain the following expression for the natural frequencies:

$$\omega = c_{CT} \frac{\pi}{L}(m + \frac{1}{2}),$$

where m is an arbitrary integer.

CHAPTER 11

11.1. Suppose the y axis of a rectangular system of coordinates is directed along the C_2 axis (as in Fig.11.1). Then $n_x = n_z = 0$ and $n_y = 1$, whence it follows that the Christoffel tensor can have the components $\Gamma_{ij} \equiv c_{iklj} n_k n_l = c_{iyyj}$. Since for classes C_2 and C_{2h} the components of the moduli of elasticity tensor c_{26} and c_{24} are zero (see Table 1.1), only the components $\Gamma_{yy} = c_{22}$, $\Gamma_{zz} = c_{44}$, $\Gamma_{xx} = c_{66}$, and $\Gamma_{xz} = \Gamma_{zx} = c_{46}$ differ from zero. In matrix form, the tensor has the form

$$\Gamma_{ij} = \begin{bmatrix} c_{66} & 0 & c_{46} \\ 0 & c_{22} & 0 \\ c_{46} & 0 & c_{44} \end{bmatrix} \qquad (11.34)$$

For this case, Eqs. (11.7) take the form

$$(\rho c^2 - c_{66})\, u_x - c_{46} u_z = 0$$

$$(\rho c^2 - c_{zz}) u_y = 0$$

$$(\rho c^2 - c_{44}) u_z - c_{46} u_x = 0$$

whence it follows that a purely longitudinal wave can propagate along C_2 with a velocity $c^{l}_{[0/0]} = \sqrt{c_{22}/\rho}$, and two purely transverse waves with different velocities.

11.2. Suppose, as in problem 11.1, that the y axis is directed along the axis of symmetry. Then, since $c_{24} = c_{26} = 0$ for all crystals with point symmetry higher than monoclinic, the tensor Γ_{ij} will have the form (11.34). Moreover, if there is an axis of symmetry of order higher than 2, we have $c_{44} = c_{66}$, and for elastic shear waves propagating along y and polarized along x and z, Eqs. (11.7) are identical, and consequently, the velocities of these waves are also identical. Hence, the velocity of propagation of a transverse wave along an axis of symmetry of order higher than 2 is independent of the polarization of the wave.

11.3. We will choose the y axis of a rectangular system of coordinates to be parallel to the fourth- or sixth-order axis of symmetry. Then $n_y = 0$ and $n_x^2 + n_z^2 = 1$. It is easy to show from definition (11.8) that the tensor Γ_{ij} has the form

$$\Gamma_{ij} = \begin{bmatrix} \Gamma_{xx} & 0 & \Gamma_{xz} \\ 0 & \Gamma_{yy} & 0 \\ \Gamma_{xz} & 0 & \Gamma_{zz} \end{bmatrix}$$

In this case $\Gamma_{yy} = c_{44}$. Hence, a transverse wave, polarized along y, is a pure mode, and its velocity is $c^{\tau} = \sqrt{c_{44}/\rho}$ irrespective of the direction of propagation in a plane perpendicular to the y axis, i.e. the fourth- or sixth-order axis.

11.4. As shown in problem 11.3, the velocity of the transverse wave considered is independent of the angle of incidence θ_1 and is equal to $c^{\tau} = \sqrt{c_{44}/\rho}$. In both crystals, the displacement vectors of the waves can only be parallel or perpendicular to the sixth-order axes of symmetry and, consequently, at the boundary there can only be two waves, reflected and refracted, with the same polarization as the incident wave. As a consequence of this the problem reduces to the problem of reflection and refraction at the boundary between two isotropic media. Using Eqs. (7.39) and (7.40) for the reflection and transmission coefficients we obtain,

$$\rho_I = \left(\frac{\sqrt{c_{44}^{(1)}/\rho_1}\cos\theta_2 - \sqrt{c_{44}^{(2)}/\rho_2}\cos\theta_1}{\sqrt{c_{44}^{(1)}/\rho_1}\cos\theta_2 + \sqrt{c_{44}^{(2)}/\rho_2}\cos\theta_1} \right)^2,$$

$$d_I = \frac{4\sqrt{c_{44}^{(1)}c_{44}^{(2)}/\rho_1\rho_2}\cos\theta_1\cos\theta_2}{\left(\sqrt{c_{44}^{(1)}/\rho_1}\cos\theta_2 + \sqrt{c_{44}^{(2)}/\rho_2}\cos\theta_1\right)^2}$$

where, according to Eq. (7.37) $\sin\theta_2 = \sin\theta_1\sqrt{c_{44}^{(2)}\rho_1/c_{44}^{(1)}\rho_2}$.

11.5. In crystals with symmetry T_d the piezoelectric moduli $f_{xyz} = f_{yxz} = f_{zxy} = f$ are not zero [8]. Hence, only the combinations $2fn_yn_z$, $2fn_xn_z$ and $2fn_xn_y$ can occur in relation (11.33). Since $n_x = \cos\phi$, $n_y = \sin\phi$, and $n_z = 0$ for a wave propagating at an angle ϕ to the x axis, the combination $2fn_xn_y = f\sin\phi$ is non-zero. Using (11.33), we can obtain that $\Gamma_{xz} = \Gamma_{yz} = 0$ and $\Gamma_{zz} = c_{44}^{E} + f^2\sin{}^2 2\phi/(\varepsilon_{il}^{u}n_in_l)$, i.e. the transverse wave is a pure mode, but unlike the result obtained in problem 11.3 its velocity depends on the direction of propagation.

REFERENCES

1*. Rayleigh, Lord. The Theory of Sound. McMillan, London (1945).

2. Morse, P. Vibration and Sound. McGraw-Hill, New York (1948).

3*. Skuchik, E. Foundations of Acoustics. Moscow (1976), Vols. 1 and 2.

4*. Rzhevkin, S. N. A Course of Lectures on the Theory of Sound, Moscow (1960).

5*. Isakovich, M. A. General Acoustics. Moscow (1973).

6*. Landau, L. D. and Lifshitz, E. M. Theory of Elasticity. Pergamon Press, New York (1959).

7. Amenzade, Yu. A. Theory of Elasticity. Moscow (1976).

8*. Nye, J. F. Physical Properties of Crystals. Oxford University Press (1957).

9. Mikhailov, I. G. and Shutilov, V. A. Diffraction of light by large-amplitude ultrasonic waves. Akust. Zh. **3**, No. 2, 203 (1957).

10. Zarembo, L. K. and Shklovskaya-Kordi, V. V. Velocity of propagation of finite-amplitude ultrasonic waves in liquids. Akust. Zh. **5**, No. 1, 47 (1960).

11. Gitis, M. B. and Mikhailov, I. G. Propagation of sound in liquid metals. Akust. Zh. **12**, No. 2, 145 (1966).

12*. Bergman, L. Ultrasound and its Applications in Science and Technology. Moscow (1956).

13*. Mikhailov, I. G. Solov'ev, V. A., and Syrnikov, Yu. P. Foundations of Molecular Acoustics. Moscow (1964).

14. Kornfel'd, M. I. Elasticity and Strength of Liquids. Moscow (1951).

15. Konstantinov, B. P. Absorption of acoustic waves accompanying reflection from a solid boundary. Zh. Tekh. Fiz. **9**, No. 3, 226 (1939).

16. Zarembo, L. K. and Chunchuzov, I. P. Characteristics of the acoustic field in a viscous medium near the boundary of a beam. Akust. Zh. **23**, No. 3, 466 (1977).

* References marked with an asterisk are especially recommended.

17. Zel'dovich, Ya. B. and Raizer, Yu. P. Physics of Shock Waves and High-Temperature Hydrodynamic Phenomena. Academic Press, New York (1966-1967).
18. Ostroumov, G. A. Foundations of Nonlinear Acoustics. Leningrad (1967).
19*. Zarembo, L. K. and Krasil'nikov, V. A. Introduction to Nonlinear Acoustics. Moscow (1966).
20. Beyer, R. Nonlinear acoustics. In: Physical Acoustics, edited by Mason, W. Academic Press, New York (1965), Vol. II, Part B, pp. 231-264.
21. Mikhailov, I. G. and Shutilov, V. A. Distortions of the form of a finite-amplitude ultrasonic wave in different liquids. Akust. Zh. **6**, No. 3, 340 (1956).
22. Shklovskaya-Kordi, V. V. Acoustic method for determining the internal pressure in liquids. Akust. Zh. **9**, No. 1, 107 (1963).
23. Mikhailov, I. G. and Shutilov, V. A. Nonlinear acoustic properties of water solutions of electrolytes. Akust. Zh. **10**, No. 4, 450 (1964).
24. Burov, V. A. and Krasil'nikov, V. A. Direct observation of the distortion of the form of intense ultrasonic waves in liquids. Dokl. Akad. Nauk SSSR **118**, No. 5, 920 (1958).
25. Mikhailov, I. G. and Shutilov V. A. Diffraction of light by large-amplitude ultrasonic waves. Akust. Zh. **4**, No. 2, 174 (1958).
26. Shutilov, V. A. Optical studies of the form of large-amplitude ultrasonic waves in liquids. Akust. Zh. **5**, No. 2, 231 (1959).
27. Mikhailov, I. G. and Shutilov, V. A. Diffraction of light by the harmonics of an ultrasonic wave distorted during propagation in a liquid. Akust. Zh. **5**, No. 1, 77 (1959).
28. Niedemmann, E. A. and Zankel, K. L. Study of ultrasonic waveform by optical methods. Acustica **11**, No. 4, 213 (1961).
29. Naugol'nykh, K. A. Absorption of finite-amplitude waves. In: Intense Ultrasonic Fields, edited by Rozenberg, L. D. Moscow (1958), Part I, pp. 5-48.
30. Gol'dberg, Z. A. Propagation of finite-amplitude plane waves. Akust. Zh. **3**, No. 4, 322 (1957).

31. Rudenko, I. V., Soluyan, S. I., and Khokhlov, R. V. Bounded quasiplanar beams of periodic disturbances in a nonlinear medium. Akust. Zh. **19**, No. 6, 871 (1973).

32. Zarembo, L. K. Temperature dependence of the absorption of finite-amplitude waves in viscous liquids. Akust. Zh. **3**, No. 2, 163 (1957).

33. Andreev, N. N. Some second-order quantities in acoustics. Akust. Zh. **1**, No. 1, 3 (1955).

34. Gol'dberg, Z. A. Sound pressure. In: Intense Ultrasonic Fields, edited by Rozenberg, L. D. Moscow (1968), Part 2, pp. 49-86.

35. King, L. V. On the acoustic radiation pressure on spheres. Proc. Roy. Soc. (London) A **147**, 212 (1934).

36. Yosioka, K. and Kawasima, Y. Acoustic radiation pressure on a compressible sphere. Acustica **5**, No. 3, 167 (1955).

37. Gor'kov, L. P. Forces acting on a small particle in an acoustic field in an ideal liquid. Dokl. Akad. Nauk SSSR **140**, No. 1, 88 (1961).

38. Kanevskii, I. N. Steady forces arising in an acoustic field: Review. Akust. Zh. **7**, No. 1, 3 (1961).

39. Lamb, H. Hydrodynamics, Dover Publications, New York (1945).

40. Dorr, W. Anziehende und abstossende Krafte zwischen Kugeln im Schallfeld. Acustica **5**, No. 3, 163.

41. Mednikov, E. P. Acoustic Coagulation and Settling of Aerosols. Moscow (1963).

42. Zarembo, L. K. Acoustic flows. In: Intense Ultrasonic Fields, edited by Rozenberg, L. D. Moscow (1968), Part 3, pp. 87-128.

43. Zarembo, L. K. and Shklovskaya-Kordi, V. V. Visualization of the acoustic flow at the boundary between two immiscible liquids. Akust. Zh. **3**, No. 4, 373 (1957).

44. Gabrial, A. M. and Richardson, E. G. A study of acoustic streaming in liquids over a wide frequency range. Acustica **5**, No. 1, 28 (1955).

45. Roi, N. A. Appearance and development of ultrasonic cavitation. Akust. Zh. **3**, No. 1, 3 (1957).

46. Pernik, A. D. Problems in Cavitation. Leningrad (1966).

47. Akulichev, V. A. Pulsations of cavitation bubbles. In: Intense

Ultrasonic Fields, edited by Rozenberg, L. D. Moscow (1968), Part 4, pp. 129-166.

48. Sirotyuk, M. G. Experimental studies of ultrasonic cavitation. In: Intense Ultrasonic Fields, edited by Rozenberg, L. D. Moscow (1968), Part 4, pp. 167-220.

49. Flynn, H. G. Physics of acoustic cavitation in liquids. In: Physical Acoustics, edited by Mason, W. Academic Press, New York (1964), Vol. 1, Part B, pp. 58-172.

50. Zel'dovich, Ya. B. Theory of formation of a new phase. Cavitation. Zh. Eksp. Teor. Fiz. **12**, No. 11-12, 525 (1942).

51. Rozenberg, L. D. Focusing ultrasonic radiators. In: Physics and Technology of Intense Ultrasound. Sources of Intense Ultrasound, edited by Rozenberg, L. D. Moscow (1967), Part 3, pp. 149-206.

52. Akulichev, V. A. Hydration of ions and the cavitation strength of water. Akust. Zh. **12**, No. 2, 160 (1966).

53. Mikhailov, I. G. and Shutilov, V. A. Simple method for observing cavitation. Akust. Zh. **5**, No. 3, 376 (1959).

54. Akulichev, V. A. and Il'ichev, V. I. Spectral indication of the appearance of ultrasonic cavitation in water. Akust. Zh. **9**, No. 2, 158 (1963).

55. Khoroshev, G. A. Collapse of vapor-air cavitation bubbles. Akust. Zh. **9**, No. 3, 340 (1963).

56. Sirotyuk, M. G. Behavior of cavitation bubbles in the presence of high ultrasonic intensities. Akust. Zh. **7**, No. 4, 499 (1961).

57. Nolting, B. E. and Neppiras, E. A. Cavitation produced by ultrasonics. Proc. Phys. Soc. B **63**, P. 9, 674 (1950); B **64**, P. 12, 1032 (1951).

58. Cole, R. Underwater Explosions. Princeton University Press, Princeton, N. J. (1948).

59. Rozenberg, L. D. Cavitation. In: Intense Ultrasonic Fields, edited by Rozenberg, L. D. Moscow (1968), Part 6, pp. 221-266.

60*. Kanevskii, I. N. Focusing of Acoustic and Ultrasonic Waves. Moscow (1977).

61. Smirnov, V. I. A Course in Higher Mathematics. Pergamon Press, N. Y. (1964), Vol. 3, Part 2.

62. Anderson, V. C. Sound scattering from a fluid sphere. J. Acoust. Soc. Amer. **22**, No. 4, 426 (1950).

63. Zarembo, L. K., Krasil'nikov, V. A., and Shklovskaya-Kordi, V. V. Distortion of the form of a finite-amplitude ultrasonic wave in liquids. Dokl. Akad. Nauk SSSR **109**, No. 3, 485 (1956); Propagation of finite-amplitude ultrasonic waves in liquids. Akust. Zh. **3**, No. 1, 29 (1957).

64*. Brekhovskikh, L. M. Waves in Layered Media. Academic Press, N. Y. (1980).

65. Dianov, D. B. Radiation of ultrasonic waves through plane-parallel layers. Akust. Zh. **5**, No. 1, 31 (1959).

66. Tartakovskii, B. D. Acoustic transitional layers. Dokl. Akad. Nauk SSSR **75**, No. 1, 29 (1950).

67. Kul'bitskaya, M. N. and Shutilov, V. A. Ultrasonic studies of glasses. Akust. Zh. **22**, No. 6, 793 (1976).

68. Shaw, R. R. and Uhlman, D. R. Effect of phase separation on the properties of simple glasses. II. Elastic properties. J. Non-Cryst. Solids **5**, No. 3, 237 (1971).

69*. Strelkov, S. P. Introduction to the Theory of Oscillations. Moscow (1951).

70. Olson, G. Dynamical Analogies. Van Nostrand, Princeton, N. J. (1958), 2nd edition.

71. Gitis, M. B. and Khimunin, A. S. Diffraction effects in ultrasonic measurements. Akust. Zh. **14**, No. 4, 489 (1968).

72. Furduev, V. V. Electroacoustics. Moscow (1948).

73. Sneddon, I. N. and Berry, S. D. The Classical Theory of Elasticity. Encyclopedia of Physics, Vol. 6, Berlin-Gottingen-Heidelberg.

74. Viktorov, I. A. Physical Foundations of the Applications of Ultrasonic Rayleigh and Lamb Waves in Engineering. Moscow (1966).

75. Surface acoustic waves: Devices and applications: Publications index. Trudy In-ta inzhenerov po elektronike i radioelektronike (TIIER) **64**, No. 5, 323 (1976).

76. Dransfeld, K. and Salzman, E. Excitation, detection, and attenuation of high-frequency elastic surface waves. In: Physical Acoustics, edited by Mason, W. and Thurston, M. N.,

Academic Press, New York (1970), Vol. 7, pp. 219-272.
77. May, J. Guided wave ultrasonic delay lines. In: Physical Acoustics, edited by Mason, W. Academic Press, New York (1964), Vol. 1A, pp. 418-483.
78. Truell, R., Elbaum, C., and Chick, B. B., Ultrasonic Methods in Solid State Physics, Academic Press, N. Y. (1969).
79. Novozhilov, V. V. Foundations of the Nonlinear Theory of Elasticity. Graylock Press, N, Y. (1950).
80. Zarembo, L. K. and Krasil'nikov, V. A. Nonlinear phenomena accompanying the propagation of elastic waves in solids. Usp. Fiz. Nauk **102**, No. 4, 549 (1970).
81. Khatkevich, A. G. Characteristic directions for elastic waves in crystals. Kristallografiya **9**, No. 5, 690 (1964).
82. Fedorov, F. I. Theory of Elastic Waves in Crystals. Plenum Press, N. Y. (1968).
83. Aleksandrov, K. S. Acoustic Crystallography: Problems in Modern Crystallography. Moscow (1975), pp. 327-345.
84. Khatkevich, A. G. Classification of crystals according to their acoustic properties. Kristallografiya **22**, No. 6, 1232 (1977).
85. Merkulov, L. G. and Merkulova, V. M. Lectures on the Physics of Ultrasound. Taganrog (19786).
86. Musgrave, M. J. P. Crystal Acoustics. Holden-Day, San Francisco (1970).
87. Tucker, J. and Rampton, V. Microwave Ultrasonics in Solid State Physics, North-Holland Publications, Amsterdam (1972).
88. Mason, W. Physical Acoustics and the Properties of Solids. Van Nostrand, New Jersey (1958).
89. Aleksandrov, K. S. and Ryzhova, T. R. Elastic properties of crystals. Kristallografiya **6**, No. 2, 289 (1961).
90. Shubnikov, A. V., Flint, E. E., and Bokii, G. G. Foundations of Crystallography. Moscow (1940).
91. Belov, N. V. Structural Crystallography. Moscow (1951).
92. Zheludev, I. S. Physics of Crystalline Dielectrics. Moscow (1968).
93. Sirotin, Yu. I. and Shaskol'skaya, M. P. Foundations of Crystallography. Moscow (1975).
94. Merkulov, L. G. and Yakovlev, L. A. Characteristic features of

the propagation and reflection of ultrasonic rays in crystals. Akust. Zh. **8**, No. 1, 99 (1962).

95. Borgnis, F. E. Specific directions of longitudinal wave propagation in anisotropic media. Phys. Rev. **98**, No. 4, 1000 (1955).
96. Neighbours, J. R. and Schacher, G. E. Determination of elastic constants from sound-velocity measurement in crystals of general symmetry. J. Appl. Phys. **38**, No. 13, 5366 (1967).
97. Koga, I. and Aruga, M. Theory of plane elastic waves in a piezoelectric crystalline medium and determination of elastic and piezoelectric constants of quartz. Phys. Rev. **109**, No. 5, 1467 (1958).
98. De Klerk, J. Elastic constants of α-ZnS. J. Phys. Chem. Solids **28**, No. 8, 1831 (1967).
99. Mayer, W. G. and Parker, P. M. Method for the determination of elastic constants of trigonal crystal systems. Acta Cryst. **14**, P. 7, 725 (1961).
100. Fisher, E. S. and McSkimin, H. J. Adiabatic elastic moduli of single alpha-uranium. J. Appl. Phys. **29**, No. 10, 1473 (1958).
101. Verma, R. K. Elasticity of some high-density crystals. J. Geophys. Res. **65**, No. 2, 757 (1960).
102. Krupnyi, A. I., Al'chikov, V. V. and Aleksandrov, K. S. Calculation of the elasticity tensor of a monoclinic crystal with the help of a computer. Kristallografiya **16**, No. 4, 801 (1971).
103. Aleksandrov, K. S. Determination of the elastic moduli of a monoclinic crystal by the impulsive ultrasonic method. Kristallografiya **3**, No. 5, 623 (1958).
104. Parker, P. M. and Mayer, W. G. Method for the determination of elastic constants for some crystallographic groups. Acta Cryst. **15**, P. 1-4, 334 (1962).
105. Cady, W. Piezoelectricity. McGraw-Hill, New York (1946).
106. Mason, W. Piezoelectric Crystals and Their Applications in Ultrasonics, Van Nostrand, Princeton, N. J. (1950).
107. Soroka, V. V. Propagation of elastic waves in piezosemiconductors. Izv. Vyssh. Uchebn. Zaved., Fiz., No. 8(88), 129 (1969).

108. Abramovich, A. A., Khromova, N. N., and Shutilov, V. A. Application of lithium iodate crystal as a wide-band transducer in an ultrasonic impulsive-phase interferometer. Akust. Zh. **22**, No. 2, 278 (1976).

109. Gulyaev, Yu. V. Surface electroacoustic waves in solids. Pis'ma Zh. Eksp. Teor. Fiz. **9**, No. 6, 202 (1969).

Bibliography

J.D. Achenbach, *Wave propagation in elastic solids*, Amsterdam, 1973. *Acoustic surface waves*, A.A. Oliner (ed), New York-Berlin, 1978.

B.A. Auld, *Acoustic fields and waves in solids*, New York, 1973.

L.L. Beranek, *Acoustics*, New York, 1954.

E.R. Berg and G.D. Stork, *The physics of sound*, New York, 1982.

L. Bergmann, *Der Ultraschall und seine Anwendung in Wissenschaft und Technik*, Stuttgart, 1954.

A.B. Bhatia, *Ultrasonic absorption. An introduction to the theory of sound absorption and dispersion in gases, liquids and solids*, Oxford, 1967.

A.P. Dowling and W.J.E. Flowcs, *Sound and sources of sound*, New York, 1982.

W.M. Ewing, W.S. Jardentzky and F. Press, *Elastic waves in layered media*, New York, 1957.

G.L. Gooberman, *Ultrasonics. Theory and application*, London, 1968.

D.J. Gorman, *Free vibration analysis of rectangular plates*, Amsterdam, 1981.

K.F. Hertzfeld and T.A. Litovitz, *Absorption and dispersion of ultrasonic waves*, New York, 1965.

M. Jessel, *Acoustique théorique*, Paris, 1973. High-Intensity Ultrasonic Fields, L.D. Rozenberg (ed), New York, 1971.

J.A. Hudson, *The excitation and propagation of elastic waves*, New York, 1980.

A. Kalnins, *Vibrations of plates and shells*, Bethlehem, 1979.

Y. Kikuchi, *Ultrasonic transducers*, Tokyo, 1969.

L.E. Kinsler and R. Frey, *Fundamentals of acoustics*, New York, 1962.

A.E.H. Love, *A treatise on the mathematical theory of elasticity*, London-New York, 1927.

G.W. Mackensie, *Acoustics*, New York-London, 1964.

W.P. Mason (ed.), *Physical acoustics. Principles and methods*, New York-London, 1964 - ... .

J. Mercier, *Traité d'acoustique*, Paris, 1962.

J. Miklowitz, *The theory of elastic waves and waveguides*, Amsterdam, 1978.

C.S. Morawetz, *Lectures on nonlinear waves and shocks*, Berlin , 1982.

P.M. Morse, *Theoretical acoustics*, New York, 1968.

P.M. Morse, *Vibration and sound*, New York, 1948.

J.W.S. Rayleigh, *Theory of sound*, New York, 1945.

H. Reismann and P.S. Pawlik, *Elasticity. Theory and application*, New York, 1980.

E.G. Richardson, *Ultrasonic physics*, Amsterdam-New York, 1962.

E. Skudrzyk, *The foundations of acoustics. Basic mathematics and basic acoustics*, Vienna-New York, 1971.

E. Skudrzyk, *Vibration of complex vibratory systems*, The Penn.State.Univ. Press, 1968.

R.W. Stephens and A.E. Bate, *Wave motion and sound*, London, 1950.

G.W. Stewart and R.B. Lindsay, *Acoustics*, New York, 1930.

G.S. Verma, *Fundamental aspects of physical acoustics*, Oxford, 1980.

R.V. Williams, *Acoustic emission*, Bristol, 1980.

J.M. Ziman, *Electrons and Phonons*, London, 1960.

SUBJECT INDEX

Absorption, at beam boundaries, 69
  Stokes, 64
  ultrasonic, 60-66, see also Coefficient, absorption
Acoustic dipole, 199
Acoustic flow, 129, 144-148
  nonlinear, 148
Acoustic impedance, see Impedance, specific acoustic
Acoustic lens, 191, 192
Acoustic power, 57
Acoustic Q factor, 236, 237
Acoustic refractive index, 190, 220
Acoustic resistance, see Impedance, characteristic acoustic
Acoustic transparency of plate, 216, 217, 222
Amplitude,
  of forced oacillations, 239
  linear relations, 54, 55
  nonlinear relations, 88
  of oscillations, 52
  resonance, 241
  of sawtooth wave, 97, 109
Aerosols, 198
Anisotropy factor, 307, 313
Axes, acoustic, 303, 312
  crystallographic, 307-309

Band, transmission, 242
Bernoulli forces, 140
Bessel-Fubini solution, 98, 113
Bjerknes forces, 139
Boyle-Marriotte law, 35, 83

Cavitation, index, 170
  nuclei, 150, 154
  strength of liquid, 153
  threshold, 154, 155
  ultrasonic, 149
Christoffel's equations, 304, 310, 338
Coagulation, ultrasonic, 141
Coefficient, absorption*, 63-66
    amplitude*, 63
    differential, 103, 104
    energy*, 69, 70
    of harmonics, 114
    of waves,
      infinitesimal-amplitude*, 63
      finite-amplitude, 111, 118
      shear, 73, 74
  attenuation, 67, 233
    amplitude, 68
    energy, 69
    spatial, 67
    temporal, 67, 233
  of bulk viscosity, 61
  of internal friction, 236
  nonlinear, 69
  reflection*, 175
    amplitude, 213, 275, 278, 280-285
    pressure, 175-176, 213, 214
      energy, see Intensity reflection coefficient
      intensity, 177, 214, 220, 267, 275
      velocity, 175-176
  of resistance, 232
  scattering, 209
  of shear viscosity, 61
  of thermal expansion, 36, 86
  transmission, 175, 214, 217, 280-285
    of plate, 214, 217, 222
    pressure, 176

*Terms marked with an asterisk are used repeatedly throughout the book, and only the pages on which they are defined are indicated.

energy, 177, 243
velocity, 176
Compressibility, 36
adiabatic, 36
isothermal, 36
Constants,
electromechanical coupling, 339
Lamé, see Lamé constants
stiffness, see Elastic moduli
Continuity equation,
see Equation,
continuity
Crystals, acoustic axes of, 303, 312
crystallographic axes of, see Axes, crystallographic
cubic, 19, 306-320
directions in,
characteristic, 303
crystallographic, 308
isonormal, see Isonormal directions
transversely isotropic, 312
elastic moduli of, see Elastic moduli
groups of, 18, 19
hexagonal, 19, 318, 320-324
monoclinic, 18, 325, 325-336
piezoelectric, 335
rhombic, 18, 323, 325, 331-332, 333-334
systems, see Systems, crystallographic
tetragonal, 18, 321, 323, 328, 329-330
triclinic, 18, 327
trigonal, 19, 318, 321, 325, 326-327
velocity of sound in, 304

Damping, time constant, 67, 234
of ultrasonic waves, see Coefficient, attenuation
Decrement,
logarithmic damping, 68
Deformations, finite, 9, 225
one-dimensional, 2
homogeneous, 2
three-dimensional, 6-8
Density, ultrasonic energy, 56
absorbed, 69
average, 56
ultrasonic energy flux, see Intensity ultrasonic
Dielectric constant, 327
tensor, see Tensor, dielectric constant
Diffraction, of light by ultrasound, 94, 101
of ultrasonic beam, 248
Directivity function, 204
Dispersion, geometric, 293
of sound, 42
Distance, critical, 91
discontinuity, 91
stabilization, 103

Effective scattering cross section, 205
Elastic moduli*, 17
adiabatic*, 20, 30-37
of crystals, 18, 19
cubic, 316-318
hexagonal, 32-324
monoclinic, 335-336
rhombic, 333-334
tetragonal, 329-330
trigonal, 326-327
dynamic, 20, 36-37
isothermal, 20, 36-37
linear*, 22
matrices of, 16-20
second order*, 22
static, 20, 36-37
tables of, 18, 19, 24, 37, 263, 264, 316-318, 323-324, 326-327, 39-320, 333-334, 335-336
tensor of, 17
third order, 34, 295
Electric induction, 336
Electromotive force, 237
Emulsion, 198

Equation, Christoffel's,
see Christoffel's
equations
of conservation of
energy, 177, 276
continuity, 8, 32, 77, 157,
218
Helmholtz's, see
Helmholtz's equation
Herring-Flynn, see
Herring-Flynn equation
Laplace's, see Laplace's
equation
Maxwell's, see Maxwell's
equation
of motion, 14,32, 77, 163,
219, 237, 260
Navier-Stokes, see Navier-
Stokes equation
Nolting-Neppiras, see
Nolting-Neppiras equation
of piezoelectric effect, 336
Poisson's, see Poisson's
equation
of state, 33-37, 56, 77,
84, 336
Tait's, see Tait's equation
wave*, 38
Equivalent circuit, 229, 232,
238, 258
Erosion, cavitation, 156
Euler variables, 30, 77
Expansion, volume, 18

Far zone, see Zone, far
Field, ultrasonic, 55
of scattered waves, 198
Focusing, of ultrasound
191, 192
Forces, Bernoulli's, see
Bernoulli's forces
Bjerknes, see Bjerknes
forces
friction, 141, 232
pressure, 49, 227, 256
Stokes, see Stokes formula
Formula, Poiseuille's, see
Poiseuilli's formula
Rayleigh's, see Rayleigh's
formula
Stokes, see Stokes formula
Stokes-Kirchhoff, see
Stokes-Kirchhoff formula
Fraunhoffer zone, see Zone, far
Frequency, characteristic,
of clamped plate, 224
of unilaterally loaded
plate, 226
circular*, 48
cyclical*, 48
resonance, 229
of gas bubble, 138,
155, 161
of circuit, 230
of mechanical system,
229
of plate, see Frequency,
characteristic
Fresnel zone, see Zone, near

Geometric dispersion, see
Dispersion, geometric
Gulyaev-Blyushtein waves,
339

Harmonics, of finite-
amplitude wave, 95, 99,
102, 113, 114
of sawtooth wave, 97
of spherical wave, 200-
204
Helmholtz's equation, 48
Herring-Flynn equation, 165
Heterogeneous media, 198
Hooke's law, 16, 20, 241,
261, 291
generalized, 16, 336
Huygėns-Frésnel principle,
245

Impedance, specific acoustic*,
50, 180
total
characteristic*, 50
of cavitating liquid, 170
total*, 50
specific*, 43, 50
of circuit, 241

Poisson's equation, 35, 83
ratio*, 25
Potential, electric field, 328
velocity*, 31
Power, acoustic
see Acoustic power
of ac current, 58, 227, 255
of radiation from pulsating sphere, 257
scattered, 205
Pressure, acoustic*, 49
average, 249
critical, 153, 154
hydrostatic, 139
radiation, 125
Langevin, 129
Rayleigh, 129
surface tension, 139, 150
threshold, 168

Q factor, 235, 236
of acoustic system, 236, 237
of circuit, 235
of plate, 236, 237

Radiation efficiency of pulsating sphere, 258
Radiation resistance, see Impedance, characteristic
Radiator, piston, 244-247
of spherical waves, 255-259
zeroth-order, 255
Radius vector*, 9
Ray, ultrasonic, 191
Rayleigh's formula, 158, 244
Rayleigh's scattering law, 204
Rayleigh waves, 285
Reflection, at normal incidence, 172-187, 285
at oblique incidence, 188-195, 266-285
total internal, 194, 271, 276
of ultrasound, 172-195, 266-285
Resonance, 239
Reynolds number, 106
instantaneous, 111
Riemann's method, 80

Scale, decibel, see Scale, logarithmic
logarithmic, 59
Scattering, coefficient, see Coefficient, scattering
cross section, see Effective scattering cross section
parameter, see Parameter, scattering
of ultrasound, 198-210
backward, 204
coherent, 209
diffuse, 209
geometric, 199
incoherent, 209
Mie, see Mie scattering
Rayleigh, 200-204
resonance, 208, 209
secondary, 209
Sonoluminescence, 156, 161, 169
Sources, coherent, 197
Stiffness, equivalent, 230
effective*, 262, 263
Stokes force, 141
formula, 141
Stokes-Kirchhoff formula, 64
Stress, mechanical*, 10
tensor, see Tensor, stress
viscous, 60, 64
Surface tension, 150
Suspension, 198
System, crystallographic,
cubic, 19, 306-320
hexagonal, 19, 309
monoclinic, 18, 309, 325, 335-336
rhombic, 18, 309, 331-332, 333-334
tetragonal, 18, 309, 321, 328, 369
triclinic, 18, 309, 335
trigonal, 19, 309, 318,

mechanical, 241
of pulsating sphere, 257
total, 50, 257
Index, acoustic refractive, 190, 220
isentropic, 84
cavitation, see Cavitation
Induction, electric, 336
Intensity, ultrasonic*, 57 101-102, 197
of electric field, 336
of spherical wave, 254
of sawtooth wave, see Waves, sawtooth, intensity
Internal energy, 21, 295
Isonormal directions, 303, 305, 307

Kirkwood-Bethe equation, 165
Kronecker delta*, 8

Lamb waves, 291
Lamé constants, 20, 23, 261
Laplace's equation, 150
Laplacian, 32, 39, 250, 261
Layer, acoustically transparent, 214, 217-223
inhomogeneous, 218-223
half-wave, 213
quarter-wave, 214
Law, Boyle-Mariotte, see Boyle-Mariotte law
Ohm's law, see Ohm's law
Rayleigh, see Rayleigh scattering law
of reflection and refraction, 190, 269
Lorentz-Lorenz equation, 53
Love waves, 288

Mach number, 78
acoustic, 78
Mass, associated, 258
equivalent, 230
Matrix notation, 7, 11, 17
Maxwell's equations, 337
Method, dilatometric, 156, 170
of electroacoustical analogies, 226-228
Riemann's, see Riemann's method
shadow, 192
of successive approximations, 115, 298
Mie scattering, 200
Miller indices, 308
Modulus, of hydrostatic compression*, 26
nonlinear, 34, 295
shear*, 26
bulk elastic, 26
effective, 61, 230
Young's, see Young's modulus
Motion, equation of, see Equation, of motion

Navier-Stokes equation, 106
Near zone, see Zone, near
Nolting-Neppiras equation, 165
Nuclei, cavitation, see Cavitation, nuclei

Ohm's law, 230, 231
Oscillations, characteristic, 223-226
damped, 231-237
forced, 237-243
free, 228-231
normal, 228
period of, 48
resonance, 239

Parameter, gas-content, 160
nonlinear, 82
scattering, 199
Parameters, equivalent, 229, 230
Piezoelectric constants, 336
Piezoelectric effect, 335
Poiseuille's formula, 144

321, 325, 326-327
electrical, 231, 233, 237
mechanical, 228, 232, 229
with distributed parameters, 229
with lumped parameters, 229
oscillatory acoustic, 227, 232, 237

Tait's equation, 36, 84
Tensor, dielectric constant, 337
strain*, 6
invariants of, 8, 295
stress*, 11,
tensile stress, 33
Transformation of ultrasonic waves, 172, 266, 284

Vector, displacement*, 1, 301
position, see Radius vector*
unit normal*, 48, 301
wave, see Wave, vector*
Velocity, particle*, 52
of sound* (ultrasound), 41
in rods, 292
local, 81
of wave track, 196

Waveform, degree of distortion of, 92, 112
Wavelength,
of standing wave, 181
of quasistanding wave, 196
of traveling wave*, 48, 181
Wave number*, 48
Waves,
Gulyaev-Blyushtein, see Gulyaev-Blyushtein waves
Lamb, see Lamb waves
longitudinal*, 28, 260
Love, see Love waves
velocity of, 290
piezoelectrically active, 338
piezoelectrically inactive, 338
quasilongitudinal, 302
quasitransverse, 302
Rayleigh, see Rayleigh waves
velocity of, 287
sawtooth, 91, 97 102, 108
absorption of, 110-111
amplitude of, 97, 109
depth of front of 112
intensity of, 115
shear*, 260
penetration depth of, 74
shock, 91
transformation of, see Transformation of ultrasonic waves
Wave vector*, 48, 301
Width of resonance line, 243

Young's modulus*, 24

Zone, far, 247
near, 247, 249

FUNDAMENTAL PHYSICS OF ULTRASOUND

by V.A. Shutilov
(with supplemental material provided by E.V. Tcharnaya)
Translated from the Russian by Michael E. Alferieff

"The quality of the book is superb . . . I know of no other book which deals with fundamental physics of ultrasound that provides such a lucid commentary of each basic idea developed."
—William D. O'Brien, Jr., University of Illinois at Urbana–Champaign, USA

Based on lectures by the author, this volume is designed as a textbook on general ultrasonics. The text provides comprehensive coverage of the propagation of ultrasonic waves in media with different elastic properties and under conditions close to those encountered in scientific and practical applications of ultrasound. As well as classical material and data from original sources, the book includes experimental data on the velocity and absorption of ultrasound in liquids and gases and data on the velocity of sound in isotropic solids and crystals. Each chapter is complemented by problems and their solutions, making this a superb textbook in virtually any university setting.

About the authors

Professor Vladimir Alexandrovich Shutilov graduated from the Faculty of Physics, Leningrad State University and worked there all his life, first as lecturer, senior lecturer, professor and from 1980 as head of the Department of Molecular Physics. He published more than 250 scientific papers, mainly on ultrasonic spectroscopy and solid-state quantum acoustics. In the early sixties, he began research on a new phenomenon – acoustic nuclear-magnetic resonance, for which he obtained his doctorate in 1974. He was a member of USSR Academy of Sciences committees on problems of ultrasonics, radiospectroscopy and the physics of ferroelectric and dielectric materials. He died in 1985.

Yelena Vladimirovna Tcharnaya is lecturer in the Faculty of Physics, Leningrad State University. She has published more than 60 scientific papers in the fields of solid-state acoustics, quantum acoustics and the acoustic properties of crystals.

Related titles of interest

Applications of Ultrasonics to Molecular Physics *V.F. Nozdrev*

Physics Reviews, *edited by I.M. Khalatnikov*

Section A in the Soviet Scientific Reviews series

Nonlinear Physics: From the Pendulum to Turbulence and Chaos
R.Z. Sagdeev, D.A. Usikov and G.M. Zaslavsky

Electromagnetoelasticity of Solids: Piezoelectrics and Electrically Conductive Materials *V.Z. Parton and B.A. Kudryavtsev*

GB **Gordon and Breach Science Publishers**
New York London [...] **Melbourne**